Evolution of Sameness and Difference

Evolution of Sameness and Difference

Perspectives on the Human Genome Project

Stanley Shostak

University of Pittsburgh
Pennsylvania, USA

harwood academic publishers

Australia • Canada • China • France • Germany • India • Japan • Luxembourg • Malaysia • The Netherlands • Russia • Singapore • Switzerland •

Amsteldijk 166
1st Floor
1079 LH Amsterdam
The Netherlands

Front cover: Ernst Haeckel's monophyletic family tree (pedigree) of organisms of 1866. Haeckel, 1914.

British Library Cataloguing in Publication Data

Shostak, Stanley
Evolution of sameness and difference : perspectives on the
human genome project. 1.Biological diversity 2.Evolution (Biology)
3.Ecological heterogeneity
I.Title
576.8

ISBN 90-5702-540-X

To *Daniel* and *Russell*
who began my education on
sameness and difference

CONTENTS

xi PREFACE

1 CHAPTER ONE TILLING BIOLOGY'S PASTURES: LIFE'S SAMENESSES AND DIFFERENCES

3 Differences and Samenesses
5 Life's Unity
6 Samenesses that Unite Living Things
16 Differences that Separate Life from Nonlife
21 Life's Units
22 Organisms
28 Cells
31 Genes
33 DNA
36 Critique of Life's Apportionment to Units
37 A Critique of Biology in the 20th Century
38 Biology's Ontological Blindspot
39 Biology's Epistemological Blindspots

43 CHAPTER TWO "NORMAL" BIOLOGY: FINDING SAMENESS IN LIFE'S MOLECULES

45 Biological Sameness
46 Molecular Samenesses
48 Individuating DNA
55 The Human Genome Project
58 Genomic Samenesses
59 Mapping
65 Genomic Sequences
68 Looking for Genes in the Genome
69 Analyzing Polypeptide-Encoding Sequences
74 Gene Families and Classes
83 Interpretation of Gene Families
83 Homology
85 Cladistics: Confusing Molecular Families with Molecular Descent
89 Summing up: Categories of Sameness

93 CHAPTER THREE BIOLOGY'S SHIFTING PARADIGMS: CONTENDING WITH DIFFERENCE

94 A Brief History of Differences in Biology
100 The New Look of Difference
100 Cladistics Methods
103 Drawing Evolutionary Inferences from Molecular Differences
105 Noncellular Life: Viruses and Retrotransposons
107 RNA Viruses
110 DNA Viruses
111 DNA/RNA and RNA/DNA Switching Viruses (Pararetroviruses and Retroviruses) and Retrotransposons
115 The New Look at Cellular Life
116 Finding the Root of Difference
118 Putting an Age on Difference
120 Domains in Life's "Universal" Phylogenetic Tree
120 Bacteria (Also Called Eubacteria)
126 Archaea (Also Called Archaebacteria)
129 Eucarya
142 Where Domains Meet
142 Shared Across the Three Domains
143 Shared Across Two Domains
145 Critique
146 "Been There; Done It"
147 Lineages Do Not Necessarily Come and Go from Points

149 CHAPTER FOUR DEVOLUTION: RECONCILING SAMENESS AND DIFFERENCE

151 Early Earthlives
152 A Geochronometric View of the Early Earth
153 Origins
160 Between Chemistry and Biology
163 RNA-Based Genomes
164 The RNA World
167 Early Intronic RNA
168 Genetic and/or Operational Codes
172 DNA-Based Genomes
174 Genes: Stable Conveyers of Heredity
178 Chromosomes: Repositories of Genomes
181 Superfamilies of Genes
186 Earthlives
186 Devolution of Noncellular Life
191 Devolution of Cellular Life
203 Critical Issues for Sameness and Difference

207 CHAPTER FIVE ALTERNATIVES: EVOLUTION VERSUS DEVOLUTION

209 Evolutionary Theory: Biology's Horseless Carriage
210 What Is Evolution?
211 Pervasive Ambiguity
214 Virtual Evolution
216 Genes and Mutation: Evolution's Fuel
217 Genes
219 Mutations
222 Devolution
227 Heredity: Evolution's Transmission
227 Genetic Continuity
228 Devolution
231 Natural Selection: Evolution's Engine
232 The Roots of Natural Selection
237 Reprise

239 ENDNOTES

277 BIBLIOGRAPHY

325 INDEX

PREFACE

Evolution of Sameness and Difference (*ESD*) is the offspring of my last book, *Death of Life* (*DOL*),[1] inheriting many of its traits — strengths, I hope, and weaknesses, I'm afraid. *DOL* began as a critique of modern molecular biology. As a practicing biologist,[2] I had come to realize that molecular biologists, in general, blatantly superimpose certain values on life and make a concerted effort to reinforce and promulgate those values throughout professional biology. Those values are desperately out of phase with life as I know it, and harmful to its study. In my opinion, *DOL* established that biologists, especially molecular biologists, have abandoned the study of life in favor of pandering to prevailing prejudices and forces of the marketplace. *DOL* also explored the philosophical implications of biologists forsaking the intellectual pursuit of life.

ESD is not a clone of *DOL*; it is interested in alternatives to the present ideological postures of biologists, postures that became apparent while I wrote *DOL*. The current work is thus a book written backwards. Its conclusions were compelling long before the way to reach them was at all clear. My objective in writing this book was to figure out how to reach those goals.

I had an enormous amount to learn before my ideas could come across in anything approaching an informed opinion, and many nagging doubts had to be settled before committing *ESD* to paper. *ESD* reflects this process, hopefully providing enough clues for molecular biologists to follow the path, signposts to the "road not taken." My hope and expectation is that the next, if not the present, generation of molecular biologists will be sufficiently curious to explore some of the roads I have traveled in *ESD*.

The process of leaving a trail so others may retrace it is not easy — as is well known by every Hänsel and Gretel who ever tried leaving their marks behind. A trail suitable for academic scientists must be parsimonious but robust, made of the same stuff as science and equally durable. The choices are limited; and in my ruminations (on the perils of bread crumbs), I opted for philosophical markers and historical milestones with the conviction that firm and rigorous data and analysis are also subject to addition and correction.

Historical and philosophical data, like those of any academic discipline, are articulated in their own language — filled with tradition and embroidered by fashion — language used to cast for problems and haul in solutions of the particular discipline. Biology has, of course, its own language, which is not necessarily consonant with that of history and philosophy. Finding consonance between these languages was a major problem for *ESD* that was only solved when my friend Marcia Landy walked me through the work of Gilles Deleuze on difference and repetition.[3] Deleuze's work resonated with biological concepts. Biology itself resided on adjacent plateaus, and the language of difference and repetition was as adaptable as it was congenial.

As it turned out, *ESD* became an intellectual palindrome: As much as the book recounts the evolution of biological samenesses and differences, it traces the fortunes of sameness and difference in the evolution of biological concepts, chapter by chapter. However, each chapter is also intended to stand on its own, as a module, containing all the ideas and definitions necessary for its independent comprehension. Furthermore, because the book is intended as a guide, each chapter is copiously endnoted and an extensive bibliography is attached to the whole.

I must say a word here on "paths not taken" in *ESD* in order to clarify my position on some thorny issues and, if possible, avoid some misunderstandings. In writing this book, I rejected a sociological critique of sameness and difference. I did so not without regrets.

For decades, liberal and progressive thinkers have appreciated that the argument of sameness stands as a bulwark for democratic principles and a compelling instrument for reform — thus the language of sameness in the Declaration of Independence, the Constitution of the United States and the Bill of Rights. This same argument forced the abolition of slavery and the extension of the franchise to women. The mitochondrial Eve may even create a stronger sense of "brotherhood" than has all the well-meaning rhetoric.

Feminist biologists also, originally, stirred up the sea of sameness. The physiologist Ruth Hubbard, although relying on the metaphor of relationships, envisioned a grand scope of samenesses in life:

> A belief that all living forms are related and that there also are deep connections between the living and nonliving has existed during much of recorded human history. Through the animism of tribal cultures, which endows everyone and everything with a common spirit; through elaborate expressions of the unity of living forms in some Far Eastern and Native American belief systems; and through Aristotelian notions of connectedness runs the theme of a web of life that includes humans among its many strands.[4]

I support the politics of sameness, but I also appreciate the ease with which it can be abused and has been undermined historically, philosophically and biologically. The problem is that any political or social agenda based on sameness is only as strong as its critique of difference. Of course, one might argue on behalf of equality based on sameness: "Don't get rid of a foundation, however rotten, until you have a stronger one." I do not propose building a stronger foundation. My work here is demolition. Liberal and progressive politics will survive my biological critique of sameness because they are not ultimately built on falsifiable premises, and I have no need to worry on their score. What is more, there is nothing whatsoever I can do about being misused and misrepresented by saying that, biologically, ideas of sameness are wrong. I can only make my point as clearly as possible: To criticize the notion of sameness is not to attack equality among human beings.

In brief, *ESD* asks why biologists say some of the things they say about data and believe some of the things they believe without data. It lays responsibility for the failure of biologists to say what they mean and mean what they say on unexamined notions of sameness and difference and a penchant to stuff differences into opaque boxes of sameness. *ESD* intends to open the lids of these boxes and expose biology to data it has ignored.

ESD begins with an exploration of biology's historical and philosophical terrain, opening leads that are followed elsewhere in the book. What is sameness? Difference? What forces have concatenated to empower them? What shapes the prevailing sense of sameness and difference and places them on a plateau of recognition? All clues point to evolution, and this book goes on to focus evolution's modern incarnation in molecular biology's quest for sequence data and other fallout from the Human Genome Project. Chapters 2 and 3 examine how data are obtained, stored, managed and manipulated, from cloning and the polymerization chain reaction to DNA and protein sequences to construction of gene superfamilies and phylogenies. At the same time, *ESD* shows, through incongruities in data and biases in interpretation, how molecular biologists and systematists use and legitimize evolutionary hypotheses rather than test them and propose alternatives. Chapter 4 introduces devolution in an effort to show the catachresis of evolution, the untruth of truth. The book ends with the proposition that devolution can play the role of alternative to evolution *sans* sameness and difference.

Devolution is not the new theory of life I might hope to see elucidated in *my* lifetime. If this book succeeds in introducing devolution, biologists may discover the ditch in which evolutionary biology is currently stuck. I only hope to push biology and get it moving in desirable directions. I hardly imagine that the examination of sameness and difference threatens it the way relativity and quantum mechanics threatened physics.[5] The analysis of evolution performed in *ESD* can only help biologists pursue their intellectual commitment to life and the consequence of its study. An unexamined science is not worth living.

Finally, I would be shamefully remiss not to acknowledge all the help I have had from friends and colleagues. First, thanks go to my editor Sally Cheney, who has been patient throughout the extended duration of this project and encouraging whenever it was necessary. Once again, I am in debt to head librarian Drynda Lee Johnston, Langley Library, University of Pittsburgh, and her able assistants, who patiently and diligently located and procured all the esoteric literature and some mundane pieces I required from around the world. I must thank William Coffman, my colleague in the Department of Biological Sciences at the University of Pittsburgh, who read the entire work, and propelled me to reboot my machine where I had taken intellectual shortcuts. I have only gratitude and admiration for Marcia Landy, film scholar and English Department member at the University of Pittsburgh, who read every chapter as it rolled out of the printer and forced me to meet her intellectual standards whenever mine flagged.

NOTES

1. Shostak, 1998.
2. My own research interests in the past have been asexual reproduction, development and evolution in Cnidaria. A smattering of my research reports appears in the bibliography. I should confess at the outset that I have never published research reports in the field of molecular biology and so my credentials are tainted, but I won't let that stop me.
3. Deleuze, 1994.
4. Hubbard, p. 88, 1990.
5. Only the stationary "aether" had to be abandoned, not mechanics or electromagnetic dynamics in almost all ordinary applications.

CHAPTER ONE

Tilling Biology's Pastures: Life's Samenesses and Differences

Yet how can it be denied that the eternal return is inseparable from the Same? Is it not itself the eternal return of *the Same?*

Gilles Deleuze (Difference & Repetition, *p. 125*)

In all fields of biology, comparisons are cultivated in the soil of empirical data, while contrasts are neither sown nor reaped. Differences as well as similarities could be nurtured and essential "trace elements" provided by the same data, but intellectual monoculture prevails in biology, and differences are torn up as weeds and culled out as damaged goods before reaching the marketplace of ideas. Biologists argue that testing samenesses is the only efficient way to use data, but biologists have been yoked historically to comparisons and do not seem ready to plow their fields any differently today.

Biologists were granted title and began cultivating their own pastures in the 19th century when fear of revolutionary fever in the peasantry was still high among the urban intelligentsia. Following the rise of expertise, the powers that be easily interposed biologists between clergy and peasant as the new experts on all matters concerning life. But having gained privileges from the State, the issue at stake was biology's contribution to the maintenance of social order. Of course, biologists also had self-interest to serve and found professionalism an increasingly powerful route to the control of perquisites and patronage.

The ear of the throne was open to the rhetoric of orderly existence and life's hierarchical structure. Biologists could duel (Cuvier/Geoffroy debates[1]) over what data were important and what were not, but power rested in what comparisons were legitimate and what were not. The route to power for the incipient biologist thus resided in ex cathedra statements of life's samenesses.

Samenesses are judgments built on values and, like other values, attributing them dispenses value (a "rose" has value; a "weed" has none; children acquire value along with the name of their father). Moreover, having the authority to make judgments, even in science, is not a hollow favor but a signifier of power, and biologists jealously guard their prerogative to attribute sameness. Power has its privileges: only members can draw and erase sameness.

The assignment of sameness is grounds for decision making (declaring that simian immunodeficiency disease is sufficiently similar to AIDS to serve as a surrogate for the human disease). The assessment of sameness sways industry to produce products (prophylactic, diagnostic and therapeutic) and services (from mere labor saving devices to the most powerful means of war). The fixing of sameness determines how life shall be controlled (which species shall be saved and which considered dispensable). Thus, sameness decides how life is lived, whether evolution or extinction, whether a disease shall be cured or allowed to spread, whether abundance shall be created or destroyed. Biology has power, and biologists can no long claim that, as a science, biology is neither good nor bad in and of itself; only its uses have moral value, and those are beyond biology.

Despite the importance of sameness and difference, biologists pretend to take no notice of them. Biologists broadly assume that certain samenesses and differences are inherent in living things just as other samenesses and differences are inherent in nonliving things. Biologists also assume that repetition brings sameness, and the failure of repetition brings difference. Unabashed positivists, biologists are persuaded that because life repeats itself, biology is possible; because life does not repeat itself, biology is necessary. With common sense as their motto, biologists see sameness and difference as residing at the descriptive end of reductionism and regard them as the stuff science is made of and not as issues for critical examination. Instead of taking a hard look at the premises of sameness and difference, biologists have traditionally reinforced them and piled sameness and difference upon each other, even when the exercise drags biology to arbitrary and contradictory positions.

Two premises of sameness dominate thinking in modern biology and are especially problematic: (1) **Life is fundamentally the same around the**

Earth and has been throughout its history; (2) **Life is parceled out in units of living things and has been since its creation**. The first premise, life's unity, is built on samenesses that unite life on Earth and differences that distinguish life from nonlife. Regrettably, a lot falls through the gap between the living and nonliving (viruses, bacteriophage, prions). The second premise, life's confinement to units, would seem to rely on difference to separate life into isolated, living things but actually depends on sameness among these units to reinforce their identification, even though such units are not uniformly defined and may be organisms, cells, genes or deoxyribonucleic acid (DNA).

Chapter 1 will show that difference is not beyond scrutiny, and sameness is not above criticism. The chapter begins with an exegesis on sameness and difference and goes on to examine, via historical reconstruction, the sources and evolution of biology's premises about life. The chapter ends with a critique of biology in the 20th century.

DIFFERENCES AND SAMENESSES

Differences seem to emerge from the failure of things to meet criteria for sameness. Aristotle might have classified such differences as the opposition of extremes or contrarieties. In the era between Leibniz and Hegel, "difference remain[ed] subordinated to identity, reduced to the negative, incarcerated within similitude and analogy."[2] Hegel, might have identified an extreme form of difference, the reality of difference that germinates from opposition and blossoms into contradiction.

The very idea of difference is loaded with preconceptions. For example, difference is frequently said to be a quality of an individual ("I'm different!"), yet it is defined by comparisons among individuals ("Different from whom?"). Difference, therefore, depends on the preconceptions of the individual and of individuals without which difference would cease to exist.

Difference also depends on the assumption of a class which cuts off some individuals from membership while binding other individuals together. Otherwise, difference would be meaningless (except as imagery or poetry). Furthermore, differences presuppose either or both a spatial and temporal arrow (which may not be double headed), since matter cannot occupy the same place at the same time. Time's arrow may point to change (movement between alternative states), and space's arrow presupposes causality (things that are made different). A statement of difference, therefore, is a compound of concealed statements about class, time, space, the possibilities of change and causality. As a result of this concealment, mainstream biologists constantly confuse differences occurring over time with evolution (the accumulation of progressive alterations in structure and function), and differences occurring over space with adaptation (the optimization of fit between an organism and its environment).

Sameness is a difficult concept because of all the things taken to be the same: "the Same, the One, the Identical and the Like."[3] Sameness is also a difficult concept because it requires narrow agreement, evoking quantitative comparisons rather than merely the qualitative comparisons evoked by difference. Of course, sameness is the dialectical opposite of difference and, like other dialectical opposites (such as man and woman), the two are defined as much by each other as by qualities of each one. Indeed, the sameness of sameness and difference is their most seductive feature. They are not only intimate but inseparable, and the preconceptions of difference also cover sameness: class, the individual and individuals, time and space, change and causality.

Especially in science, sameness is problematic, since to say, "They are the same," may not be a statement of absolute sameness but a relative statement requiring an additional explanation regarding the way things are the same. Sameness is not counterintuitive, but sameness is sometimes attributed by default (when differences are too difficult to identify). For example, to identify people as adults is to assert their sameness (that they have passed through childhood, reached a plateau of growth and obtained sexual maturity) while ignoring their difference (where they reside on a scale of aging [What is their status apropos of breeding? Are they post-reproductive "adults" or in some way permanently sterile?]).

Quantitative criteria of sameness are not a problem as such, but what they represent (and cover up) can be problematic. For example, the much celebrated sameness of the internal organismic environment (the *milieu interior* and favorite subject of physiology for over a century) is achieved through dynamic change (a relatively steady state) not a static constancy. It is a statistical sameness, a sameness only in the sense that one obtains the same values when measuring a variable at two places or times, but nothing, other than the quantities measured may be the same, including, in all probability, the very things impinging on the instruments of measurement.[4]

In the last analysis, biological sameness and difference are yoked in tautology: the samenesses that bind living things together also differentiate them from nonlife and from each other; the differences that individuate living things from each other also bind them into classes of similar living things. Sameness like difference can be as refined as amino acid sequence in polypeptides or as gross as the *Baupläne* of superphyla. The antipoles can neither be integrated nor disentangled, and biologists explain neither life's samenesses nor its differences.

LIFE'S UNITY

The unity of life, or the idea that all living things possess similar attributes, might be dismissed as a mere tautology if it were not so universally accepted as a biological principle. Historically, it would seem, features of life in human beings were identified in other things which were, as a consequence, considered "living" as well. Presumably, ever since human beings started to domesticate animals, characteristics, such as warmth, the requirement for nutrition, and motion suggested attributing life to nonhuman animals. Aristotle went further and attributed life to plants, where the same qualities were present "if only hidden and not evident."[5] The attribution of sex and soul, the capacity for desire, pleasure, pain, sadness, etc., to plants was agreed with considerable hedging, although respiration and waste removal were more easily granted to plants. Possibly the most general notion of life was that of "organicism," the notion that life alone sustained life, while nonlife did not. Organicism reinforced attributing life to nonhuman animals and plants since they provided resources for the sustenance of human life. The notion of organization, embodied in "organism" and of individuals, or unique systems, as the identifying features of life were finally added in the Renaissance.

Nonliving things were probably not identified historically by any criteria for nonlife but simply by differences with living things. The nonliving category consisted of attributes left over after identifying the attributes of living things. In the 17th century, vitalism attempted to secularize life, and efforts to define differences between life and nonlife were even suspended. By the close of the 18th century, some incipient biologists struggled to link natural history to natural laws, but, while chemistry blossomed under Lavoisier's custodianship, the flower of biology wilted (preserved in the arboretum of the bourgeoisie). The century that would climax in revolution, both armed and social, saw the efforts of materialists to confront the ineffable turned back by a determined effort to co-optate vitalism,[6] and even reintroduce God's volunteerism.

In the end, "All things must change so that everything can remain the same,"[7] and what had once been the exclusive province of God, became the province of a transcendent vitalism. Living matter was liberated from the tyranny of capricious gods, and biologists acquired the authority and right to discover the laws of life, but unity of life remained the foremost of those laws, even if it defied the natural laws of physics and chemistry.

Samenesses that Unite Living Things

Unity of life was not always as conceptually entrenched as it is today. Despite the commitment of 18th century natural philosophers to unity, epitomized by the Great Chain of Being, in the 19th century, evidence for biological change drove natural historians to ponder difference in new ways.[8] Irreconcilable differences among living things were reexamined for the possibility of transmutation, and great debates were held over whether life was continuous or discontinuous.[9] Sameness was implicit in structure, whether the structure of organisms or their parts,[10] while difference was implicit in function, whether meeting demands of the environment or those of other parts of the organism. Irreconcilable difference was thus on a collision course with unalterable sameness.

The rise of a concept of life's disunity was only forestalled when Charles Robert Darwin (1809–1882) suggested how structural and functional change could be derived through linear relationships filtered through environmental constraints. Darwin thus held out the promise of uniting heredity and evolution. The problem was, "How can the very same mechanism that perpetuates the biological accomplishments of the past (heredity) also create those accomplishments (evolution)?"

Heredity: Sameness that Descends

Repetition is the key to concepts of heredity, repetition of form and the similarity of offspring and parents (as well as offspring and offspring). Heredity was "the way things are": The lioness gave birth to the lion because that's what lionesses did and could do no differently. Ordinary folk knew better, however, and speculated on how one child had "his grandmother's eyes" and another child had "his father's hands." Heredity was, indeed, "up for grabs," and the idea of discrete characters, available for mixing and inheritance through lineages was well entrenched, at least in popular culture.

Plant breeders and botanists had also grasped the concept of discrete heredity units predetermining life, and they did so long before the 20th century. The issue had been resolved with the discovery of the equality of male and female principles of reproduction demonstrated by reciprocal crosses. Joachim Camerarius (also Camerer, 1534–1598) had performed the first reciprocal crosses between the pistil and pollen of different flowers, and, a century later, Joseph Köllreuter (1733–1806) demonstrated the equivalence of characteristics transmitted by pollen and pistil. A hundred years later, the Czech patriot, Abbot of Brno and gardener, Gregor Mendel (1822–1884), performed similar experiments, and the weight of modern genetics would rest on his accomplishments (although he was largely ignored in his own time and only appreciated posthumously after being "rediscovered").

Modern concepts of heredity began just prior to 1900 with the exhumation by botanists of Mendel's laws of heredity. While searching the older literature for new ideas, Hugo De Vries (1845–1935), among others, came upon Mendel's report of 1866. Youthful enthusiasm drove De Vries to hand over his prize to the Cambridge zoologist, William Bateson (1861–1926)[11] who excitedly announced the "rediscovery of Mendel's laws," heralding in the new age of genetics. While few in the academy had been interested in the ramblings of a country priest, Gregor Mendel gained currency once his discoveries were translated to the elevated language of scientific laws and redoubtable reductionism.

Mendel's concept of immutable hereditary factors was quickly confirmed in several hereditary situations (but not others) using rudimentary data on traits and arcane algebraic formulas, but these studies hardly anticipated the future science of genetics. To begin with, Wilhelm Johannsen (1857–1927) coined "gene" in 1903. At the time, he was interested in restoring mystery to materialism and countering the rampant reductionism overtaking concepts of heredity and development swarming around him. He proposed "gene" to reassert the enigma of Mendelian quantities influencing inherited qualities. Regrettably, the word was quickly appropriated by those who would use it for the very purpose it was intended to combat.

Indeed, the phenomenal success of genetics in the 20th century did not spring so much from geneticists as from the synergism of genetics with cytology and embryology. Cytologists had already reasoned that cells required a self-perpetuating hereditary material (alias idioplasm; alias germ plasm) and had already traced it to the cell's nucleus. Meanwhile, the study of nuclear control over development had proceeded apace in the laboratories of experimental embryologists.

Chromosomes, those heavily staining bodies that appear in nuclei prior to cell division were the candidates of choice for a hereditary material, since they demonstrated the mathematical fluctuations expected of Mendelian factors. Oscar Hertwig (1849–1922) deduced that the number of chromosomes characteristic of a species's body cells (the somatic cell line) was reduced by half in egg and sperm (the germ line) and doubled by fertilization. Theodore Boveri (1862–1915) then demonstrated, with the help of statistical inference, that normal development in sea urchin embryos required the presence of complete sets of chromosomes, suggesting thereby that chromosomes were different from each other.[12]

Boveri's idea was conveyed to William Sutton (1877–1916) who passed it on to his mentor, Edmund Beecher Wilson (1856–1939) who brought Sutton's speculations to the attention of his colleague, Thomas Hunt Morgan (1866–1939), who would ultimately receive the Nobel Prize in Physiology or Medicine for the chromosomal theory of inheritance and his confirmation of Mendel's work. Morgan initially found the concept of chromosomes acting as conveyers of hereditary factors appallingly reductive and set out to discredit it experimentally. By 1926, however, he was only interested in supporting the chromosomal theory, and he did so in such a compelling way that he determined how heredity would be studied for decades.

Like Bateson who advised his followers to "value exceptions," Morgan sought to study heredity through exceptions. Borrowing Hugo De Vries's word, "mutation" (change), individuals with an exceptional appearance, or phenotypes, would thereafter be known as "mutants." Morgan's group of young investigators quickly accumulated several mutants in the fruit fly, *Drosophila melanogaster*, that were reared in the famous "fly room" at Columbia University. Breeding them revealed that the mutant phenotypes were inherited in Mendelian fashion. The idea that mutants were caused by mutations in normal genes required a little more refining, but that could wait.

Alfred Sturtevant and Charles Bridges, working with Morgan, showed that mutations segregated disproportionately in "linkage groups" that could be traced to chromosomes, via their morphology. Many mutations, or so it would seem, resided on the same chromosome. Moreover, not only could more than one mutation affect the same trait, but different mutations could virtually occupy the same position on a chromosome. The superposition of mutations broke the geneticist's faith in their difference, and suggested that normal genes could also be variants of mutations. The normal gene and each of its mutations became known as alleles and their mutual location on a chromosome, their locus. Genes had become not only the undisputed agents of hereditary (sameness) but the authorized source of hereditary difference (mutation).

Evolution: Samenesses that Ascend

The miracle of evolution is its ability to move through imperceptibly small steps that unite inertia and progress. Biological inertia is the concept that everything remains the same as long as nothing changes. Darwinian evolution argued on behalf of inertia with amazing persuasiveness, demonstrating that evolution, turned backward upon itself, restored similarities to living things. Indeed, the very presence of difference among living things assumed the existence of similarities in once common ancestors. Life's unity thus was due to life having been wrought from a self-same source. This concept of life's singularity received a major boost from 19th century churchmen who grasped that unity provided a compromise between the materialism of evolution and Christian creationism. The churchmen's concept flourished in the wake of Darwin's breakthrough, and dominated theories of evolution, both popular ones[13] and those largely ignored.[14] Ironically, inertia is reinforced today by concepts of change, such as mutation and DNA clocks, which leave undisturbed the so-called molecular "homologies" or samenesses among molecules.

Progress has a more checkered career in biology than inertia. Biologists rarely elaborate upon the concept of evolutionary progress,[15] but it generally means change toward an optimal state. Optimality is, likewise, broadly defined and can refer, without prejudice, to the fit of an organism to its environment or the conformation of an enzyme to its substrate. Of course, optimality changes with conditions (environments and metabolites), but life either perseveres in its drive to perfection, despite altered circumstance, or ceases to exist. According to the Darwinian doctrine of natural selection, if two otherwise similar individuals exist at the same time, a difference between them which gives a slightly better fit for a trait in one of the individuals (under prevailing circumstance) will allow that individual to leave more offspring than the other individual and the trait to appear in more individuals in future generations. Notwithstanding the vast samenesses of individuals, the theory of natural selection derives progress exclusively from competition among individuals possessing small differences.

Darwin must be credited with inaugurating the modern age of scientific speculation on evolutionary progress, although he acknowledged his debt to Georges Cuvier (1769–1832) for the concept of unity of type, and to Etienne Geoffroy Saint-Hilaire (1772–1844) for the concept of homology. Cuvier's unity of type was predicated on sameness, a prelude to today's concept of *Baupläne*,[16] which acknowledged irreconcilable differences among organisms in different branches of life. Geoffroy's homology featured similarity in the relationship of structures that nevertheless appeared different from each other. Darwin, thus, speculated on both samenesses and differences, but he drew upon yet two other sources of inspiration that inflated the sameness of ancestors and the difference of descendants.

The embryologist Karl Ernst von Baer (1792–1876) greatly admired by Darwin, had earlier proposed[17] that, in the course of development, embryos moved from categories of the general to the specific, of homogeneity to heterogeneity, of the simple to the complex. Darwin, likewise, proposed that evolution proceeded by selecting specificity, heterogeneity and complexity above generality, homogeneity and simplicity. Finally, although he vigorously denied any such debt,[18] Darwin borrowed from Jean-Baptiste Pierre Antoine de Monet, otherwise known as Chevalier de Lamarck (1744–1829), the notion that life progresses via evolution.

The great, if unwitting, iconoclast, Lamarck promoted the concept that living things possessed an inherent push toward perfection quite independent of God and of the laws of physics and chemistry.[19] His view of transmutation presumed upon this drive to deliver ever more perfect organisms, fitting even more perfectly with their environments. Darwin, attempted to distance himself from such a drive, but failed wherever he came upon adaptation.

Evolution's need for a materialist foundation was supposed to be supplied from below. Félix Pouchet, in France, and H. Charlton Bastian, in England, found in spontaneous generation a constant source of new life ready to evolve. Louis Pasteur quickly shot down spontaneous generation,[20] however, and to the victor (along with Robert Koch), was left the spoil, the germ theory of disease. Life's simplest units took on new complexity, but not the ability to power evolution.

In the late 19th century, Heinrich Ernst Phillip Haeckel (1834–1919) recast the unity of life as a new form of Monism or Monadism. It had, and continues to have enormous influence on evolutionary theory. As a modern Prussian, he adamantly opposed any form of theism or holy intervention, but he also opposed deism or rule by arbitrary laws, even those of science. Haeckel was a master of rhetoric, but his acute prose style proved to be a double-edged sword, and his unitary principle became an extension of vitalism, equally transcendental and as rigid as any theology. The consequences of monadism extended well beyond anything even Haeckel could have imagined:[21] all the way to modern cladistics and its insistence on simple branching and monophyletic origins of all life's forms.

The idea of life's unity in the 20th century rose on the star of evolution, but the trajectory was not smooth. Prominent biologists in the early 20th century were impatient to explain evolution with the help of newly emerging genetics and hoped to find hereditary changes that would create new species in a single blow. Hugo De Vries's (1845–1935) discovery of "mutations" in plants seemed to provide exactly what was necessary (and desired), although the peculiarities of strains among his evening primroses soon discredited "species by mutation" scenarios.[22] Morgan also advanced a concept of evolution by mutation, but most of the mutations he accumulated in the "fly room," appeared to be deleterious, and hardly the instruments of adaptation.[23] Darwinian natural selection required the gradual acquisition of small, quantitative, beneficial variations and could not be brought into line with either sort of qualitative mutations.

Ultimately, Johannsen's term, "gene," would mediate the dispute if not resolve it. The gene became a useful instrument for explaining evolution, but genes had to be wrenched into several contortions (if not distortions) and redefined in order to make them compatible with data. First, the "gene" had to be modified to include the small effects required by Darwinian evolution. This was accomplished in 1909 when H. Nilsson-Ehle, among others, showed that some genes (distributed as Mendelian factors in breeding experiments) influenced quantitative traits and contributed stepwise, additive effects to the appearance of individuals. Geneticists then generalized upon these genes and invented polygenes (also called multiple factors), or genes with equal, additive, quantitative effects. The problem of reducing the "steps" into a smooth curve was solved by adopting Johannsen's distinction between the collection of genes in an individual (which he called the genotype), that was independent of the environment, and the collection of traits exhibited by the individual (which he called phenotype), that was influenced by the environment as well as by the genotype. Thus, the environment could theoretically smooth out the stepwise effects of polygenes, and polygenes, acting like any good gene, could undergo mutation and sort out in sexual reproduction. Henceforth, the study of gradual, quantitative change would be based on competition predicated on mechanism of violent departures (mutation) and built on the study of qualitative heredity (genetics). Ultimately, genotypic changes during evolution would be accomplished with qualitative genes acting as surrogates for quantitative genes, and a theory for the modification of species would be modeled after the inheritance of unchanged genes.

The genetics of evolutionary theory is known as neoDarwinism (also the synthetic theory of evolution or populational microevolution).[24] The Merlins whose magic transformed Darwinism into neoDarwinism were Theodosius Dobzhansky (1900–1975), Ronald Aylmer Fisher (1890–1962), J(ohn) B(urdon) S(anderson) Haldane (1892–1964), Ernst Walter Mayr (b. 1904), George Ledyard Stebbins (1906–1993) and Sewall Wright (1889–1988). Fisher and Haldane demonstrated with mathematical rigor that selection was much more powerful than mutation for driving evolution (and provided evolution with an alternative to necessity as well as to purposeful design). The founders were unanimous on the importance of natural selection, but they differed regarding the sort of accidents that provided the sources of variation upon which natural selection might act. Stebbins advanced the cause of polyploidization events in plant evolution, and Mayr saw evolutionary consequences pouring into populations through genetic bottlenecks. Wright, the dean of American population geneticists, bequeathed to evolutionary theory two lasting concepts: the concept of "drift" or the effect on gene frequency of small population size, and the "adaptive landscape" that was stochastic to the core while honeycombed with purposeful hill climbs. Dobzhansky, who published, *Genetics and the Origin of Species* in 1937, smoothed out differences and made neoDarwinism sensible for the average biologist.[25]

Of course, not everyone was satisfied. Ernst Mayr[26] saw reaction to neoDarwinism as pitting biologists in two camps, the naturalists and geneticists. He, of course, led the naturalists who were (and are) willing to admit to a creative side to natural selection, an ability to improve the fit between a species's organisms and their environment. The geneticists, on the other hand, led by Morgan and his followers, reduced by their own reductionism to the hereditary operation of genes, saw (and see) natural selection as a warmed over Lamarckism and insisted that heredity was altered only by chance mutations.

In the détente that descended, the drive to perfection as such was officially disqualified as a principle of evolutionary theory, but many biologists continued to wonder how organisms scaled the evolutionary landscape toward ever higher peaks of adaptation. Life's units would seem to be perched on every peak with no one able to say how they got there. Proteinaceous enzymes, for example, would seem to be exquisitely precise and efficient at their catalytic tasks. One is not permitted to attribute intention or directed guidance to their evolution, but each amino acid in a polypeptide sequence is thought to have been pushed over a "protein sequence space"[27] into its place in an extant enzyme. "Hence, the process of enzyme evolution is thought to be hill climbing on the landscape by adaptive walks using mutation and selection."[28]

Generation: The Mechanism of Sameness

In less than thirty years after the "rediscovery of Mendel's laws," Mendelian factors had gone from variables in algebraic formulas to mutations, to "beads" on a chromosomal "string"[29] to "genes."[30] At that point, biology bogged down and vacillated between different concepts of genes: Were they ferments, enzymes, proteins? The mythical complexity of protein was well suited to a self-perpetuating and self-actuating gene, but this same complexity frustrated efforts to unravel the physical-chemistry of genes. Then, pressed by more pressing concerns coincident with World War II, those interested in genes turned to infectious bacteria where the results of some obscure experiments suggested that inheritance was conveyed by deoxyribonucleic acid (DNA).

By the end of World War II, the climate had changed and DNA had become an acceptable candidate for a hereditary material. James Dewey Watson (b. 1928) and Francis Harry Compton Crick (b. 1916) then entered the picture, setting out to win a Nobel Prize[31] by solving the riddle of heredity. Their premise was that heredity was basically a structural problem (how molecules made copies of themselves). Thus, by figuring out the structure of DNA (called solving the structure of DNA in the argot of crystallography) they would be led, ipso facto, to how DNA made copies of itself. The tale of how they obtained data and insights is now part of biology's folklore.[32] Suffice it to say that in the Spring of 1953, Watson and Crick discovered that DNA consisted of antiparallel strands of nucleic acid wound in a double helix with the portions of the molecules known as nitrogenous bases facing inward and matched in complementary pairs: for every adenine (A) in one strand there was a thymine (T) in the other strand, and for every cytosine (C) in one strand there was a guanine (G) in the other. In the words of Watson and Crick, and the most "coy"[33] sentence to be found in the biological literature: "It has not escaped our notice that the specific pairing we have postulated immediately suggests a possible copying mechanism for the genetic material."[34]

Watson and Crick's double helix and their concept of complementary, or canonical base pairing (also known as Watson/Crick pairs) pointed biology toward its long sought-after explanation for heredity: DNA could act as a "copying mechanism" in molecular reproduction. According to the concept of "replication," or DNA dependent DNA synthesis, the double helix opens up, providing access to the complementary nitrogenous bases of free monomers (called nucleotide precursors); these line up against the original strands; following "zippering up" by enzymes, exactly replicated, double helices are synthesized containing one "template" strand from the old molecule and one new strand of former monomers. Since DNA was equated to genes, by this time, DNA replication became equated with gene duplication.

Watson and Crick were true wizards when they transmuted DNA's physical-chemistry into the duplication of genes.[35] The twosome did not stop there, however. The structure they discovered also suggested how DNA could perform other activities of genes: how it could determine protein structure, mutate, and remain immune from directed environmental influences. This emerging "DNA paradigm," offered the ultimate explanation for all of life's properties. As for heredity, everything that an organism is (or a species for that matter) must be encoded in its DNA. This encoded information was replicated from one generation to the next through a perfect match of complementary base pairs in identical double helices, and transcribed to the instructions for making an individual through the matching of equally complementary but slightly different base pairs in ribonucleic acid (RNA). As for evolution, everything a species (or an organism for that matter) can become must be encoded in its DNA. But it is in generation, the production of new organisms, that the DNA paradigm reached its apotheosis.

Just five years after elucidating the structure of DNA, Crick[36] promulgated his sequence theory, that hereditary information flowed from DNA to RNA and from RNA to protein. All else is commentary. The cell's proteinaceous enzymes, the machinery used by cells for making whatever else they required, were determined by DNA. Once the base was in place (DNA), the superstructure (the organism) was a foregone conclusion. Ever since Crick delivered his Bull, all problems of generation found their way to Crick's sequence theory. Humanity could henceforth provide everything for the health and well-being of organisms by figuring out the specifics of the sequence theory and finding ways of compensating when the sequence somehow got off track, whether developmentally or physiologically. Indeed, the late Peter Medawar, Nobelist and molecular biologist, would not deign to argue "with anyone so obtuse as not to realize that this complex of discoveries is the greatest achievement of science in the twentieth century."[37]

The triumph of DNA was so complete that it not only ended the debate over the identity of the gene but changed irreversibly the gene's place in life. No longer would it be merely the unit of heredity; from now on, it would be the starting place for all features of life. DNA utterly monopolized the imagination of biologists on all questions of heredity, evolution, generation, regulation and maintenance, and sensitivity and movement in living things, etc. Biology moved away from its traditional concerns with structure and function toward concerns with information and information processing and, at the same time, away from an analysis of life to the control of life. Under the guidance of physicists- and chemists-turned biologists with an attachment to bacteriophage, of politicians with a need to please voters, and entrepreneurs with ambitions for new product lines, the study of DNA changed from a branch of structural biochemistry into molecular biology, the premier science of the late 20th century. An esoteric problem of basic biology inspired an industry, and biotechnology created more jobs for biologists than had existed in the entire history of biology. Bioengineering also generated more profits than even Watson and Crick had foreseen in their wildest imaginings.

Critique of Samenesses that Unite Life

Today, the replication and transmission of DNA through cells is widely considered equivalent to heredity itself, and sequences of nucleotides in an organism's DNA are frequently equated with the organism's legacy to future generations. Even human beings do not escape this extreme reductionism, and the Human Genome[38] Project was sold to governments worldwide as the ultimate way to document knowledge about human beings. Moreover, the potential of DNA to determine RNA (and, hence, the polypeptides of protein [see Chapter 2]) has allowed the molecular biologists to proclaim DNA the underlying principle behind the unity of life. Responsibility for any similarity among macromolecules in living things is laid on DNA's doorstep,[39] and virtually every chemical or physical parameter of molecules is drawn into the DNA doctrine.

Inevitably, problems appear when claims are as sweeping as those made for DNA. The flip side of attributing differences in DNA sequences to change is attributing samenesses in DNA sequences to inertia. What evidence suggests that similarities among sequences of nucleotides in DNA are due to a failure of the macromolecule to change in the course of time? No evidence whatsoever may be offered to justify the molecular biologists' assertion of biological inertia; no data rule out the possibility that the similarities among the macromolecules are not themselves products of evolution; hardly any mainstream molecular biologist has even considered the possibility that life had many origins[40] and that similarities among macromolecules might have resulted from the merging of living forms and the sharing of genes.

The cornerstone of claims for DNA's role in life's unity is replication, but replication has not turned out to be the simple, physical-chemical problem it seemed on first blush, and mysteries remain. First of all, any nitrogenous base can pair with any other nitrogenous base via the same hydrogen bonds that bind bases in the canonical, Watson/Crick complementary pairs: adenine–thymine (A–T); cytosine–guanine (G–C). Second, nitrogenous bases do not bond together automatically in solution. Although some pairings (G–C) between the bases in nucleic acids form more hydrogen bonds than others, and are thus more stable, if left to their own devices, the nitrogenous bases of monomeric nucleotides pair as a function of concentration and not necessarily in the prescribed formulas (A–T, G–C). As it turns out, the formation of Watson/Crick base pairs during replication is an enzymatically mediated and regulated process much like other processes in living things.

Replication is performed by a monstrously large enzyme complex or holoenzyme that performs several activities as it moves over the double

DNA helix. Enzymes in the complex include topoisomerases which change the surface configuration of the helix, helicases that unwind the helix while causing supercoiling elsewhere, swivel enzymes or gyrases that relieve stress produced by unwinding and relax the strands, and stabilizing proteins or single stranded binding proteins that protect the single stranded DNA from enzymatic degradation.

This massive involvement of proteins in replication completely undermines claims for DNA's independence during replication. Of course, replication is important or the cell would not go to so much trouble to do the job correctly, but replication is not the automatic processes it was once thought (hoped?) to be. Furthermore, the process of replication is not only qualitative: the cell places a premium on replicating its content of DNA quantitatively as well. Indeed, when it comes to the ends of chromosomes, the telomeres, the cell employs enzymes, telomerases, that restore any loss possibly suffered by these ends during replication. Inasmuch as 95% of the mammalian genome is not genetic, in the usual sense (concerned with the transcription of DNA into RNA and hence into polypeptides), one would think that molecular biologists might pay more attention to the quantitative aspects of replication than merely to its qualitative aspects. The question remains open: "Why does the cell need so much DNA and why does it replicate it quantitatively?"

No matter how well buttressed and fortified by molecular similarity, belief in the unity of life is based on nothing stronger than consistency, and "consistency is no proof." Molecular biologists even defy the Popperian hypothetical-deductive method of science, and drive viruses and bacteriophage as well as viroids and prions[41] from the pale of life's demesne rather than admit that their uniqueness negates the unity of life. The dissemination and uncritical acceptance of life's unity by contemporary biologists requires a "suspension of disbelief" if not an act of faith.

Differences that Separate Life from Nonlife

Many a warhorse has vaunted differences between life and nonlife over the centuries only to fall by the wayside under the crushing burden of relentless discovery. With the rise of the DNA paradigm, heredity, evolution and generation joined the ranks of fallen comrades, but enough has already been said about these features of life. Regulation, or the ability to return to a prior state from a disturbed state, and purposefulness, revealed by sensation and movements, remain to be discussed here.

Regulation

Regulation, or the ability to maintain some sort of steady state in an adult, has been acknowledged as a principle of life since Aristotle (despite recognition of deterioration on one hand and rejuvenation through birth on the other). Geological uniformitarianism in the 19th century, or the idea that "what goes up (somewhere) comes down (if only somewhere else)" did not threaten life's monopoly on regulation. In the hands of the profoundly religious (Sir) Charles Lyell (1795–1875), uniformitarianism was brought into line with life: It was deprived of cataclysmic convulsions, and geological changes were seen to result from more gentle forces yet in operation, if given enough time. This "kinder, gentler" uniformitarianism handed down from a benevolent God reinforced nonmaterialist regulation in living things, and, by the late 19th century, Claude Bernard (1813–1878)[42] and his French colleagues, elevated regulation to the top of life's purposes. At least in advanced animals, everything happening in an organism's internal environment led to stability despite changes in the organism's external environment.

The issue surrounding regulation was not so much that it occurred but what determined it. Throughout the 18th century and well into the 19th, regulation was attributed to some form of biological memory, the individual somehow remembered unconsciously what was "good" for it. In any individual, part of that memory was acquired during a lifetime, but a greater part was inherited. The concepts of "use and disuse" and the "inheritance of acquired characteristics," chiefly, associated with Lamarck, were considered intuitively obvious, common sense, and uncontroversial. Whatever affected the biological memory of the individual could also be passed on to the biological memory of the individual's offspring.

Lamarck never fleshed out memory, and his ideas on the subject did not resemble those of feedback mechanisms in engineering and physics with a materialist bent. Lamarck's First Law of "use and disuse" applied to changes induced during the lifetime of an animal: "Every new need, necessitating new activities for its satisfaction, requires the animal ... to make use of entirely new parts, to which the needs have imperceptibly given birth."[43] "Use" was firmly grounded in observation, exemplified by the blacksmith's biceps, enlarged as a consequence of repeatedly wielding a heavy hammer against an anvil. "Disuse" was grounded in analogy rather than observations. Lamarck drew his concept of "disuse" from reversals accompanying aging (although he attempted to exclude "senility"[44]) and seems to have anticipated a concept of vestigial organs: "[T]he result will be the use of some one part in preference to some other part, and in some cases the total disuse of some part no longer necessary."[45]

His Second Law,[46] "inheritance of acquired characteristics" referred to a permanent change in biological memory induced by "use and disuse." The law required that the environmental challenge originally inducing a structural change would persist or be repeated for several generations, but Lamarck does not seem to have had any notion of selection. Ultimately, structures created by the necessity of meeting environmental challenges for several generation were to become fixed even in the event of a relaxation of the relevant environmental pressure. Transformations, or habitual changes in animals, thus, were induced by changes in environments.

Lamarck's theory of transformation was offered as a solution to the problem of apparent breaks in the continuity of the Great Chain of Being. "[N]ature passes from one system to another without a break, if they are closely allied; it is indeed by this faculty that she succeeded in fashioning them all in turn, passing from the simplest to the most complex."[47] Breaks in the Chain occur because "*Progress in complexity of organisation exhibits anomalies here and there in the general series of animals, due to the influence of environment and of acquired habits*."[48] Taken together, "use and disuse" and the "inheritance of acquired characteristics" explained how organisms on one link in the Great Chain of Being could be so utterly transformed in just a few generation as to appear to break the Chain.

Today, "use and disuse" and the "inheritance of acquired characteristics" are unfashionable, but they are not totally absent as a primer for current concepts of biology.[49] For example isozymes (alternate versions of enzymes often produced by cells under different physiological functions; also allozymes in different species) have a Lamarckian flavor, and some theories on the control of "control genes" (genes whose products influence the activities of other genes), are distinctly Lamarckian.

Neither alternative enzymes nor control genes undermine the molecular biologists faith in DNA as the root of all causes, however. Genes may control one process at a time, but organisms are not made one process at a time nor do they operate that way. The antidote to neoLamarckism, therefore, is to conceive of genes's role in more sweeping processes than those performed merely one at a time. In the end, only DNA provides biologists with "a party line. For the first time, too, biologists were in possession of a chemical structure that said something about function."[50]

Purposeful Motion and Sensation

Once upon a time, the difference between a living thing and a rock was simply that a living thing could kick a rock, but a rock could not kick a living thing. Furthermore, the living thing could choose to kick the idle rock and would sense the insentient rock at the moment of impact. Renaissance science challenged these distinctions. In the first place, the law of equal and opposite reactions meant that the rock kicked back every bit as hard as the living thing kicked it. Action was one consequence of the application of force; reaction was another. Thus, movement within living things came under renewed scrutiny as a subject for physical explanation, and a compromise had to be found between the laws of nature and the laws of life.

Of all the accomplishments of the 17th century mechanists, one of the most far-reaching was William Harvey's (1578–1657) inference that blood flowed through the heart, a flow that was further elucidated when Jan Swammerdam (1637–1680) described the movement of blood corpuscles in capillaries (although not published until 1737, more than a half-century after his death), thus completing the loop of circulation. René Descartes (1596–1650) would have been delighted with these physical explanations for motion in living things. Later, in 1827, while studying nuclei in plant cells, Robert Brown (1773–1858) discovered "Brownian motion," the random movement of minute, suspended particles caused by contingency and the kinetic energy of molecules.

Even if physical law explained motion in living and nonliving things, the same could not be said about the sources of motion. The difference between living and nonliving things was that motion in nonliving things was external (conditioned by local circumstance) and followed the application of external force, while motion in living things was internal and self-generated or inherent. In the halcyon days of teleology, in the 19th century, the explanation for the movement of living things was well-being and survival. Organisms moved, even if no more than to raise their leaves to the source of nourishment. Threats from without, likewise, produced escape movements. Threats from within produced regulatory movements.

Descartes would have accepted purposeful behavior and given a nod of approval to the regulation of a separate internal environment (as he did to the separation of mind and body). Life exhibited purpose, epitomized by sentience and climaxing in sapience; nonlife did not. Harvey would have agreed, since he demoted the heart to the status of a warehouse for blood upon discovering that the heart had no feeling.[51] Not until the 20th century would sentience come under the gun of physical laws, and then only secondarily as an extension of genetics to mental capacities and molecularization of the neural sciences.

Today, an endless stream of genes flows from the backwaters of purposeful motion and sensation to the pulsating sea of biotechnology. Genes are now widely acknowledged to be the root causes of alcoholism, aggressiveness, manic depression, schizophrenia, even if little pans out in practice.[52] Attempts continue to correlate dyslexia or reading disability, attention-deficit hyperactivity disorder (ADHD) and autism with chromosomal markers linked to autoimmune disorders and genes (i.e., "quantitative trait loci") in the human leukocyte antigen complex and in the major histocompatibility complex of chromosomes[53] Other studies go beyond pathology to link patterns of human behavior to genes, for example, associating a region of the X chromosome with a gene for male homosexuality.[54] The situation is, however, quite messy, since conclusions evaporate as more evidence accumulates and subjects in one category of behavior are moved to another category after the original publication.

Critique of Differences that Separate Life from Nonlife

The idea of features of life that distinguish it from nonlife is philosophically flawed inasmuch as life's exclusive features are derived tautologically. Even well defined concepts such as regulation, motion and sensation are nothing more than self-reinforcing tautologies. Moreover, ultimately, all the differences of living things that separate them from nonliving things fade at the point where life came from nonlife — whether from "ashes" or abiogenesis — and rudimentary living forms emerged from the physics and chemistry of the early-Earth atmosphere.

Saying that life came from nonlife does not, of course, make it happen, and theories on life's origins broadly fail to lay out a convincing scenario.[55] The assumptions of scientists doing research on early-Earth seem to beg the question of life's origin rather than answer it. Habits of thought, based on life as it is seen or experienced on Earth today, are not likely to give any hint of what went on before there was a recognizable form of life. Prevailing views of life, such as ideas that separate it from nonlife, are probably obstacles to thinking about prebiotic and protobiotic chemistry. Instead of "primitive" life, one must try conceiving of unincorporated processes and materials, however unrecognizable in terms of life. Instead of life's unity, one might imagine independent lives, with separate origins, cooperating instead of competing; mixing instead of separating. Instead of progress, early life may have been chaotic; instead of purposeful, accidental; instead of passing on its achievement, melting difference; instead of reproducing, merging; etc.

Viruses and phages, episomes, transposomes, prions and mycoplasmas, considered nonliving in the formalism of life's unity, share samenesses with living things. The separation of life from nonlife by difference, coupled to a chauvinism about human life, has done a disservice to the search for biological facts. Only a minute ecologically restricted subset of viruses and phage is known at all, while the greatest mass of viruses may infect marine phytoplankton, and a similarly great mass of unknown phage probably infect the free-living bacteria, especially those residing deep in the earth's crust. Indeed, RNA viruses are ubiquitous and the single-stranded RNA viruses are probably the most abundant parasites on Earth.[56]

Conceptions of life based on differences with a nonliving world also do disservice to the search for a unifying theory of life. Even elementary thermodynamic principles inform us that life is inseparable from its abiotic environment, but biologists routinely portray life as an isolated system, as separate from nonlife as black and white. The distortion spreads to all levels of complexity: at the physiological level where individuals exchange oxygen and carbon dioxide with their environment; at the biogeological level where oxygen and carbon dioxide achieve quasi-steady-state kinetics.[57] Indeed, erasing the present, static charcoal sketch of differences between life and nonlife might reveal an underlying portrait of life in fully saturated colors.

Finally, molecular biologists embarrass themselves by disparaging Lamarckian memory, in all its various incarnations, because it does not live up to Watson/Crick reductionism. Fabricated machines are, after all, designed with built-in feedback mechanisms because engineers have some precise ideas of how the machine is supposed to run. Organisms too exhibit regulation, but arguments of design are forbidden in the forum of molecular materialism. One might rule out arguments of transcendent design, but what is material here is the demystification of design itself.[58] For example, the most pedestrian theory of evolution proposes that environments select organisms. If particular environments recur, they may have an impact on life in the present or past. If, thus, the future designs life, the final state to which an organism's performance is regulated is not only set in advance, but living things exhibit Lamarckian memory!

LIFE'S UNITS

In the 19th century, the unit of life was personified by the cell. At the dawn of the 20th century, it was epitomized by the gene, and today it is embodied by the gene's molecular idealization, deoxyribonucleic acid, alias DNA. The atomization of life had a long and profound history, traceable to Lucretius. In the 19th century, the heyday of the cell theory, the organism was seen as a profoundly complex class in which fundamentally similar, but differentiated cells took charge of the overall activities constituting life in the organism. The waxing of DNA brought about the waning of cells, and today cells are seen as normally more controlled than controlling.

Today, mainstream biologists assume that life will ultimately be explained by engineering and informational principles operating within and among the parts of individuals. In the current, reductive (analytic) atmosphere, genes of DNA are widely thought of as the unit of life, the fundamentally differentiated part of living computing machines whose functions and interactions result in life. Life is thus reduced to DNA's functions beginning with replication = reproduction.

These attractive concepts are not entirely satisfying, however, since they elide concepts of individual organisms and classes of cells and genes, turning individuals into a class of repeating machines, while cells and genes are turned into differentiated parts of organisms. The apportionment of life to individuals is not that easily reconciled with the apportionment of life to parts, and problems arise from trying to do so. Above all, individuals only function when all their parts are present and in working order, whereas a class is more open than an individual and can exist, function and be made by any quorum of its membership. Since existence and function presume the operation of an individual at some level, the question of how life is apportioned can be answered by looking at how life is made.

Organisms

Aristotle (384–322 B.C.E.) had no difficulty with the idea of life apportioned to individuals, but he had a great deal of difficulty accounting for the creation of their parts. According to Aristotle,[59] life developed in individuals. Their parts were never preformed entities continuing from one generation to the next. In his tripartite division of animals, oviparous animals reproduced by laying eggs, and ovoviviparous animals reproduced by retaining eggs in the uterus; but viviparous animals (such as ourselves) reproduced by giving birth (i.e., without eggs whatsoever). His trail of mammalian generation was soaked in blood, a theory he borrowed from Indian healers and Hippocrates. Menstrual flow was supposed to supply the soil in which the male planted the seed of generation. "Blood relatives" were thus, related through the mother's flow, but this relationship was superficial. After all, the menses of one month had no relationship to that of another month, and offspring were thus, created *de novo*.

In the second century, Galen of Pergamum (131–201) picked up on the suggestion of the Alexandrian physician, Herophilus of Chalcedon, and hinted at a role for ovaries (testicles or diminutive testes), suggesting that the uterus attracted a female semen on a par with male semen which met in the uterus thereby creating a conceptus (Lat. *concipere*, to receive).[60] This concept of a mammalian egg did not acquire legitimacy, however, until the Renaissance. Andreas Vesalius (1523–1562) of Brussels, the first anatomists in the great succession at Padua, illustrated follicles and luteal glands. His student, Gabriello Fallopio (1523–1562) later described the uterine tubes bearing his name, and his student, Hieronymus Fabricius ab Aquapendente (1537–1619), drew further attention to the ovary, although he did not suggest that embryos came from eggs.[61] Finally, his student, William Harvey (1578–1657; the same Harvey of blood) declared in 1651 that reproduction was unified by eggs:

> We, however, maintain (and shall take care to show that it is so) that all animals whatsoever, even the viviparous, and man himself not excepted, are produced from ova; that the first conception, from which the foetus proceeds in all, is an ovum of one description or another, as well as the seeds of all kinds of plants.[62]

Harvey called the production of animals from eggs and plants from seeds "epigenesis,"[63] and imagined the origin of an individual's parts by unfolding and progressive change in form.

Epigenesis: Beginning with Sameness

Harvey's contribution to embryology may be somewhat exaggerated by historians. His priority to the egg is questionable, since Volcher Coiter (1534–1576) had already suggested that eggs arose in the ovary and embryos formed in the egg (a suggestion for which he is belatedly recognized as the parent of embryology[64]). What is more, Harvey equated the mammalian uterus to the shell of the hen's egg. and he was totally confused by fertilization. As for chickens, he suggested that male semen (liquid seed) performed its role in reproduction by being absorbed through the walls of the hen's cloaca (vagina, in the case of mammals), becoming active as it moved through the female's bloodstream.[65] Harvey's reference to blood as a vehicle for the conveyance of the male element to the egg cast "blood relatives" in a new mold, quite different from Aristotle's. Furthermore, Harvey did not write the phrase for which he is given credit, "Ex ovo omnia." It was the work of the unknown engraver of the frontispiece of

Generatione Animalium.[66] What Harvey contributed, however, was the conclusion that reproduction proceeded from the development of an egg, rather than the individual as a whole who would be the parent of the offspring. That egg would ultimately be identified as a cell, but first it had to be characterized correctly as an egg.

The Mammalian Egg

The search for the mammalian egg returned to the ovary if only because the eggs of oviparous species (the hen egg) came from ovaries, not the uterus (where the hen egg received its shell). Reijnier de Graaf (1641–1673) drew attention to the mature mammalian follicle, now bearing his name, and considered it the mammalian egg itself. He had good reason: De Graaf had observed ovulation following coitus in the rabbit, and he observed tiny spherical bodies (undoubtedly blastocysts) in the oviducts of rabbits three days after mating. A day later, slightly larger bodies entered the uterus, commencing pregnancy. The Swiss physiologist, Albrecht von Haller (1708–1771), however, failed to find anything like de Graaf's cystic embryos in the uterus of artiodactyles (where they would have appeared as mucous strings), and thus, Harvey's theory of a uterine, mammalian egg continued to prevail against de Graaf's observation of an ovarian follicle.

Finally, in 1827, the Estonian born, Karl Ernst von Baer (1792–1876) reported on his efforts to trace the "vesicles of reproduction" in dogs.[67] Repeating de Graaf's methods and observation, von Baer, at first, identified the mature follicle as the mammalian egg, but taking his clue from the size of the follicle in the ovary and the "egg" in a bitch's uterus, von Baer backtracked and concluded that the "embedded vesicle" within the follicle behaved "with regard to the coming embryo, as the real egg."[68] Calling it the "fetal egg," von Baer proceeded to demonstrate its presence in mature follicles of a variety of mammals from pigs to porpoises as well as human beings. Thus he had closed the gap in reproduction left open by Aristotle: mammals had eggs.

The Male's Role

What then was the male's role in reproduction? Botanists were, once again, well ahead of zoologists in answering a question in natural philosophy. Joachim Camerarius had discovered that, following the removal of anthers, castor oil plants produced empty "seeds," incapable of germination, and that, following removal of stigmas, no seeds developed whatsoever. In

England, Nehemiah Grew (1641–1721) detailed the microscopic anatomy of hermaphroditic flowers. Discrete male and female elements, thus, were required for reproduction.

Controversy over the male's role in animals climaxed in the late 17th century. In 1677, Anton van Leeuwenhoek (1632–1723) reported the presence of animalcules or zoa in semen, later to be called spermatozoa,[69] and while the observation was confirmed by others, most anatomists dismissed the zoa as just another infusorian, a parasite of semen, immaterial to fertilization. This cavalier attitude toward Leeuwenhoek's discovery is understandable, first as a consequence of the popularity, at the time, of eggs as the sole instruments of reproduction (see below), and second as a result of unfortunate sighting of preformed creatures within the animalcules. Several late 17th century microscopists claimed to see the homunculus or miniature adult within the animalcules of semen, suggesting that the embryo got into the egg through the male element. In the 18th century, Lazaro Spallanzani (1729–1799), among others, conducted experiments on artificial insemination with filtered and unfiltered semen.. The results in frogs and even a pet bitch suggested that semen with zoa acted as an activator of generation, and even though a role for animalcules in insemination became a distinct possibility, it was considered only a remote or indirect cause.

The English anatomist, Martin Barry, was probably the first to report seeing spermatozoa inside an egg's envelope, and George Newport observed the disappearance of a spermatozoon into a frog egg, suggestive of a role in fertilization. Later, at the marine station at Roscoff, Oscar Hertwig (1849–1922) discovered that one of the two nuclei in fertilized sea urchin eggs was derived from the spermatozoon, the other from the egg. Moreover, the two nuclei in sea urchin eggs fused prior to the first cleavage division. Although this situation in sea urchins is unusual (nuclei within fertilized eggs approach each other and may juxtapose each other quite intimately, but they generally do not fuse), Hertwig had solved the riddle of fertilization: the spermatozoon triggers development when it is taken up by the egg at fertilization and contributes half of the nuclear content of the future embryo. The egg's nucleus contributes the other half.

Preformation: Beginning with Difference

The alternative to an "unfabricated" individual is a prefabricated one, and the antithesis to epigenesis is preformation. Toward the end of the 17th century, Platonic preformation awakened from centuries of slumber, and the concept of organisms enlarging from miniature versions of themselves yawned in the face of an exhausted epigenesis. The classical microscopist, Marcello Malpighi (1628–1694) was the century's foremost architect of preformation (although his foremost biographer, disputes the claim[70]). Malpighi was a committed empiricist, devoted to the proposition that observation is preferable to hypothesis, and determined to expose his methods and results to criticism from the broadest possible audience by having his reports published by the Royal Society in London, but he also believed that development was attributed to preexisting structures in an already differentiated embryo. He also believed in the Holy Roman Church and was keenly aware of the difficulty faced by the Christian position on creation, namely, that all the generations of Man had somehow to be condensed in the mother of us all, Eve.

One possibility for condensing generations was the celebrated idea of *emboîtement* or encapsulation, according to which, embryos were stacked within embryos (like Russian dolls). *Emboîtement* required the reduction of each succeeding generation. Hence, every generation began its development from increasingly miniature versions of itself. The idea was neither novel nor did it require much ingenuity to imagine how it might work: Miniaturization did not exempt working parts from function, even if "the law [of nature] does not apply to the very small."[71] Furthermore, in the era of the first microscopes, one was hardly compelled to believe that the lower limits of visibility were necessarily the lower limits of life. Organisms stored within organisms since the time of Creation might also be invisible.

Malpighi set out to test *emboîtement* observationally. He chose hen eggs for his "system" and examined developing chicks with the aid of the newly invented microscope (a bead of glass capable of magnifying illuminated objects brought close to it). Thorough and accurate, his observation (published initially in 1672 with illustrations added in 1675)[72] were the most complete of his time, and in every case he found miniature chickens growing on the yolk — even in unincubated eggs! He never found a fertile egg that did not have recognizable chicken parts in miniature.

Malpighi's confidence in his own observations was thoroughly justified. Arguments by others over temperature were not as impressive as what he had seen himself; the failure of others to see tiny embryos in eggs was hardly more convincing than evidence of his own eyes. Furthermore, the results of "lifeless" experiments with temperature were not worthy of consideration. Preformation was by far the best explanation for the positive information available (even if contrary to the authority of Aristotle).

Determinism Versus Indeterminism

Preformation was the dominant, biological ideology of 18th century natural philosophy, and Charles Bonnet (1720–1793) of Geneva[73] was its undisputed master (with apologies to Albrecht von Haller). Preformation was not what it started out as, however, since it had begun to concede physical preformation to a sort of representation resembling contemporary ideas of encoded information.[74] Preformationism, thus, had matured into a form of determinism in which differences continued to be present at the beginning of development but these differences were no longer equivalent to the thing in itself.

Epigenesis also matured, if only at the close of the 19th century. The impetus came from Hans Driesch[75] (1867–1941) when he succeeded in producing complete, miniature larvae from the separated blastomeres of early sea urchin embryos. The results showed that neither the fertilized egg nor the blastomeres were predetermined. Moreover, because each blastomere would ordinarily have produced only part of a larva, the blastomeres must interact to produce an embryo as a whole in the undisturbed situation. The fertilized sea urchin egg and its early blastomeres, thus, were indeterminate, waiting upon and relying on signals that gave them directions for later development.

Possibly the most peculiar aspect of determinism and indeterminism is that the proponents for one side of the controversy frequently produced results that supported the other side. In the 19th century, Wilhelm Roux (1850–1924) was a determined determinist who nevertheless aided the cause of indeterminism. He is best known for his demonstration that when one of the first two blastomeres of a frog egg is killed, the other blastomere develops into only half an embryo. Development is thus predetermined. Roux also reported, however, that half embryos may regenerate missing structures, providing ammunition for the guns of epigenesists. Similarly, Hans Spemann, who demonstrated indeterminacy through a variety of isolation and transplantation experiments in amphibian embryos,[76] also showed that some influences (those of the so-called gray crescent) were embedded in embryonic tissue and were thus predetermined.

Mammalian embryos seem to play tic-tac-toe with determinism and indeterminism. On the one hand, the general failure of isolated blastomeres at the eight-cell stage to give rise to embryos suggests that the determinants of development are sufficiently restricted at that time to limit development.[77] On the other hand, the successful cloning of an adult sheep breast cell via transfer of nuclei to an enucleated ovine egg[78] suggests that the full spectrum of developmental genes are present in adult-cell nuclei.

The ultimate question posed by determinism is whether it may be reversible and, therefore, indeterminant. Naked cambium cells of a variety of adult plants (from banana and carrot to sunflower and tobacco) are capable, under the right culture conditions, of giving rise to complete adult plants demonstrating that differentiation of cambium cells is reversible. Likewise, a variety of embryonal carcinoma cells of mice, originating in adult testes, are capable of seeding mouse embryos and differentiating into many types of adult cells, including gametes. On the other hand, in Amphibia, the proportion of embryonic nuclei supporting complete development upon transfer to enucleated eggs diminishes with the age of the donor embryo and drops off precipitously following gastrulation. With all the effort and repetitions that have gone into these experiments, it is hard not to conclude that at least in some cells, irreversible nuclear differentiation accompanies early embryonic development.[79] In other cells, it would seem, nuclear differentiation, if it occurs, is reversible

Cells

In the Renaissance, as Galileo turned his lenses upward toward the heavens, the classical microscopists turned their lenses downward toward life's unseen boundaries. Robert Hooke (1635–1703) and Nehemiah Grew looked at botanical material, and the Bolognese, Marcello Malpighi wrote his greatest work on the microscopic anatomy of developing chicks, although he published the historic illustrations as a brief addendum to a work on plants. Virtually everything that could be held in suspension went before the microscopist's lens, and everything imaginable was suspended. Anton van Leeuwenhoek of Delft, for one, looked at "infusions" (samples of water which, with their contents, were left standing prior to examination) and discovered "animalcules." He pushed resolution to (or beyond?) theoretical limits and discovered bacteria, and yeast as well as the animalcules of semen.

The idea of cells of one form or another was tossed about in the 18th century but largely as a clay pigeon to be shot down by professional nature philosophers. Cells were entirely too unformed for the taste of Charles Bonnet and the dominant school of preformationists. Support for preformationism came from Bonnet's detailed observations on parthenogenesis in aphids, while the epigenetic cell languished in a botanical model of vessels capable only of moving fluid. This model was transported to zoological material by Caspar Friedrich Wolff (1738–1794), among others, however, and their search for the origins of vessels led to a description of "globules" and "granules" that would later be recognized as cells (although they were probably nuclei) present in embryonic "leaflets" (i.e., today's germ layers).[80]

The 18th century's cell bore little similarity to today's cell. According to the testimony of the parent of biology, Lamarck, cells formed the membranes and investments that contained the moving fluid, hence the essence of life. Cells thus resembled an updated version of Aristotle's blood, "nature's method of gradually creating and developing these [membranous] organs out of this [cellular] tissue."[81] The legacy of 18th century notions of cells remains with us, for example, in the name, "parenchyma" (taken from botany), for a tissue poured into a cellular bed or stroma. Lamarck's ideas were not a point of departure but a hint of where cells would go in the ongoing struggle over how living matter was put together.

Marie Francois Xavier Bichat (1771–1802),[82] adopted the principles of atomism and mechanism from chemistry and conceived of living things constructing organs (the equivalent of molecules) from tissues (the equivalent of atoms). Increasing awareness of the complexity of microscopic life was required before the cell theory (or theories, as the case may be) could be formulated.

The first glint of cells as builders is found in the work of Lorenz Oken (born Ockenfuss, 1779–1851), preeminent German nature philosopher. By the weight of his authority as a morphologist, he succeeded in forever dividing the infusorians into animals made of "single mucous vesicles" and animals made of "multiple mucous vesicles" or "agglomerations" of vesicles.[83] Oken's insight and separation of infusorians into two classes was not the naive gesture of a disinterested party: Oken intended to inscribe these vesicles with distinctly romantic values, to incorporate a concept of progress, attendant on complexity, to legitimize "lower" and "higher" designations for organisms and, however mildly, to suggest evolution.

The issue of whether living matter was cellular was settled early in the 19th century by fiat. In 1824, Henri Dutrochet (1776–1847) asserted that living things were comprised of cellular units. Robert Brown named nuclei in 1831, and Felix Dujardin (1801–1862) characterized cytoplasm in 1835. In 1839, Evangelista Purkynjê (1782–1869) named the composite of cytoplasm and nucleus protoplasm,[84] the stuff of life, and cells were thereupon defined as separated units of protoplasm.

The 19th century "cell" was biology's great leap into reductionism, its foremost effort, until the mid-20th century, to explain properties of the larger through recourse to activities of the smaller. Theodore Schwann (1810–1882) and his friend Mathias Jakob Schleiden (1804–1881)[85] took the leap and theorized that a proper analogy was between cells and atoms and, hence, tissues and molecules. The two theoreticians proceeded to promulgate a cell theory that, regrettably, added enormously to the difficulty of understanding how cells constructed organisms. According to Schwann, the idea "that there exists one general principle for the formation of cells ... may be comprised under the term cell *theory*."[86] Schleiden and Schwann were utterly confused about how cells were constructed, suggesting that they were somehow crystallized from granular material, the nucleus preceding the cell proper, but, building on Robert Remak's (1815–1865) analysis of cleavage (cell division) in frog embryos, the revolutionary pathologist, Rudolf Virchow (1821–1902) "got it right when it counted," and pointed cytology toward mitotic division as the chief mechanism of cellular propagation. As a microbiologist, pathologist and physician profoundly concerned with hygienics, he drew together the threads of unicellular and multicellular life. Virchow, the Prussian liberal in the revolution of 1848 and later (1880–1893) member of the Reichstag, argued from an egalitarian perspective, that unicellular organisms were nomadic, while cells in multicellular organisms achieved their destiny through cooperation, through a division of labor and by disciplined differentiation.

The larger issue of whether tissues were made of cells would persist into the 20th century, with the neuroanatomists Santiago Ramón y Cajal (1852–1934) defending cells and Camillo Golgi (1844[43]–1926) carrying the fight against them to the Nobel podium.[87] The issue would not be settled until Ross Granville Harrison (1870–1959) invented tissue culture in order to see if precursors of nerve cells could actually spin out the long processes associated with nerves on their own. They did, and from that point onward, cells could do anything living organisms could do.

Cells quickly jumped into the role of individuals, and organisms became societies. The question was, "What kind?" The Prussian conservative and nationalist, Ernst Haeckel had already importuned biologists with his conviction that the multicellular masses had to be rescued from anarchy by a Kaiser capable of disciplining them in the embryo and imposing cellular dependency in the adult organism. Haeckel's success as author and lecturer was not only due to his overt politics (which he fervently advanced to a particularly receptive audience), but to his cryptic politics in the form of rationalizing change as progress. His passion was to promote a program for social change parallel to his version of biological evolution.[88]

The concept of cells changed dramatically in mid-20th century. Haeckel's cell, which had been all but ready to march off to war in defense of the Fatherland (organism), had been defeated. Moreover, the singularity of cells was fractured by speculations of Lynn Margulis,[89] among others, that the eukaryotic cell's mitochondria and chloroplasts, and possibly other organelles, were derived from bacteria. If cells of eukaryotic organisms had acquired their organelles piecemeal and not derived them from themselves, then cells were not quite the equivalent of atoms. Finally, the ascendancy of the DNA model of the gene knocked cells off their pedestal as the preeminent units of life.

Genes

The idea that life was governed by stable, controlling elements was alluded to in 17th century debates over vitalism (i.e., laws governing living matter), materialism (i.e., laws governing nonliving matter [sometimes incorporated into living matter]), and a host of related issues (epigenesis and preformation, spontaneous generation and generational continuity).[90] Concepts of control systems won the day, in the late 19th century, through the eloquence of the vitalist, Claude Bernard[91] and his French (not especially collegial) colleagues.[92] Bernard was concerned with the regulation of an organism's internal environment and saw the purpose of life's organization as maintaining constancy in this environment despite changes in the organism's external environment. With the exception of Darwin's concept of transmutation, virtually all qualities of life were thought to flow from the constancy of the *milieu interior*, at least in large animals. Today, biology is redolent with concepts based on resistance to outside forces, the most far-reaching of which concepts is the concept of the gene.

Genes reached their present, august heights in several giant steps.[93] First, August Weismann (1834–1914) brought the concept of a *milieu interior* down to cells, suggesting that a germ plasm in the cell's nucleus represented the location of life's regulating elements, while the cytoplasm represented an environment subject to regulation. Motivated by a fanatical anti-Lamarckism, Weismann[94] portrayed the germ plasm as passing from cell to cell and from generation to generation quite immune to externally controlling influences. At the same time, however, the germ plasm controlled everything about the organism including development and the maintenance of the organism's internal environment. Germ plasm was both self-perpetuating (through cell division and from generation to generation) and capable of guiding all life's processes: from a cell's differentiation to an organism's life expectancy. Germ plasm was transmitted to new generations through the so-called, "germ line" of eggs and sperm,[95] while a trophoplasm, consisting of one or another part of the germ plasm, was introduced into the so-called "somatic line" of ordinary body cells accompanying or following cell division. Cells of the somatic line were doomed to develop and die at the behest of trophoplasm, while cells of the germ linc escaped the "mortal coil."

Weismann's celebrity is due to his having argued passionately and repeatedly that the germ plasm (nucleus) influenced trophoplasm (cytoplasm), but the latter did not influence the former. Weismann is thus credited with proposing biology's first "central dogma."[96] Sheltered by trophoplasm and immune to the environment, the germ line passed germ plasm but could not be influenced by whatever environment an individual experienced during its life time, whether positively or negatively.

Contemporary biology has been kind to Weismann despite claims against his priority to the ideas and the failure of the germ line in many organisms to be immortal, entirely separate from the somatic line, and indifferent to the environment. Moreover, evidence for cytoplasmic and maternal inheritance show that development is not limited to nuclear control. Furthermore, and even more damning, the overwhelming burden of evidence shows that in organisms such as ourselves (eukaryotic organisms) the hereditary material of the nucleus, namely nuclear DNA, does not enter the cytoplasm whatsoever. Instead, messengers of ribonucleic acid (RNA) convey information from the nucleus to the cell's cytoplasm (see below). Weismann's doctrine of an isolated germ plasm dominates modern biology, nevertheless, because, in the words of the late developmental biologist and historian Jane Oppenheimer (1911–1996), "like that of so many influential figures in the history of science, [his] was the expression of doctrine to a century philosophically ripe for its acceptance."[97]

The event that ripened the 20th century was the "rediscovery of Mendel's laws of heredity." Heralding in the new century, Mendel's idea of independent, hereditary factors seemed virtually interchangeable with Weismann's germ plasm (without trophoplasm), and, under the direction of Thomas Hunt Morgan, Mendel's laws proved a bonanza to geneticists prospecting for ways of studying heredity free of environmental influence. Ironically, Herman Joseph Muller's (1890–1967)[98] discovery (made while working with Morgan) of the ability X-rays to induce mutations had to be rationalized as random and undirected environmental influences in order to maintain the concept of an insular gene.

Trophoplasm had a much more checkerboard history than germ plasm and the gene. The issue of how development and physiology were controlled had been recast in terms of genes in 1909,[99] but then only in the midst of confusion over the gene's identity. Initially, the equation of hereditary material with enzymes (ferments) clouded the issue, and then, for decades, thick smoke rose from heated debates over the breadth of genetics.[100] On one side, Morgan and other advocates of a corpuscular gene, argued that genes were transmitted through reproduction (germ line) and were undisturbed by their passage through organisms; on the other side, although not committed to Weismann's trophoplasm (or germ plasm for that matter), William Bateson, Richard Goldschmidt (1878–1958), Charles Manning Child (1869–1954) and others advocated a more physiological view of heredity, arguing that inheritance took place through development and could not be separated from it.[101] A partial solution to the problem began to take shape with discoveries suggesting that polypeptide synthesis in cells was effected by ribonucleic acid (RNA). RNA would later bridge the gap between genes of DNA and cytoplasmic proteins, but the larger issue of inheritance beyond genes (cytoplasmic inheritance) would not be breached (and has yet to enter mainstream biology).

The history of an environmentally isolated hereditary material capable of influencing development and physiology was completed in the later half of the 20th century, when Francis Crick molecularized the germ plasm and gene, claiming in his version of the "central dogma" that DNA was stable and immune to outside influences. At the same time Crick's central dogma isolated DNA, it reiterated the role of hereditary material as the sole arbiter of hereditary information: heredity flowed outward from DNA and never inward to DNA.

Crick's central dogma paralleled his sequence theory (DNA > RNA > protein [see above]). The central dogma's further implications were also clear: since life was regulated ultimately by DNA, failures of regulation must be caused by the wrong kind of DNA and, therefore, could be corrected by substitution with the right kind of DNA. Not since Charles Darwin first suggested that life could be molded by hybridization and domestication and that evolution could be directed through differential breeding has so powerful an idea crossed the threshold of biology. Today, mainstream biologists's hope and expectation for shaping life rock in the cradle of Crick's central dogma.

DNA

Possibly the greatest scientific discoveries of 20th century were those which explained (to one degree of confidence or another) how anything as monotonously the same as DNA does all the different things attributed to nuclear genes. What is more, discoveries showing how DNA mimicked genes followed Watson's and Crick's breakthrough with explosive rapidity. The same sort of Watson/Crick pairing of nitrogenous bases in DNA, which had suggested how DNA could replicate, also suggested how DNA could determine RNA and hence influence the amino acid sequence in the polypeptide chains composing protein. Within twenty years, biologists had unraveled the mysteries of DNA replication, of DNA's transcription to RNA and RNA's translation to polypeptide.[102]

The first discovery was that RNA was not one species of molecule. A unique RNA, now called transfer RNA (tRNA) was discovered because, due to its small size, it could not be pelleted in high speed centrifuges with the bulk of the RNA. *In vitro* experiments later showed that tRNA was a mobile carrier of reactive amino acids to sites of polypeptide synthesis. There, the tRNA complexed with large, ribonucleoprotein particles (i.e., particles made of both RNA and protein) now called ribosomes (and known to consist of a separate small and large subunit). A third type of RNA was predicted by default: it had to be there if only because (1) ribosomes were too stable to account for the initiation of polypeptide synthesis following the initiation of RNA synthesis and the cessation of polypeptide synthesis in the absence of RNA synthesis; (2) tRNA was entirely too small to program the synthesis of most polypeptides. The third type of RNA was therefore predicted and quickly found. Called messenger RNA (mRNA), it carried the information from nuclear DNA to cytoplasmic ribosomes for assembling a polypeptide chain during "translation" or RNA dependent polypeptide synthesis.

A more difficult part of the scenario was discovering how RNA directed the specific type and order of amino acids in the polypeptide chains of proteins. One possibility, that the different types of RNA were synthesized in different ways was eliminated with the discovery that all the forms of RNA were synthesized on DNA templates. The process, known as "transcription," or DNA dependent RNA synthesis, utilizes the same RNA polymerase in prokaryotes to synthesize all the RNAs, while in eukaryotes, three RNA polymerases are employed.[103] Enzymatically enforced base pairing results in the formation of RNA strands complementary to the transcribed strand of DNA except that the sugar ribose is present instead of deoxyribose and the nitrogenous base uracil (U) is substituted for thymine (instead of adenine pairing with thymine [A–T], adenine pairs with uracil [A–U]; guanine continues to pair with cytosine [G–C]).

Ultimately, the specificity of translation was attributed to the transcription of the sequences of bases in a strand of DNA (the sense strand) to the complementary sequence in mRNA, but several additional processes occurred along the way. First, freshly transcribed RNAs, or transcripts, are prepared for their role in translation by "transcript processing" and for survival during transport out of the nucleus (in the case of eukaryotic cells) by "transcript modification." Processing involves small, nuclear subunits of RNA-protein particles (called snurps) in which the RNA eliminates portions of the transcript called introns that are not translated as a result. Other portions of the transcript called exons are enzymatically linked together, and most of these are translated. Modification involves capping the head of the molecule and adding a tail, the consequences of which allow the molecule to safely move into the cytoplasm where it can condense with ribosomes and tRNA preparatory to translation.

In the cytoplasm, tRNA is charged with its amino acid by unique enzymes (called aminoacylsynthetases) that recognize a specific portion of the tRNA, called the functional RNA code. At least 20 tRNAs must be recognized by an equal number of specific enzymes to account for the 20 amino acids incorporated into polypeptides, but usually more than one tRNA and enzyme condense with the same amino acid. What is remarkable is that the tRNA with a specific RNA code also has a specific group of three nitrogenous bases called an anticodon elsewhere on the molecule and, while the anticodon has no relationship to the functional RNA code, the anticodon is also specific for precisely the same amino acid, albeit in a different way. The amino acid is incorporated into a polypeptide as a function of the complementarity of its anticodon with a codon in mRNA.[104]

The ribosome is the machinery for polypeptide synthesis, providing sites where an incoming tRNA can transfer its amino acid to the amino acid at the growing end of the polypeptide chain. Since each of the amino acids incorporated into polypeptides is carried by a specific tRNA, the specificity of the amino acids in a polypeptide is a function of the particular tRNA binding its amino acid in a precise order. That order is prescribed by the sequence of nitrogenous bases in the mRNA, specifically the order of groups of three nitrogenous bases or triplets known as codons. Each codon in mRNA is complementary to the anticodon in a specific tRNA.

At the commencement of translation, a group of three nitrogenous bases in the mRNA, called the start codon (AUG), signals the beginning of a polypeptide chain with the amino acid methionine (in eukaryotes). This start codon also designates the "open reading frame," or sequence of codons in the mRNA which, in turn, dictate the specific anticodons in tRNAs for the assembly of a polypeptide. The triplets in mRNA prescribing amino acids constitute the "genetic code."

DNA thus became the ultimate unit of life. The cell and organisms were merely DNA's way of reproducing itself. Everything about DNA quickly became recast in the model of organisms, and sociobiologists, such as Richard Dawkins, could dilate endlessly on the merits of selfish DNA as opposed to altruistic DNA.

Critique of Life's Apportionment to Units

Opening the box of life's units has not been easy. The idea of the individual organism is so well entrenched in the Western concept of life that biologists hardly contemplate life without it. By and large, funding for biological research, from basic to health-related and agricultural-enhancing research, is sought under the aegis of one model of an individual or another, whether the organism, the cell, the gene or DNA.[105] Nevertheless, organisms are not machines if only because they are so vastly more complex than any machine we have built that they defy comparison.

Cells, likewise, are not machines, nor are they easily pressured into classes. Even in their heyday, cells were not quite the universal "atoms" of life they were supposed to be. The most glaring problem of the cell theory became obvious in the difficulty in defining cells consistently: Unlike eukaryotic cells, bacterial and archaeal cells have neither nuclei nor cytoplasmic organelles, membranous organelles such as the golgi apparatus, filamentous organelles of the cytoskeleton, symbiogenetic organelles including mitochondria and plastids. Furthermore, not even all eukaryotic cells have the same sort of nuclei and may or may not have cytoplasmic organelles of the symbiogenetic variety (especially mitochondria and chloroplasts), and many so-called eukaryotic cells are or contain cytoplasmic compartments with more than one nucleus. These "cells" are syncytia, plasmodia or coenocytes depending (most of the time) on whether they are derived by either the fusion of cells or the failure of the cytoplasmic compartment to divide following the doubling of their nuclei. Moreover, the eukaryotic cell is not a cell at all in the sense of an individual, since it was occupied by bacteria in the form of cellular organelles, and possibly was comprised totally of different individuals.

Genes are considered individuals, it would seem, only because the concept of a biological individual isolated from its environment spilled over into the concept of the gene isolated from the cell. The central dogma had to triumph over the idea of the inheritance of acquired characteristics to make life safe for the concept of the gene. Indeed, Morgan was even more antiLamarckian than Weismann in his passion to separate genes from environmental influence. Regulation (both developmental and remedial) was built into the individual organism as much as into the gene and both were rendered incapable of being influenced by characteristics acquired through use or lost through disuse. For the most part, the central dogma remains victorious.

Nevertheless, neoLamarckists continue to search for "hopeful monsters" in Goldschmidt's[106] sense of "mutants" with major restructuring affecting many organs at once, Bateson's "saltational change of species," and, today, in Baltscheffsky's[107] "anastrophies," or constructive, sudden and/or drastic changes in biological evolution. "Use" now refers to mechanisms producing *useful* features in such organisms, and "disuse" refers to mechanisms altering, reducing or eliminating features (called secondary loss) in these new organisms. "Inheritance of acquired characteristics" refers to the evolution of traits especially through nonDarwinian devices such as genetic drift, cooperation, hybridization and genetic transfer.

Furthermore, the central dogma has problems of its own. Conspicuously, DNA can be manufactured upon templates other than DNA and even synthesized enzymatically.[108] Indeed, probably the most common animal viruses on Earth are RNA viruses. Among these, retroviruses replicate DNA from RNA with the help of a so-called reverse transcriptase. What is more, DNA is made from an RNA template during the replication of human hepatitis delta virus, and DNA at the ends of eukaryotic chromosomes (known as telomeres) is replicated via enzymes known as telomerases which contain an RNA template.

DNA is not a class of identical individuals. It is more dynamic than this static view can accommodate. Not only is it found in different secondary configurations, but it changes in its primary structure (the monomers making up its chains). Ironically, these mutations in DNA are generally passed off as the exceptions that prove the rule: DNA is stable. Moreover, DNA is not as physiologically inactive as it is frequently blown up to be. DNA physiology is illustrated by the development of immune cells where DNA is rearranged and entirely new sequences are created (sequences not present in the fertilized egg at all). Furthermore, DNA contains mutational "hot spots," more prone to change than other portions of DNA, and recombination "hot spots," more prone to insert sequences of foreign DNA. Finally, DNA would seem to interact with cells, since their size is influenced by the amount of DNA in the nucleus. Of course weakness in the central dogma does not put use and disuse and the inheritance of acquired characteristics back into the rank of received biology. On the other hand, the elimination of Lamarckian inheritance is not quite as secure as it was in the 1970s when the central dogma seemed invulnerable.

A CRITIQUE OF BIOLOGY IN THE 20TH CENTURY

Molecular biologists have not contemplated life's potential beyond the individual and class. They have not confronted the enigma of life's units, whether individual or class, being centered in historicity and temporality. Fortunately, some Western philosophers have.

These philosophers grasp the place of the observer in the observed and, as interlocutors, confront the subject/object dualism. In the phenomenology of Edmund Husserl and Martin Heidegger, questions about life are asked by a privileged being already possessing a sense of life. Indeed, saying "the same" is possible because we have already thought "difference." Phenomenology thus helps one avoid the traps of essentialism (searching for an ultimate explanation "which is neither capable of further explanation, nor in need of it"[109]) and of reification (transforming qualitative values to quantitative values). Moreover, one is less likely to confuse the ontological (knowledge specific to philosophy) with the ontic (knowledge of facts and technology). Ultimately, the comprehension of the problematic attached to possessing life should allow one "to comprehend [life] itself not starting from its own possibilities, but from this objective reality, as an object similar to many others."[110]

Biology's Ontological Blindspot

At the beginning of the 19th century, incipient biologists succeeded in elevating life to a subject of natural science. Because life existed, its samenesses and difference could be reduced to analytical dimensions. Biologists proposed that life is the same throughout the Earth and has been for all time. Thus, the chemistry of matter must have solved all the problems of making life on the early Earth, and biologists's job was simply to rediscover chemistry's solutions: How was adenosine synthesized from adenine? How were polypeptides synthesized from amino acids? Thus biologists confronted life, even if many problems seemed unsolvable: How can Darwinian principles account for the evolution of biochemical pathways (ABCD) requiring catalysis and coupled oxidation–reduction reactions with short-lived intermediates? How can gradualism be made compatible with the origins of "irreducibly complex systems."[111] These problems became "works in progress" for tomorrow, and biology moved on, but how would biologists ever learn if they were asking the wrong questions?

Possibly, biologists are more sophisticated than they seem and are aware of the circularity inherent in studying life through its samenesses and differences. Like anyone else working in a complex society, biologists may be "saying one thing yet meaning another." They may be appearing to rely on sameness and difference in order "not to rock the boat," so to speak, of public acceptance and support. After all, biologists are already in troubled waters just working on problems such as life's origins and may see the greater part of wisdom in using guarded phrases while speaking in the public forum. Thus, the biologist might talk of "the origin of life," while meaning "the origins of lives." The biologist may not have to say that there was no "singularity" or origin of life and incur the wrath of a culture dedicated to oneness, especially if other scientists are aware of "the facts." The loss, in this case, is the risk of creating a secret society of scientists, an elite, which alone is privileged to information, while everyone else is relegated to the dust bin of ignorance (and of opportunities that science makes available for careers and jobs). Especially in what is supposed to be an egalitarian democracy, the loss would seem to exceed the gain.

Biologists are not, of course, aliens transported from another society and culture. They may not be colonizers, dedicated to the spread of belief in one God who created one life once and for all, and they need not be hell-bent on proving some ideology come what may, but biologists are inevitably part of the problem (and not necessarily part of the solution). Biologists may be ill equipped to criticize the role of concepts such as sameness and difference in science. Anyone is liable to have difficulty expressing new and novel thoughts in a culture with embedded ways of expression and commitments to conventional thought. In Western culture, like all cultures, legitimized ways of thinking loom large and hang heavily over every discourse. Moreover, to think out of the mainstream may precipitate ideas that violate current paradigms. For example, if biologists were to admit the possibility that all life is not the same, then they would also have to admit the possibility that early-life may have consisted of many lives. Furthermore, if many lives were lived together, could they have done so in cooperation and equilibrium rather than through competition and natural selection as we know it? Sameness and difference does not admit the possibility that life evolved from mixing "lives," but it is a possibility too important to be relegated forever to footnotes and the status of "alternative possibilities" in textbooks and scientific discourse.[112]

Biology's Epistemological Blindspots

The Epistemology of Sameness

Simplicity is the essence of clarity and sameness is the ultimate simplifier. Indeed, if it were not for the potential of sameness to simplify, so much of what appears understandable would fall by the wayside as incommunicable complexity. In the epistemology of life, nothing ranks higher than sameness for communicating, especially communicating ideology with conviction. As a consequence, many patently absurd assertions about life go unchallenged in "ordinary" science and are only elevated in "revolutionary" science: molds and mosses have a thallus (even if their thalli are the same only in word); Embryophyta and Metazoa produce embryos (with nothing whatsoever to recommend the comparison); head, thorax and abdomen, appendages, jaws (mandibles, maxillae), and limb (trochanter, tibia, etc.) exist in insects and vertebrates (sharing nothing more than their names).

The same habit of thought, sucking on simplifying samenesses, draws on molecules, atoms and ions. Ca^{++} may be Ca^{++} and K^{+} may be K^{+} anywhere in the Universe, but their occurrence in living things is never in isolation, and assertions about them cannot be made independently of their associations. Whether within or without cells, context carries the consequences for what ions are and what they are doing. For example, finding low Ca^{++} and high K^{+} concentrations in the cytosol of many cells may have a host of very different causes and consequences, but differences in process are ignored, while emphasis is placed on the sameness of the phenomenon.

Perhaps the silliest claim for sameness based on exaggerating the minuscule was made by micro- and molecular biologists looking for collagen in bacteriophage.[113] Collagen is a massive, macromolecular protein composed of three helical strands, yet the criterion employed by the collagen-seekers was "the characteristic collagen-like repeat $(Gly–X–Y)_n$," where *n*, it turned out, could be as small as five.

Other macromolecular complexes are also assumed to be the same despite demonstrable differences. Ribosomes, for example, which are universally considered to contain highly homologous RNA molecules and protein elongation factors differ significantly in dimensions and in other components.

The power of sameness to elide difference and create an illusion of identity is exaggerated by repetition. When an event occurs more than once, the observer is more than twice as likely to look for a "cause" rather than assume an "accident." For example, the secondary structure of RNA molecules (the complementarity of bases in different portions of a strand) is rarely observed by physical means, but "strong evidence for base-pairing"[114] is said to exist when similar segments of RNA in different species show comparable changes in the putative partners.

At a higher level of complexity, "chromosomes" provide a perfect example of misplaced "sameness." The DNA in bacterial and eukaryotic "chromosomes" is not remotely comparable, neither in mitotic cells and budding or dividing cells nor in the chromatin of interphase (nonmitotic) cells; prokaryotic DNA is neither decorated with histones characteristic of eukaryotic chromosomes nor are the qualities of genes in prokaryotes similar to those in eukaryotes (i.e., cistrons versus unique sequences with abundant separation). Furthermore, enzymes involved in DNA replication in addition to polymerase (topoisomerases), are different even within one of life's domains to say nothing of between life's domains.[115] Biologists had an opportunity to set the record straight when they floated the name "genophor" for bacterial chromosomes but the forces of sameness sunk the genophor and "chromosome" was retained for both the eukaryotic and prokaryotic "chromosome(s)."

The power of sameness to elide difference might not be so problematic if it did not also mislead scientists and undermine scientific education. Of course, one can always say, "What's in a word?"

The Epistemology of Difference

Difference provides the edge where thoughts bend and sometimes break. Ideas in Western culture require difference to express bifurcation and arborization, whether between alternate premises or constructions, between wholes and parts, causes and effects, means and ends. Of course, statements of differences are supposed to represent only the beginning of the scientific process, while retrodiction is supposed to dispense with difference, replacing it with the mechanism somewhere (and somehow) before the end of the process. Regrettably, difference does not yield easily. Rather, it hardens thinking, making it more permanent and brittle, less pliable and still less permeable to fresh ideas.

In scientific discourse, the solidification of thought by difference is illustrated by reductionism, exemplified in biology by insistence that living things consist of working parts operating through action and interaction (to which may be added, while assessing the consonance of performance among parts and in the thing as a whole).[116] Jacques Monod, championed this point of view when he asserted that "the cell is indeed a *machine*" (emphasis original),[117] while Richard Feynman went beyond it when he reduced life to atomic forces:

> *Everything is made of atoms*. That is the key hypothesis. The most important hypothesis in all biology, for example, is that *everything that animals do atoms can do. In other words, there is nothing that living things do that cannot be understood from the point of view that they are made of atoms acting according to the laws of physics* [emphasis original].[118]

Such hard-nosed reductionism is easily traced to 19th century mechanists, conspicuously to Robert Koch (1843–1910), Joseph Lister (1827–1912) and Louis Pasteur (1822–1895), who "had done so much to show that scientific research could pay off handsomely in practical results."[119] Jacques Loeb (1859–1924) also appreciated possibilities for controlling life chemically (or, at least, activating eggs), but Loeb's early speculations were premature.[120] These possibilities only entered the realm of the practical in 1953, following the discovery of DNA's secondary structure. Then, as Francis Crick boasted, "[W]e have discovered the secret of life."[121] Within five years, the DNA paradigm adopted lines of information storage, processing and retrieval,[122] and, in theory, the physical-chemistry of DNA became capable of producing the right protein in the right place at the right time and thus determining everything about a cell and hence life itself.[123] Since life's molecularization, the biotechnology industry flourished, and gifted molecular biologists have found opportunities for profits and power (as opposed to penalties and ostracism) in life's molecules.[124]

The hardening around difference is not always so well calculated for results as it is in the biotech industry. For example, biologists use computers as instruments for Baconian inference or searching for knowledge in "cumulation."[125] The Human Genome Project epitomizes this approach, in which vast amounts of data are piled up in the expectation that knowledge will trickle down. Likewise, informatics, the "information-intensive" approach to molecular pharmacology of the Developmental Therapeutics Program of the National Cancer Institute[126] expects drug activity patterns to pop out of computer-generated "clustered correlations" maps. The odds may be bad, but as long as computers are doing the work, why not let them have their go at it? One would seem to be in deep epistemological trouble, however, were one unable to distinguish between the assumptions antecedent to a method of data analysis and inferences drawn from data analysis.

Unless one believes that "this is the best of all possible worlds," the problem of giving expression to novel ideas may require alternative ways of expression as much as alternative ways of thinking. The problem is that a culture specializing in bifurcation and arborization has little tolerance for rhizomatic and synthetic alternatives, and terms of excess and lack are incommensurate with the language of flight and strata. Today, reductionism requires scientific analysis to break down a thing (phenomenon or process) even when that thing is as irreducibly complex as a living thing, and one has no way of anticipating all the possible consequences of the breakdown.

CHAPTER TWO

"NORMAL" BIOLOGY: FINDING SAMENESS IN LIFE'S MOLECULES

According to the law of nature, repetition is impossible.

Gilles Deleuze (Difference & Repetition, *p. 6*)

Throughout most of its two-hundred year history, biology waxed with revolutionary ardor. As the 19th century progressed, compromise with the State was succeeded by accommodation with capitalists, and, although biologists did not contribute directly to the industrial revolution and the renaissance of armaments as did physicists and chemists, biologists nevertheless tied their fate to the rising star of science. Between Humboldt and Huxley, a new religion of science was founded, and a new priesthood installed. Confident of their power, the select few passed on their arcana and privilege, while piously preaching platitudes to the rapt masses.

Biologists were granted a seat in the choir, but they were not ordained full bishops until the advent of the germ theory. Biologists could then make a claim for theories (immunity) that sometimes led to the production of coveted products (vaccines). Designers of medicine, agriculture, hygiene and waste disposal all paid homage to the sagacity of biologists. The first fruits of the new profession were plucked by Koch, Lister and Pasteur, but, ever since, whole industries devoted themselves to making "better things for better living" through biology.

Biology's best route to productivity was still uncertain, however. Would it follow physics and the pursuit of nature's universal laws, although their application to practical problems was vitiated by populational and stochastic uncertainty, or would biology follow chemistry and the pursuit of atomism, discerning nature's classes and identifying nature's individuals? At least, atomism's application to problems (here on Earth) was direct, albeit local. Evolutionary theory had been biology's greatest move in the direction of universal laws, à la physics, but it did not lead to predictions with any regularity, despite the occasional evidence for convergence. Even after evolution's demotion to microevolution by early-20th century bio-mathematicians, predictability did not rise above the waves of gene pools. On the other hand, in the second half of the 20th century, trinucleotide codons in nucleic acids,[1] composing the genetic code, correctly predicted amino acid sequences in polypeptides. While no theory whatsoever explained why particular codons encoded particular amino acids, the predictions worked, and predictability, predictably decided the issue: biology would pursue its destiny by following the route of chemistry. Like chemists, biologists would devote themselves to atomizing; they would peal away complexity until they located biological units; they would individuate units and funnel them into biological classes that would brook no exceptions. The focus of these effort would be the predictability of deoxyribonucleic acid (DNA), the stuff codons are made of,[2] and of polypeptides, the stuff predicted by a sequence of codons.

Biologists had opted for a sure thing: to imitate chemists and follow their paradigm. At least it worked, and never have more biologists been gainfully employed than today. But what happened to biology's revolution? It went the way of other "scientific revolutions." In the words of Thomas Kuhn, "Mopping-up operations are what engage most scientists throughout their careers. They constitute what I am here calling normal science."[3] Kuhn could not have been thinking about the Human Genome Project itself (almost thirty years before it was conceived), but he was certainly prophetic when he explained (under the heading of technological "puzzle solving") that

> Though an outcome can be anticipated, often in detail so great that what remains to be known is itself uninteresting, the way to achieve that outcome remains very much in doubt. Bringing a normal research problem to a conclusion is achieving the anticipated in a new way, and it requires the solution of all sorts of complex instrumental, conceptual, and mathematical puzzles.[4]

Chapter 2 examines the growth of this Kuhnian, nonrevolutionary biology in the 20th century, biology's transition into a mature science, and the passage of biologists to the security of individuating DNA and polypeptides. The examination extends to the history of ascribing samenesses and differences to molecules and the role of the Human Genome Project in determining current molecular preferences. Above all, the chapter shows how the methods of molecular biology can do nothing more than advance a "self-fulfilling prophecy." The chapter illustrates how, in the pursuit of molecular classes, biologists invented techniques for data acquisition that were inseparable from the biologists's methods of comparison, thereby circumscribing their own opportunities for discovery. Samenesses became inevitable, while differences were culled without ever surfacing. The chapter ends with a brief, philosophical recapitulation of lessons drawn from research on molecular sameness.

BIOLOGICAL SAMENESS

The criteria biologists traditionally set for sameness in a class are never especially stringent. Biologists build boxes of sameness in which they hide all the difference they can stuff inside. Zuckerkandl and Pauling[5] resorted to this strategy when they made a box of molecular sameness in which to hide differences in hemoglobins.[6] Their notion that comparisons among molecules revealed something about their evolutionary history implied that inertia prevents molecules from changing. If, however, molecules did change, "a recognition of many differences between two [sequences] does not preclude the recognition of their similarity."

Data began entering the box in 1985 when Carl Woese and Norman Pace began the sequencing of ribonucleic acid (RNA) in the small ribosomal subunit,[7] and by the 1990s, data produced under the aegis of the Human Genome Project required enlarging the box considerably. Sequence data flowed into biology, while increasingly sophisticated computer programs channeled these data into well-worn comparisons. Organisms were routinely identified as a consequence of their molecules.[8] Molecules were not only identified by similarities to other molecules, but functions of molecules were attributed to them from the alleged functions of still other molecules, and scanty similarities among the molecules's parts gave rise to huge scenarios on the evolution of molecules and organisms.[9]

Zuckerkandl and Pauling's scheme became the standard for modern molecular biology and remains so today. Elsewhere in the natural sciences, Karl Popper's hypothetical-deductive method (progress by negation), has greatly changed the style and content of scientific communication, but biologist rarely ask, "How much difference is necessary to burst a box of sameness?" In the case of molecular sameness, for example, similarity as low as 30–40% in the sequences of nucleic acids or polypeptides is considered tantamount to a demonstration of evolutionary relatedness (homology)!

Indeed, the very machinery used by scientists is corrupted, by "Scanning this parameter-space at a sensitivity just above noise level (50% identity on the DNA level)."[10] Instead of science proceeding from hypotheses to testing, it proceeds from assumptions to dogma.[11]

Today's molecular biologists rely on quantitation to attribute formal sameness while eliding differences. Dubious (if not false) claims for molecular sameness (based on "eyeballing" data, or historical assumptions about descent) are now credited to chemical data, "metricized" to make it reportable, repeatable, and redolent with mathematical rationales.[12] Molecular biologists crunch[13] differences into metric conformity, align them through omissions or gaps (and commissions [assigning "X" to anything whatsoever]) and wrench them into computer generated consilience. Moreover, exaggerated samenesses are bled into generalities concerning living things and blended across a broad spectrum of life, to such grand notions as phyla, body plans (*Bauplan*), kingdoms and life's domains.

Of course, biology did not proceed in a vacuum. Governments steadily increased their hold on biology, especially since World War II. Old charitable foundations, such as Ford and Rockefeller lost their influence while others, such as Howard Hughes and the Wellcome Trust increased theirs. The pharmaceutical industry and biotechnology start-up firms erupted as major shakers and shapers of biology. Finally, and most spectacularly, governments, foundations and industry joined forces in the Human Genome Project, the greatest and most expensive enterprise ever to lay pearls before biologists.

In the last decades of the century, biology was poised to become another "big science" (like astronomy, space and planetary science, nuclear and particle physics). Ironically, bigness in biology would not be pursued in the biosphere, in the study of global cycles of exchange, but in the pursuit of smallness, in the sameness of sequences among nucleotides in DNA and amino acids in polypeptides. The preference for molecular sameness was not hard to explain, since vast sums of money were available, especially through the Human Genome Project. Moreover, the methods and machinery for handling molecules rapidly became part of biology's new arcana (its paradigm).

MOLECULAR SAMENESSES

The saturation of contemporary biology by molecular methods and computer technology was no more unexpected that it was resisted. Searching for molecular sameness had the appearance of a high-stakes horse race. Hi-tech gadgetry for ascertaining the molecular weight of macromolecules and determining purity was already in hi-gear by the 1920s, when Theodor Svedberg (1884–1971)[14] developed his analytical ultracentrifuge, and, by the 1930s, when Arne Tiselius (1902–1971)[15] constructed the "Tiselius," the grandparent of the electrophoresis apparatus, both exercises abetted by Rockefeller largesse.

All roads led to proteins during these golden days of physiology, when biochemists unraveled metabolic pathways almost daily. The pursuit of molecular sameness did not take on its contemporary form, however, until the 1940s when prevailing prejudice dictated that the *grand prix* or Nobel Prize would go to those who uncovered the structure of the gene. It was genes, after all, that would unlock the structure of other biological components of life, or so it was thought.

In the pre-World War II era, while the identity of genes was uncertain, enzymes or proteins were considered the leading candidates for the gene. In the 1940s, Frederick Sanger (b. 1918) undertook the arduous task of identifying 51 amino acid residues in the protein hormone, insulin (discovering, while he was at it, the source of antigenic difference in the insulins of pig, sheep, horse and whale insulin). In 1958, he was awarded the Nobel Prize in Chemistry for the success of this efforts.

Since then, important advances have been made in purifying proteins and characterizing their polypeptide sequence.[16] Automated "Edman degradation" relies on the chemical removal of one amino acid at a time and its identification by liquid chromatography. The process is laborious (limited to about 50 residues per day), however, and results do not identify changes in amino acids due to most posttranslational modifications (such as the phosphorylation of residues). The much faster technique of protein ladder sequencing, utilizing matrix-assisted laser-desorption mass spectrometry to identify amino acids, allows the identification of phosphorylated residues but loses resolution in chains above 35–65 amino acids.[17] But direct, amino acid sequencing was by no means the center of activity in polypeptide analysis.

By the mid-1950s, as a direct result of James Watson's and Francis Crick's unmasking the predominant secondary structure of DNA,[18] the gene was no longer identified with protein but with deoxyribonucleic acid (DNA).[19] Sanger and others turned their attention to identifying the thousands (and thousands upon thousands) and millions (and millions upon millions) of nitrogenous bases in DNA. A great deal of technological progress would soon make DNA sequencing feasible.

The palace coup in biology[20] which followed the cracking of the genetic code in the 1960s, was "not incommensurate with any other ideas then accepted in chemistry or biology,"[21] and all the surprises that followed were readily accommodated within existing paradigms. Biology's vanguard rolled rapidly to the threshold of bioengineering, developing various strategies to accomplish its objectives (cloning, developing DNA libraries, the polymerase chain reaction [PCR], sequencing, assembly and mapping [see below]) without ever once questioning the prevailing faith in the dogma of sameness.

Individuating DNA

Biologists were never adverse to bringing humanity under biology's control, but DNA offered the first practical means for reaching that end, if only DNA itself could be brought under the biologists's control. The approach taken was that of the chemist: to individuate DNA and place it in classes.[22] By the early 1970s several methods had developed to reach this goal. The key insight leading to these methods came from research on restriction enzymes.[23] These are enzymes produced by bacteria which function in breaking down the "foreign" DNA of infectious viruses (bacteriophage) while not harming the "domestic" variety of DNA (hence "restricting" infectivity). Restriction enzymes (over 200 were known by 1997) cut through the double helix of DNA at restriction-enzyme specific sites, frequently consisting of particular sequences of four to six base pairs. Complete digestion of a particular DNA with a restriction enzyme produces the same fragments of DNA, fractured at the particular restriction-enzyme specific sites. These fragments are then identified by their number of base pairs, which is to say, their length, with the help of electrophoretic separation on an agarose or polyacrylamide gel calibrated with markers of known lengths.

Further identification of fragments produced by enzymatic digestion is achieved with the aid of probes, single-stranded fragments of nucleic acid (either RNA or DNA) typically manufactured to be highly radioactive or otherwise labeled. Probes identify particular sequences in single stranded DNA as a consequence of the complementarity of base pairs. Under appropriate conditions,[24] a strand of DNA consisting of any sequence of bases will anneal or hybridize (in the usual, antiparallel fashion) with a strand consisting of the complementary sequence of bases. The label attached to the probe permits one to detect the hybridized, two stranded molecule, for example, by autoradiography.[25] In DNA or Southern[26] transfer (or blotting), fragments of DNA are separated (if indistinctly) by electrophoresis on an agarose gel, and the DNA duplexes are separated into single strands (denatured) and then transferred, or captured in place, on a nitrocellulose or nylon filter (through the expedient if messy device of drawing a salt solution through the filter and soaking it up on paper towels). DNA probes can then identify the "blotted" DNA fragments.[27]

Cloning

In the decade of the mid-1970s to mid-1980s, cloning became the method of choice for handling fragments of DNA. Indeed, the method is still so central to the DNA mystic, that genes and their mutants are not considered thoroughly characterized until they are cloned. Cloning is the production of a large number of host cells containing a particular fragment of foreign DNA. Host cells are usually bacteria, most frequently of the species, *Escherichia coli*, but sometimes yeast or tissue culture cells of various types. Cloning requires getting a cell to take up and replicate the desired DNA without treating it as "foreign" or reacting negatively to its presence (by lysing). These problems were solved through the development of recombinant DNA technology that allowed one to combine DNA from different sources.

The ability to combine DNA is achieved most conveniently by taking advantage of a property of some restriction enzymes to leave "sticky ends" on the fragments of DNA produced by digestion. "Sticky ends" are generated by restriction enzymes that cut DNA's double helix in a staggered fashion through specific "restriction-enzyme sites." The bases protruding from each end are complementary to bases protruding from the other end and may reannealed with their opposite number (hence "sticky ends"). Recombinant DNA is DNA reconstituted from the DNA of different sources, made by digesting DNA from these sources with the same restriction enzyme and allowing the different fragments to anneal. A ligase is then employed to form covalent bonds between abutting ends, thereby forming a continuous strand.

Pieces of DNA intended for cloning are generally prepared by partial digestion with restriction enzymes. Random restriction-enzyme specific sites are cleaved, while other sites are left intact. Fragments selected for cloning are generally the large ones segregated by pulsed-field gel electrophoresis.[28] These pieces are inserted into the DNA of a cloning vector, usually a bacterial plasmid, a bacteriophage or cosmid (a bacterial plasmid capable of being packaged *in vitro* in a λ phage coat[29]) to make a hybrid (chimeric) DNA.

Molecular biologists have been remarkably inventive in solving the problems of cloning. In theory, the plasmid or phage DNA provides a vehicle for entering a host cell and protection against host-cell defenses once inside, but, in practice, host cells exhibit different preferences for various vectors and may require some "persuasion" (such as opening transient holes in the host cell's surface) before taking up recombinant DNA. Furthermore, precautions must sometimes be taken against the host's turning on the "foreign" DNA or some of its own genes and producing products leading to cell lysis. Several tricks may be necessary to identify compatible host/vector partners, preventing the destruction of the recombinant DNA by the host cell's defenses, and even preventing the lysis of the host cell by the recombinant DNA's defenses. Other tricks, involving antibiotics and coloring agents, are also employed to kill off host cells infected with the vector alone and to certify that the hybrid DNA was incorporated by host cells.

Once inside the host cell, the hybrid DNA may insert itself into (recombine with) the host cell's own DNA (chromosome), or the hybrid DNA may remain independent as a plasmid (depending on the cloning vector and host cell). In either case, the recombinant DNA is replicated during the host cell's cycle of reproduction, and host cells pile up in a colony or clone with their content of replicated, recombinant DNA. The cloned DNA, can then be harvested (retrieved),[30] usually with the help of the same restriction enzymes used to create it.

Of course, recombinant DNA technology and cloning have greater potential than merely individuating DNA. Molecular biologists claim that virtually any piece of DNA from any species can be inserted into the DNA of virtually any other species whatsoever. Given all the potential risks of recombinant DNA technology inherent in moving DNA, including the introduction of runaway pathogenic genes, a year-long moratorium on "genetic engineering experiments" was brokered by Paul Berg (b. 1926)[31] who chaired an international committee that drafted guidelines evaluating protocols on transferred DNA. These guidelines were widely adopted, and while relaxed today, are still broadly in force. In 1978, Berg was the first investigator to transfer genes between cells of two mammalian species.

DNA Libraries

Two categories of DNA are generally cloned: genomic DNA (or chromosomal DNA), and complementary (or copy DNA). Genomic DNA is supposed to represent the DNA of a species (a virus or bacteriophage, mitochondria or chloroplast, prokaryotic cell or a eukaryotic cell nucleus), and the sum of all the cloned genomic fragments constitutes a genomic library. Genomic DNA libraries can be enormous. Moreover, subclones, made from small fragments of particular interest or as a convenience for harvesting known sequences of DNA, can add greatly to libraries. For example, the genome of mice and man, 99% of which was cloned by 1996, occupy libraries of 800,000 cloned fragments.

Complementary DNA, better known as cDNA, is a DNA "facsimile" of RNA. cDNA is made with the help of an enzyme isolated from retroviruses known as reverse transcriptase,[32] that synthesizes DNA on an RNA template. cDNA clones may be tailored to polypeptide-encoding, messenger RNA (mRNA) by taking advantage of a unique feature of mRNA in eukaryotic cytoplasm and a peculiarity of replication. The unique feature of mRNA is a tail of multiple adenines (poly[dA]) at one end, specifically, the 3′ end. The peculiar feature of replication is its requirement for a primer of ribonucleic acid. Clones of cDNA replicated from mRNA can thus be prepared by utilizing a primer of multiple thymidines (oligo[dt]). The sum of all the cDNA clones from a particular source constitute an expression[33] or cDNA library. In eukaryotes, especially, the RNAs present in the cytoplasm at any one time may be different from those present at some other time. cDNA libraries, therefore, are identified by stages of development, differentiation or function. Moreover, because the DNA expressed in different parts of an organism differ, cDNA libraries are also identified by tissue, organ, and cell type.

In the early 1990s, at the urging of Sydney Brenner (b. 1927), the Human Genome Project placed a premium on human cDNA clones as the most efficient route for creating a human gene catalog. At that time, less than 2000 uniquely human genes were actually identified of the 50,000 to 100,000 polypeptide-encoding regions anticipated in the human genome. Moreover, polypeptide-encoding sequences would seem to comprise something in the vicinity of 3% of the three thousand million (billion) base pairs (megabase pairs [Mbp]) in human DNA. cDNA libraries offered the possibility for accessing vastly more genes and suggested an alternative to the "needle in a haystack" approach to finding genes in genomic DNA. Soon thereafter, Craig Venter, then at the National Institutions of Health (NIH), but more recently at The Institute for Genetic Research (TIGR), developed the concept of "expressed sequence tags" (ESTs) to identify cDNA clones.[34]

The strategy paid off abundantly, and cloning human cDNA and sequencing its ends soon became the rage. As EST sequences were pumped out, they were deposited in the National Center for Biotechnology Information (NCBI) EST database, from whence "database ESTs" (dbESTs) moved to the GenBank database. Public cDNA collections swelled and by 1992, an entire division of GenBank was created for dbESTs.[35] By 1996, 65% of all GenBank entries were dbESTs for a total of more than 600,000 dbESTs of which, about 450,000 were uniquely human.[36] In addition, the database for the fruit fly, *Drosophila*, or FlyBase, contained information on 32,000 alleles of nearly 10,000 genes, with accession numbers of more than 9000 nucleic acids and 3000 protein sequences. The mouse database had sequences for more then 48,000 ESTs, thanks to the Washington University Genome Sequencing Center, funded by Howard Hughes Medical Institute.[37]

Many statistical tricks have been played on all these data to normalize data by organs and stages of development and reduce redundancy, thereby allowing access to "deeper" sequences of lower abundance.[38] Recently, large scale comparisons of human genes by optimal alignment programs among DNA sequence of 163,215 3′-ends of ESTs and 8516 3′-ends of known genes selected from GenBank yielded "49,625 clusters, which is a reasonable estimate of the number of human genes sampled so far."[39]

PCR

For many purposes, the alternative to cloning DNA is the polymerase chain reaction (PCR).[40] Kary B. Mullis (b. 1944)[41] "invented" PCR to escape the tedium of identifying DNA fragments by Southern blotting (see above). A host of others and the "environment at Cetus," an early biotechnology firm where all the principal investigators were employed, made PCR work and turned it into the most popular method (with all its variations) for amplifying and identifying DNA fragments.[42] Indeed, for sequencing analysis, PCR is also so much cheaper and convenient than cloning that cloned fragments generally provide templates only at early stages of sequencing. PCR is then brought in for the purpose of amplifying desired regions of DNA.

Like other *in vitro* replication techniques, PCR requires four reagents: (1) single strands of DNA for templates; (2) the four "building blocks" (triphosphodeoxyribonucleosides[43]) of DNA; (3) RNA primers, complementary to a portion of the DNA-template; (4) a DNA polymerase to extend the primers with nucleotides complementary to the nitrogenous bases in the template and link them into a patent strand. Where PCR is unique is its ability to amplify a target DNA segment in genomic DNA by using flanking primers and reiterating replication, thereby increasing exponentially the amount of DNA (at least until the polymerase is saturated).

Each iteration consists of separating DNA into single strands (denaturing the DNA), annealing to primers, and replicating. In practice, temperature is elevated to achieve DNA denaturation and then lowered to allow the strands to anneal with RNA primers. Primers generally consist of twenty to thirty ribonucleotides chosen because of known complementarity to sequences at both ends of the DNA (corresponding to restriction-enzyme specific sites and sequences in cloning vectors). Alternatively, "degenerate" primers are employed, containing sequences found widely in DNA and capable of beginning replication virtually at random. Nowadays, the technique is performed with *Taq*, a peculiarly thermostable DNA polymerase originally from a hyperthermophilic prokaryote (the archaean source, *Thermus aquaticus* YT1, is now replaced by a recombinant source) that normally lives at temperatures as high as 95°C. Because the enzyme does not lose activity at relatively high temperatures, it survives the heat-denaturing step and in 25 cycles of replication amplifies the original DNA some four million times. An error rate in the vicinity of two incorrect bases per ten thousand bases synthesized may reduce specificity, however, especially if an error occurs in early iterations. PCR requires multiple samples, therefore, to identify errors on the basis of comparison.

PCR has vastly simplified the logistics of DNA research by reducing the dependence on clones. Now, once a unique set of primers are available, fragments of DNA as long as 2 kilobases (kb) can be amplified by PCR. Indeed, the only thing that has further reduced the reliance on cloning is complete genomic sequences (see below).

Gross Characterization of Fragments

A variety of techniques have been invented for identifying clones within libraries and cloning specific fragments of DNA. These techniques use the methods of classical microbiology (replica plating from a master plate) along with techniques of high-powered biotechnology. In the colony hybridization technique, the DNA from bacterial colonies is transferred to nitrocellulose membranes, denatured, and tested with known probes. Those probes that hybridize with clones identify the clones. For example, a probe representing a cDNA fragment might be used to find a clone in a genomic library containing the corresponding polypeptide-encoding region.

Fragments are also identified by their length in terms of the number of base pairs. Restriction-enzyme fragments of DNA are characterized not only by their site-specific ends, but by the distance between them measured as number of base pairs (bp[44]). Following enzymatic digestion, fragments are separated electrophoretically on agarose or polyacrylamide gels, and the fragments's length is determined by comparison with the migration of markers of known size.

According to the principle of complete additivity, moreover, the length of fragments should be equal to the sum of lengths of sub-fragments prepared by further restriction-enzyme digestion. In the end-labeling technique, radioactive phosphorus is added at one or the other end of a DNA strand (the 3′ or 5′ end) prior to digestion, and the distance between these ends and enzyme-specific fragmentation sites is determined by electrophoresis. By using different restriction enzymes, some fragments are inevitably found extending to the right of overlapping regions and other fragments to the left. Simple restriction or physical maps are thus built, consisting of linearized restriction-enzyme specific sites with known distances in base pairs between them.

Sequencing Individuated DNA

Ultimately, the identify of DNA fragments rests in their primary structure, their sequence of base pairs. In the 1970s, techniques for end-labeling fragments with radioactive phosphorous were adapted for "reading" sequences of nitrogenous bases as a function of distance from ends.[45] Alan M. Maxam and Walter Gilbert used physical-chemical reactions for breaking up DNA at the level of specific nitrogenous bases, while Fred Sanger and others interrupted DNA replication at specific sites. Sanger and his group then deciphered the linear sequence of nitrogenous bases in fragments of DNA and deduced the sequence in larger pieces of DNA from similar, and presumably overlapping sequences. By filling in gaps with sequences from other bits and pieces, Sanger and his group assembled the sequences of complete DNA macromolecules. By 1977, the sequence of 5386-bp in bacteriophage ϕX174 DNA and 17-kbp (kilobase pairs) in a mitochondrial DNA were ascertained, leading to Sanger's second Nobel Prize for Chemistry in 1980.[46] These techniques would be improved in the 1980s, beginning with the efforts of Leroy Hood and the automated DNA sequencing technology developed at Applied Biosystems Inc., and climaxing with "high-throughput" techniques that leave the DNA "untouched by human hands" while producing results "faster and better."

Current techniques based on Sanger's methods use "illegitimate" nucleotides to interrupt replication and introduce a specific, fluorescent end-label. The "illegitimate" nucleotides are the dideoxyribonucleotide analogues of the deoxyribonucleotide substrates normally incorporated into DNA during replication. The "illegitimate" nucleotides resemble the normal ones sufficiently to form base pairs along the template strand and polymerize with previously incorporated nucleotides, but, in the absence of a second oxygen (hence, dideoxy-[47]), the replicating sequence is unable to elongate following the addition of the "illegitimate" nucleotide and replication is prematurely terminated. The technique calls for adding small amounts of the dideoxyribonucleotides to the normal substrates so that normal nucleotides are incorporated at sites along the length of a replicating strand of DNA before the random incorporation of an "illegitimate" nucleotide terminates replication. All the truncated fragments are then removed from their DNA templates and separated electrophoretically on polyacrylamide gels. Because migration on the gel is a function of fragment length, the replicated fragments separate into a linear, stepladder-like array (of about 300 steps but potentially of 1000 steps), each step corresponding to fragments differing by only one nucleotide from fragments on adjacent steps. When their fluorescent dyes are specifically excited, each dideoxyribonucleotide "lights up," thereby identifying the nitrogenous base incorporated at each position and allowing the "reading" of the sequence.

Accuracy is amazing by any standard, but nothing human is absolute. Originally slated for a goal of one error in one thousand bases, accuracy at the level of one error in 100 bases became the standard of utility in the pharmaceutical industry.[48] "Sequence accuracy is not known precisely, but for dbEST overall, it is ~97%."[49] This estimate may be slightly optimistic, however, since, according to Dan Hartl, "*Caveat emptor*. In 8% of the [*Drosophila*-related expressed sequence] DRES isolates, the sequence of the retrieved clone was different from the sequences reported in dbEST!"[50] Errors result from various sources, but accuracy is improved by sequencing both strands of a given DNA fragment. Errors are then detected whenever the sequences fail to demonstrate base-pair matching, and additional effort can be devoted to figuring out which sequence is correct. Errors due to missing clones or inaccuracy in PCR may be detected by sequencing across adjacent fragments with overlapping sequences. Other problems continue to haunt sequences as well and improvements in stability, and reduction in artifacts (lower chimerism) of clone libraries are vitally needed.

The Human Genome Project

By the late 1980s, DNA sequencing was getting too big to be left to biologists. The implications of sequencing for achieving "better things for better living" through biotechnology were not only apparent to scientists but were the subject of popular imagination and public clamor. Britain and Japan were deeply involved in genome research, and sequencing was going on in Denmark, France, Italy, the Netherlands, the Soviet Union, and West Germany to say nothing of the United States. Congressmen in the United States stewed about foreign entrepreneurs waiting to profit from American biotechnology without paying their fair share, while foreigners viewed America's penchant for patenting and protectionism with deep suspicion, if not contempt. The world seemed to be poised on the brink of another exercise in competition to solve its problems, even if the effort produced ruinous bankruptcy instead of prosperous progress for many of those making the effort. But that was not to happen. Cooperation came to the rescue, and the Human Genome Project was launched as an international effort on a grand scale and with a degree of collegiality unprecedented in human endeavors (although many investigators would later be confronted with large-scale encroachments on their parts of the project).

The brain child of Robert Sinsheimer at the Santa Cruz campus of the University of California and Charles DeLisi, of the Office of Health and Environment at the Department of Energy,[51] the Human Genome Project was born with the assistance of Walter Gilbert (who dubbed the complete human genome "the Grail of modern biology") and James Watson, who initially headed the office for Human Genome Research at the National Institutes of Health. International agencies sprang up, notably HUGO, the Human Genome Organization, funded by the Howard Hughes Medical Institute and the Imperial Cancer Research Fund, to help coordinate interactions among groups studying various aspects of genomics,[52] and scores of databases came into being, such as CEPH, the *Centre d'Etude du Polymorphisme Humain*, to accumulate and disseminate information. Above all, the Human Genome Project, was by no means restricted to human DNA. Many other organisms, from viruses to mice, were adopted as "model" systems, ostensibly to aid workers in one or another aspect of ascertaining the human genome but actually functioning as projects in their own right to aid workers interested in these organisms.[53]

Some problems arose and some solutions were not entirely satisfactory. The National Institutes of Health (NIH) attempted to file for patents on the partial sequences of some 1200 "expressed sequence tags" (ESTs; see above), but scientists protested, charging that incomplete sequences of ESTs were unpatentable, since they lacked precise functional associations and utility. The patents were denied but, in the fall out at NIH, the Institute for Genomic Research (TIGR) was established (with corporate sponsorship from Human Genome Sciences), and Incyte Co. was incorporated (also with corporate sponsorship) to build EST databases on an industrial scale with limited access to pharmaceutical and academic groups by licensing agreements. The stakes changed, however, in 1994, when the pharmaceutical giant, Merck & Co., in conjunction with Washington University, initiated GenBank, a repository for ESTs, offering public access to the database and the possibility of acquiring clones from commercial sources. Public access was an enormous boon to research: By 1996, as many as 80% of the articles published in the *Journal of Molecular Evolution* and 85% of the articles on molecular evolution in the prestigious *Proceedings of the National Academy of Sciences* concerned comparisons of amino acid sequence deduced from sequences of cDNA.[54] By 1997, twenty-seven cDNA libraries supplied clones of cDNA inserts for sequencing.

A volcano was also erupting in bioinformatics, the technology for collecting, organizing and interpreting large amounts of biological information with appropriate algorithms, software, database tools and operational infrastructure. By 1994, the Human Genome Project could rely on three databases,[55] each with its own kind of information: GenBank had DNA sequences; Genome Data Base had chromosome mapping information, and Protein Information Resource had protein sequence and structure. These databases originally lacked the connectivity of a virtual database, however, and could not support interrogation and retrieval of related information available in a primary database. Structured Query Language (SQL) and File transfer protocol (FTP) would soon develop to serve the genome community with ready access to recently produced data.

Several aspects of the project were then ready for international cooperation: HUGO managed some aspects of the chromosome workshops; the United States and United Kingdom collaborated on sequencing the round worm, *Caenorhabditis elegans*; Britain's Wellcome Trust supported the international effort to sequence the genome of brewer's yeast and the malaria parasite, and the trust's Sanger Centre near Cambridge and the Pasteur Institute in Paris led the sequencing of the tuberculosis bacillus; the US scientists at the Los Alamos National Laboratory collaborated with Australians on the physical map of human chromosome 16; investigators at the Lawrence Livermore National Laboratory worked with Japanese on the high-resolution mapping of human chromosome 21; NIH and CEPH engaged in a joint effort on the human genetic map, and the Whitehead Laboratory of the Massachusetts Institute of Technology cooperated with Généthon on the human physical map.

The Project has been incredibly successful in many regards. It contributed enormously merely by picking up the bill for automating DNA sequencing and turning it into a high-throughput, information processing, computer driven technology. But the main, ostensible goal of the Project, the assembly of DNA sequence data for individuated fragments of DNA and their assembly into sequences for entire chromosomes and hence genomes for *Homo sapiens*, is not yet assured of success. This very ambitious goal was to be completed in fifteen years, by 2005, at an expected level of funding in the United States of $200 million annually (adjusted for inflation) by 1991. This level of funding was not achieved on schedule, but, in 1998, the National Institutes of Health (NIH) allocated roughly $218 million to its

Human Genome Research, and the Department of Energy, allocated $87 million to genomic work. Optimists believed the project was on schedule, and, as early as 1993, suggested additional goals (such as identifying polygenes through remapping and mapping within maps and further developing rapid genotyping technologies with DNA "chips," arrays using 1000 or more DNA sequences on a single sensor) with an expectation of reaching them as well within the original time constraints.[56] In the meantime, however, sequencing rates have failed to meet targets at a collective rate (for both human and model organisms of 50 Mb [megabases] per year and a cost of $0.5 per base), and urgent appeals for alternate technologies seem to go unattended.[57]

The Human Genome Project is, by no means, an eleemosynary organization. It expects to draw lessons from DNA for understanding the function of various genes and mutations, for ascertaining the action of drugs and mechanisms of resistance to drugs, for launching preemptive therapies and initiating therapeutic interventions, for assessing causal mechanisms of heritable diseases (from those of childhood to adult-onset), for improving diagnostic methods, for evaluating propensities to cancer, and for offering genetic counseling, prenatal diagnostics and screening newborns. The Human Genome Project is thus, potentially, of enormous consequences for human life on Earth and should somehow be brought under the control of human life on Earth.

GENOMIC SAMENESSES

The sequence of nitrogenous base pairs on individual chromosomes and in entire genomes is just the sequence of nitrogenous base pairs in a DNA fragment "writ large." Writing on this scale, however, requires some tricks. Indeed, the Human Genome Project has had to invest heavily in the development of techniques for welding individuated fragments of DNA into individuated chromosomes and, ultimately, individuated genomes.

Generally, a "scaffold" or "framework" is necessary on which to hang sequence information. Unfortunately, biologists concerned with the sequences of large pieces of DNA placed a premium on known genes and hence adopted the traditional language of linkage maps from geneticists. Biologists, therefore, employed a mixed metaphor to work out chromosomal sequences and called the assembly of scaffolds or frameworks "mapping."

Mapping

Mapping employs various techniques for synthesizing sequence information and assembling pieces of DNA into chromosomes. Scaffolds and frameworks are intended to provide large-scale maps for placing sequences. In general:

Genetic (or Linkage) maps > Physical maps > Genomic sequences

Several independent and merging subcategories exist for several subtypes of map.

Genetic Maps

A genetic map consists of a linear sequence of known genes, or more accurately, known mutants. In sexually reproducing eukaryotes, genetic mapping depends on the availability of genetic variation in the population and on genetic polymorphism, the coexistence of multiple alleles of phenotypically observable variants of the wild type at a genetic locus. The distance between genes is measured with classical genetic techniques, from pedigree analysis to breeding experiments. Distances are given in units of centiMorgans (cM), corresponding to frequencies of recombination.[58] In the case of human beings with few progeny per generation and few generations available for genetic analysis, mutants have been assigned to chromosomes on the basis of a LOD score, the logarithm of the odds ratio for linkage (the ratios of the probabilities that a gene is linked to other particular genetic loci as opposed to unlinked). Scores of 3 to 4 correspond to ratios of 1000:1 and 10,000:1 and odds of 20:1 and 200:1 that suspected sites are actually linked.

Since the 1980s, especially as a result of the expansion of computer tools for the evaluation of lineages among genetic markers,[59] genetic maps have provided the framework (skeleton) on which to hang DNA-based polymorphisms. The first of these polymorphisms employed to extend genetic maps was restriction polymorphisms, or variations in the lengths of restriction fragments found among individuals within a population. Because restriction polymorphisms are more common than mutant forms of genes as such, fine grain genetic maps were constructed by treating restriction fragment length polymorphisms (RFLPs) as "phenotypes" and determining frequencies of recombination between them through genetic analysis. LOD scores were used to identify linkages among RFLPs and between RFLPs and conventional mutants (including those for cystic fibrosis and Huntington's disease in human beings). The assignment of RFLPs to chromosomes was also validated by *in situ* hybridization, in which, a probe identifying the restriction-enzyme fragment or clone was literally seen to anneal to the suspected chromosome in the vicinity of the alleged RFLP site. RFLP maps have evolved by adopting several conventions. A haplotype is the particular combination of restriction sites found in defined regions of the genome. Informative meiosis is mating identified by computer programs which generate recombination between RFLP sites.

Other PCR-based markers include tandem repeat polymorphisms, single-strand conformational polymorphism, denaturing gel electrophoresis-based polymorphism, and allele-specific oligonucleotides. These are especially useful in eukaryotes where nongenetic DNA may occupy 95% (or more) of a chromosome. This nongenetic DNA, frequently characterized as mini- and microsatellites for its electrophoretic migration outside the range of other DNA, consists of monotonous, tandem repeats of short (one to six) sequences of nitrogenous bases and is variously known as simple sequences repeats (SSRs), short tandem repeats (STRs), short tandem repeat polymorphisms (STRPs), and sequence length polymorphisms (SSLPs). Call it what you will, polymorphic DNA is key to the construction of recombination haplotypes for each chromosome and "meiotic bins" (defined for each sex by loci within skeletal maps of genotypic data) in which recombination events are likely.

Perhaps the biggest surprise coming out of this work on genetic mapping is the degree to which portions of chromosomes are different from other portions. Some chromosomal regions, for example, exhibit recombinational deficiencies, possibly as a result of small inversions, while other regions are hot spots of recombination (the Chi site in *Escherichia coli*). Telomeric regions exhibit a much higher frequency of recombination than centromeric regions of chromosomes. Moreover, overall, recombination distances appear larger in female than male meiosis, even though male recombination distances are greater for some chromosomal regions.

In the case of the human genetic map, the *Centre d'Etude du Polymorphisme Humain* (CEPH) led the way initially by making available a set of data (blood group markers and protein polymorphisms, as well as DNA-based markers [RFLPs and STRPs]) for 61 families of cell lines (the CEPH reference pedigree set, publicly available at Coriell Cell Repository, Camden, NJ and the GDB and CHLC database). CEPH also invited investigators from around the world to pool data for markers (and genotype their own markers). In 1993, the National Advisory Council for Human Genome Research of the National Institutes of Health (NIH), and the Health Advisory Committee of the Department of Energy (DOE), coordinating the Human Genome Project, set a goal of constructing a 2- to 5-cM map of the human genome by 1995.[60] The project built on the "index" maps for chromosomes prepared by the NIH/CEPH Collaborative Mapping Group of 1992,[61] the European Genetic Linkage Map Project (EUROGEM), and Généthon, and exceeded its goal by 1994 with the publication of a map with an average 0.7-cM density.[62] Not surprisingly, inasmuch as similar methods are used in constructing the maps of mice and men, of 2616 loci listed in a high-resolution mouse genetic map, 917 have similar loci mapped in human beings among which are 101 segments with conserved linkage homology.[63]

Three large groups developed comprehensive genome-wide, microsatellite markers for human linkage maps: Généthon for markers with the CA[64] repeat motif; the Cooperative Human Linkage Center (CHLC), for markers with tri- and tetranucleotide repeats; the Utah group for markers with di-, tri-, and tetranucleotide repeats. In addition, groups at the National Institutes of Health (NIH)/CEPH consortium and EUROGEM work on chromosome-specific markers.

For eukaryotes, confirmation of chromosomal localization is sought by identifying a gene on a chromosome with the help of a fluorescent-labeled probe or *in situ* hybridization (fluorescent *in situ* hybridization [FISH]). Mammalian chromosomes are available in so-called mapping reagents or genetic mapping panels. In the case of human beings, cell lines from human families provide various meiotic products of parental chromosomes; radiation hybrid (RH) panels of hamster cell lines containing many large fragments of human DNA produced by radiation breakage; and in yeast artificial chromosomes (YAC) libraries consisting of yeast cells that contain individual fragments of human DNA. Sequencing other eukaryotes, including the plant, *Arabidopsis thaliana*, require the use of bacterial artificial chromosomes (BAC) libraries as well.[65]

Through the cooperation of Généthon, CHLC, NIH and 110 collaborators of CEPH, the first, comprehensive human linkage map was published in 1994.[66] Built hierarchically from STRPs to genes and scaled by the physical size of each chromosome, the map consisted of 5840 genetic loci (970 uniquely ordered), covering 4000 cM (on the sex-averaged map, although separate data are also available on female and male maps) on CEPH reference families.[67] The climax of the project, aligning the maps with standard karyotypes, was reached when several hundred probes for known markers (yeast artificial chromosomes [YACs] containing Généthon-based STRPs) were localized through FISH on appropriate chromosomal bands and distances of fractional chromosomal length relative to the terminus.

High-resolution genetic linkage maps of the mouse are also now available, based on polymorphism between laboratory strains and distantly related species (*Mus spretus*) and subspecies (*Mus musculus castaneus* and

Mus musculus molossinus), with a common set of evenly spaced (every 10 to 20 cM) anchor loci among the probes mapped in each cross. In 1993, two DNA marker maps (a gene-based [Frederick] map with 1098 loci [useful for mouse–human comparisons] and a SSLP [MIT] map containing 1518 loci) were integrated, creating a high-resolution, genetic linkage map for the mouse aligned with the centromeres. The genetic map was built with the help of the MAPMAKER computer program and likely typing errors were reexamined by a mathematical error-checking procedure. Finally SSLPs were genotyped in progeny from gene-based, interspecific back crosses, achieving approximately 2 cM resolution.

Estimates of heterozygosity, the size of the mammalian genome, and the size of genes suggest that a complete genetic map would require a map density of 0.0001 cM. Resolution at this level would require major advances in variant identification. At this point, physical maps fall into place.

Physical Maps

A physical map is a restriction map with distances between marker sites measured in base pairs or kilobase pairs (bp or kbp). The markers can be any DNA-based polymorphism present in sufficiently large numbers that their distribution throughout the genome is close enough to be measured (with existing methods). A variety of these physical markers are generated by PCR: sequence specific tags (SSTs or sequence tagged sites [STSs]), variable number of tandem repeats (VNTRs), and short tandem repeat polymorphisms (STRPs). In theory, two STS markers are considered linked if their appearance is correlated in a panel of mapping reagents. A sufficiently large collection of markers makes it possible "to assemble a 'framework' map spanning all or most of the genome. Further loci can then be mapped relative to this framework."[68]

Sequence tagged sites (STSs) are markers consisting of short stretches of unique sequences detected by PCR analysis. Specific primer pairs, complementary to opposing ends of restriction fragments, provide unique reference points. By 1996, the physical map of the mouse genome compiled by the Whitehead Institute encompassed more than 6,000 polymorphisms,[69] and the Généthon linkage map provided the "backbone" (skeleton, framework) of the first-generation physical map of the human genome. Efficiency and cost-effectiveness were improved in the case of the human physical map by selecting a single representative sequence from each unique gene. The 3′ untranslated regions (3′ UTRs) of mRNAs became the object of interest when their sequences were converted to gene-specific STSs. These 3′ UTRs rarely contained introns, allowing PCR to amplify pieces of useful size, and they tended to be more polymorphic than coding regions, making it easier to discriminate among gene family members that were very similar in their coding regions.[70] As a result, the physical map of the human genome compiled by the Washington University Genome Sequencing Center quickly included more than 48,000 EST sequences.[71]

Possibly the most significant use thus far made of physical maps has been positional cloning, a strategy used initially to identify the gene for Duchenne muscular dystrophy. Once the general location of a particular gene is known, the association of the gene with specific polymorphisms permits narrowing down the area of the gene's location until a "candidate gene" can be found in a cloning plasmid. In the case of the mouse, SSLP and other closely linked markers are now sufficiently dense that, "it is now possible to positionally clone virtually any mouse mutation." Several additional strategies are also employed to localize genes and mutants: (1) clone characterization for presence of CpG islands or other associated sequences; (2) searches for tissue-specific expression patterns of cloned sequences; (3) hybridization or exon trapping to search for coding sequences; (4) searches for single-strand conformational variants by chemical cleavage and denaturing gradient gels; and, ultimately (5), direct, large-scale sequencing and genomic searches.[72]

Sequence Assembly

Given the power of computers to work backward from data, a "shotgun" approach is commonly adopted for assembling genomic DNA. The source DNA is broken into restriction fragments and separated by pulsed-field gel, cloned as small inserts in cloning vectors (Janus vector), harvested, partially sequenced, selected to reduce redundancies, resequenced, and compared to sequences in fragments cloned with other cloning vectors (15- to 20-kb lambda clones) or "amplimers" from long-range polymerase chain reaction (PCR) fragments. The DNA sequences are "cleaned up,"[73] by expunging those sequences imported with the cloning vector and incorporated at the ends of sequence files, and stored by processes heavily dependent on computer software. The extracted (clean) data is then loaded into a relational database from which other programs can retrieve specified sequences. Pairs of sequences with a high degree of identity (in the vicinity of 97%) are recognized with the aid of programs designed for optimal alignment, and sequences are built into continuously adjoining sequences or contigs. The assembly of still larger sequences may require the addition of sequences from different sorts of libraries, and filling gaps may require PCR.

Of course, nothing is ever quite this simple, and numerous problems arise in practice. Some chromosomal regions, for example, are especially rich in guanine and cytosine (as high as 85% of the nitrogenous bases), causing curling of DNA strands and requiring the cloning of especially short fragments of DNA. Moreover, numerous checks and controls must be run simultaneously to prevent errors and correct them when they are detected in passing.

For the assembly of DNA sequences into an entire genome, TIGR ASSEMBLER may be employed to simultaneously cluster and assemble sequence fragments. A single fragment chosen as the initial contig is extended by candidate fragments with the "best overlap" in nucleotide content (utilizing the Smith–Waterman algorithm for optimal gap alignments). BLASTX, which assigns probabilities to the identity of sequences, makes links by searching the ends of each contig. Contigs whose ends matched appropriate database sequences are tentatively placed adjacent to each other. AB AUTOASSEMBLER and the Fast Data Finder reassembles overlapping sections of contigs. Visual inspection and the manual assignment of the most likely base at a position can occur at this point. Overlaps are then removed and data compressed by searching the contigs with the GRASTA program (a modified version of FASTA).

Gaps are eliminated when clones in different contigs are identified and ordered by the ASM_ALIGN program. Gaps between adjacent contigs on the order of 300 base pairs or less are closed with the help of PCR using appropriate oligonucleotide primers designed from sequences at the ends of contigs, while larger gaps are generally closed when GRASTA finds relevant DNA sequences at opposite ends of the same λ clone (or artificial yeast chromosomes [YACs]).

Additional editing may still be required before final assembly. TIGR EDITOR, a modified AB AUTOASSEMBLER, is an editing program capable of downloading contigs from databases, providing a graphical interface, and editing them with aligned sequences from TIGR ASSEMBLER. In the end, COMB_ASM is called on to splice together contigs on the basis of short-sequence overlaps. For eukaryotes, additional work is required for assigning sequences to linkage groups or individual chromosomes, although high density genetic and physical maps (see above) make it possible to locate contigs on chromosomes without the necessity of first separating individual chromosomes.

In the case of the eukaryote, *Saccharomyces cerevisiae*, sequencing of the 12 million base pairs on 16 chromosomes was accomplished through the worldwide effort, headed by André Goffeau, involving over 600 scientists from 96 laboratories, performing some 300,000 independent sequence reads (with an expected error rate of 0.03% or 3 errors in 10,000 bases, in the final sequence).[74] The European Union contingent used the Martinsrieder Institut für Protein Sequenzen (MIPS) to assemble, verify and analyze contigs, and that is the form of data freely available to the public.

Genomic Sequences

The completion of genetic and of physical maps remain ahead of genomic sequences, but by the end of 1996, genome sequences were successfully completed for 141 viruses and 51 DNA-containing organelles of eukaryotic cells (chloroplasts and mitochondria),[75] and on April 24, 1996, the complete sequence of the eukaryote, brewer's (or budding) yeast (*Saccharomyces cerevisiae*) was made available to the public.[76] By early 1998, complete sequences were also available for eleven bacteria (*Aquifex aeolicus*,[77] *Bacillus subtilis*,[78] *Borrelia burgdorferi*[79], *Escherichia coli*,[80] *Haemophilus influenzae*,[81] *Helicobacter pylori*,[82] *Mycobacterium tuberculosis*,[83] *Mycoplasma genitalium*,[84] *Mycoplasma pneumoniae*,[85] *Synechocystis* sp. PCC6803, and *Treponema pallidum*[86]), and two archaea (*Archaeoglobus fulgidus*[87] and *Methanococcus jannaschii*[88]). At present (1998), the genomic sequences of no fewer than fifty species are in the pipeline.*

The completion of genome sequences is still sufficiently rare that it generates some pride of accomplishment within the molecular biology community, but the sequences are preserved in electronic databases and only "presented" or "reviewed" in the print medium. There is, after all, hardly any utility in printing all the ACGTs detected in the sequence. As far as research biologists are concerned, the important part of genomic sequences is not found in print but stored in the electronic medium and available in silicon. In the case of the yeast genome, for example, ten electronic searches for samenesses are supported by the World Wide Web site of the Martinsrieder Institut für Protein Sequenzen yeast resource. Users can (1) inspect for genetic elements on visualized chromosomes and selected chromosomal regions; (2) search for detailed information by accession numbers (systematic codes or gene names); (3) browse through functional classes; (4) search for human homologues; (5) obtain information on functional properties; (6) download sequence data (either nucleic acid or polypeptide); (7) inspect sequence homologies and alignments (displaying FASTA scores greater than 300); (8) browse families and superfamilies of polypeptides; (9) search for sequence patterns and similarities; (10) inspect for gene redundancy.[89]

The print medium is paltry in comparison, and no protocol has been generally accepted for print publications on genomic sequences. The report generally begins with the number of base pairs in the sequence and a gross

---

*The genome of two more prokaryotes and a multicellular eukaryote, *Caenorhabdidis elegans*, were completed in the fall of 1998. See Andersson *et al.* (1998), Cole *et al.* (1998), and The *C. elegans* Sequencing Consortium (1998).

analysis of the number of base pairs on a (or each) chromosome (and plasmid, when present), followed by an assessment of relative and quantitative features of the sequence. Conventional "earmarks" are used to identify centromeres[90] and telomeres, as well as long terminal repeats (LTR), several hundred nucleotides long bracketing transposons, and LTR-containing retrotransposons bracketing sequences resembling those of retroviruses (Ty elements 5–6 kilobases long in yeast), solo and remnant LTRs (lacking Ty elements), said to be former transposon-integration sites of recent and early origins, intergenic regions, and sites with high periodic G+C (or, reciprocally, A+T) content. A+T-rich islands are also assessed as "signature elements" (such as bacteriophage lysogens or other inserted elements[91], or insertion sequences [IS][92]). The identification of genes (see below) allows the investigators to calculate the average gene size (usually in the vicinity of 1000 bp), their density or the percentage of the genome that they occupy, and the intergenic distance between them. The sites where replication begins (autonomous replication sequences [ARS] or origin of replication), and where it ends, are detected by "consensus" patterns, such as the "DnaA box," or located by their dinucleotide bias, and significant inversions of the G–C/G+C ratio (indicating an asymmetry between leading and lagging replication strands[93]). Longer "words" of nucleotide sequences are cited as evidence for the presence of ribosomal RNA (rRNA) genes and their flanking regions. Localized groups of long "words" (consisting of complex, repeated sequences) are interpreted as evidence of gene clusters.[94] Comparisons are also made among sequences in plasmids or, in the case of eukaryotes, chromosomes. Similarity among these regions is frequently considered evidence for recombination.

Genes encoding various types of "structural" or "stable" RNA (the sort that is not translated into polypeptides [rRNA, tRNA, and, in eukaryotes, small-cytoplasmic RNA {scRNA} of signal recognition particles {SRPs},[95] small-nuclear RNA {snRNA[96]}, introns, etc.]) are also identified by computer-driven scans based on conventional criteria. Several rRNA genes (the templates for the different rRNAs[97]) are typically clustered in one or more rRNA operons or rRNA organizing centers. *Bacillus subtilis*, for example, has ten such rRNA operons clustered around the origin of replication.[98] In *Borrelia burgdorferi*, the rRNA operon contains the 16S rRNA and two copies each of the 23S rRNA and 5S rRNA coding region. Interspersed among these regions are two tRNA genes (Ala–tRNA and Ile–tRNA) and four unrelated polypeptide-encoding genes (3-methyladenine glycosylase, hydrolase and two with no database match). Except for one tRNA gene (Ile–tRNA) and one of the unrelated polypeptide-encoding genes, the operon's genes are transcribed in the same direction. Thirty-one tRNAs, specifying all twenty amino acids in *Borrelia burgdorferi*, are organized into seven clusters and thirteen single genes,[99] while *Bacillus subtilis* has 84 previously identified tRNA genes and another four putative new tRNA loci found by computer scan in the complete genomic sequence.[100]

Visualizing the sequence as a whole pays off when it validates earlier hypotheses or suggests new ones. For example, the layout of genes in complete bacterial genomic sequences[101] has validated earlier evidence for "codirectionality" of replication and transcription in *Escherichia coli*.[102] Additional surprises and hypotheses[103] have flowed from the discovery of genes in clusters, in some of which, the genes retain individuality (transcribed as a unit) while in others, genes are integrated in operons,[104] or regulated sites, usually with a single promoter[105] controlling the manufacture of a single transcript encoding more than one polypeptide. A close, physiological relationship attributed to members of clusters and operons has even served as a basis for predicting the location of some genes in the genome.

Introns continue to be enigmatic. In *Arabidopsis thaliana*, where 1.9 Mb of chromosome 4 has been sequenced, introns show no distinguishing features other than the consensus donor and acceptor sites. However, the frequency of A+T (66.48%) in introns is substantially closer to that of intergenic regions (67.77%) than to exonic regions (55.96%),[106] suggesting that exons were once aliens in more homogeneous "intron-intergenic" space.

In the case of viruses, prokaryotes and yeast, with compact genomes and few introns, complete sequences simplify many research problems. Typically, an investigator needs nothing more than a fragment of DNA to discover a whole gene. The DNA is sent away to a service lab which will sequence about one hundred base pairs from one end and send the results back to the investigator within a day. A short list of the most reliable bases (sometimes a computer generated "consensus sequence," but more often an "eye ball" consensus sequence) is then sent electronically to a database for the complete genomic sequence, and the sequence of nitrogenous bases (or amino acids) in the whole gene (or polypeptide) in question is quickly returned via e-mail. Expertise only enters at the level of picking the most reliable bases in the sequence, and surprises sometimes result when whole families of proteins are represented by the original sequence rather than unique gene products.

Genomic sequences, especially those of model organisms, have also paid off in the demonstration of previously unsuspected genes.[107] Almost half of the open reading frames (ORF; see below) identified in the genomic DNA of the round worm, *Caenorhabditis elegans*, the yeast, *Saccharomyces cerevisiae*, and the bacterium, *Escherichia coli*, represent heretofore unidentified genes. A still greater surprise (which is not to say an unwelcome one) is the identification of a number of genes in model organisms resembling important, human disease-associated genes (Huntington's disease, amyotrophic lateral sclerosis, neurofibromatosis types 1 and 2, myotonic dystrophy, and fragile X syndrome) and other human genes that confer a predisposition to some familial variety of common diseases (breast cancer, colon cancer, hypertension, diabetes and Alzheimer's disease).

Identifying genes of one species with those of another species has given rise to syntenic maps, plotted for one species with the similar genes of another species. The similar sequences are called "homologous" sequences whether or not the sequences are known to encode polypeptides in both locations. Jackson Laboratory's mouse genome database, is a syntenic database that allows for rapid searches for mouse homologues of human disorders.

LOOKING FOR GENES IN THE GENOME

The main objective of genomic studies is the analysis of genes, their place, distribution, density and circumstance in genomes. The problem is, "How does one identify genes in genomic sequence data?" Ambiguity over what a gene is and conflicts between various definitions of genes compound the problem. Adding to the complexity, are "split" genes, containing intervening sequences (introns) interspersed with expressed sequences (exons). Following transcription, RNA sequences complementary to exons are "spliced" together while sequences complementary to introns are cut out. Most RNA genes are "split," and, in many eukaryotes, a host of polypeptide-encoding genes are also split. What is more, the transcripts of some of these split polypeptide-encoding genes may be assembled by alternative splicing, joining different downstream exons to the same upstream exon.

In addition, there is the problem of "RNA editing," the enzymatic alteration of coding sequences in RNA. In mitochondrial transcripts of trypanosomes and related protozoa, editing is predominantly the insertion of uridylate residues into primary transcripts and, if less frequently, deletion by U-exonuclease, effectively "reformatting" the functional mRNA.[108] In the case of plant mitochondrion-encoded genes, the 5′ end of a small (~45–70 nucleotide) guide RNA (gRNA) is complementary to pre-edited mRNA and anchors it during editing, while the more 3′ region of the same gRNA directs editing of sites within an editing block and is complementary to the final, edited messenger. Editing domains require the services of several gRNAs which may even be assembled into complexes or "editosomes" (kinetoplastid mitochondria).[109] Genes are not, therefore, simply regions of DNA whose information is placed directly at the service of the cell. The cell, it would appear, has its own mechanisms for sorting out and dealing with the information residing in DNA.

Sequences are searched for tRNAs, repeats and recognized splice sites, but the computer programs devised by contemporary molecular biologists, emphasize polypeptide encoding regions.[110] BLASTX analysis provides an initial comparison of all reading frames with known protein sequences, while Genefinder, GenMark and XGRAIL, modified for known sequences in a given species, are used to identify new genes in the same species. BLASTP and FASTA analyze and assign coding regions, and additional ORFs are identified, with start sites refined, with CRITICA. In order to check for consistency with putative protein sequences, FINDORFS may be used to extract protein sequences from cDNA sequences, and other genome analyses are carried out by MAGPIE. ORFs can be analyzed for the presence of particular types of proteins, protein families (pfam) and motifs with MacPattern. Membrane-spanning domains can be identified by TopPred, and "searches" are easily performed for "boxes" and "signatures" such as the "lipobox" in a polypeptide's first 30 amino acids and aromatic amino acids at the ends of outer-membrane proteins. Likewise, homopolymers and dinucleotide repeats are searched for using programs such as RepScan.

Analyzing Polypeptide-Encoding Sequences

Predicted (putative) protein- or polypeptide-encoding sequences (known as CDSs or, more commonly, "open reading frames" [ORFs]) comprise the "transcriptome" portion of a genomic sequence. Most genes in this portion are "annotated,"[111] with the help of prior assumptions of resemblance to known polypeptide-encoding genes in the species of interest or other species, usually without the benefit of experimental substantiation. Even "unknown" genes bearing one or another presumed "earmark" of polypeptide encryption serve as a basis of identification. In the case of the yeast genome, polypeptide-encoding sequences are classified as (1) known proteins, (2) similar to known proteins; (3) exhibiting a weak similarity to known protein; (4) exhibiting a similarity to hypothetical proteins; (5) having no similarity with any other proteins; or (6) only questionable ORFs.

Sequences stored at publicly accessible sites permit searches tailored for known gene sequences. "Sequence skimming" with low density sequences of random small DNA fragments allows one to find putative genes by matching sequences to ESTs or sequences of known genes. Generally, sequences are evaluated for ORFs (Geneplot) and a program such as BLAST picks out polypeptide-encoding units by scanning a composite protein databank (SWISS-PROT, TREMBL) or previously identified proteins in the species of interest (Link's database of NH_2-terminal peptide sequences from *E. coli*). Other searches utilize computer prediction of signal peptides, upstream matches to the Shine–Dalgarno[112] [and TATA box] ribosome binding site, etc.

Gilbert et al. offered a variation on this theme when they reported on "ancient" proteins:

> We defined a set of those "ancient" proteins that are homologous to prokaryotic proteins by comparing our intron-containing database to the prokaryotic database (GenBank 85) using BLAST. We identify genes in our database that match prokaryotic proteins with BLAST scores above 75 (probability below 0.005 for the size of the prokaryotic database). These are genes like triosefphahate [*sic*] isomerase, or the "mitochondrial" genes that have introns in nuclear-encoded proteins, or other genes of ancient ancestry. We aligned each of the prokaryotic matches to the eukaryotic gene using the Smith–Waterman alignment, and defined a new database that contains only those introns that lie in the region of the eukaryotic protein that matches the prokaryotic one.[113]

After matching genes, the degree of matching is scored: 1) if identical bases occupy a position, no substitution is inferred; 2) if different bases occupy the equivalent position, a substitution is assumed to have taken place. With the help of CLUSTAL V, a multiple alignment program, a gap may be inserted representing a loss of a base in one sequence compared to others, and parallel mutations or conversions which increase divergence may be corrected if accompanied by sequences that have not diverged.

The degree to which the polypeptide-encoding regions use particular codons (their "codon bias" or, reciprocally, their codon preference [codon adaptation index {CAI}][114]) is calculated for recognized genes, and the frequency of particular start codons is computed (the frequency of the not-quite "universal" ATG is compared to the frequencies of the less common TTG and GTG or "unusual" ATT and CTG at the start of recognized ORFs or sequences with high "GeneMark predictions" and similarity to polypeptide-encoding genes in other species[115]). Operons are identified, usually by the presence of a common transcription termination site (terminator) following coupled transcription units. Operons may contain as few as two genes or as many as twenty-five. In *Bacillus subtilis*, operons contain an average of three genes.[116] Unfortunately, promoters (specifically, the consensus sequences for sigma factors associated with promoters, such as *E.*

coli's 5′–TTGACG–n_{17}–TATAAT–3′) identify "an extremely high number of false-positive results."[117]

The analysis then shifts to its primary target: gene types according to functional and evolutionary assumptions. Sequences in a cell's genome resembling those of viruses and transposon (and transposase), chromosomal ends or telomeric regions, and other "anomalies," provide possible (as opposed to presumptive) evidence for horizontal gene transfer in genomic evolution. Some genes are found to be employed with the enhanced efficiency characteristic of viruses, for example, by having their products manufactured through "programmed ribosomal frameshifting" (α, β and γ, and τ proteins of DNA polymerase III in *Borrelia burgdorferi*[118]), while genes with "authentic" frameshifts and stop triplets (in-frame termination codons) are identified as nonfunctional "pseudogenes" or considered "junk DNA" or "decaying genes not encoding functional proteins."[119] Two types of polypeptide-encoding genes are classified as "of unknown function." These are (1) genes which fail to have sequences matching or similar to sequences in genes with known functions and (2) genes with sequences matching or similar to sequences in genes with no known functions. All other genes are considered functional, at least, for the purpose of argument.

Polypeptide-encoding genes may appear once or in multiple copies; they may not resemble other genes or their similarities to other genes may justify placing them into genetic families (see below) presumably expanded by "gene duplication"; duplicates may have evolved into isozymes, or multiple enzymes with nearly identical function, or changed drastically into virtually new genes. "Orphan" genes are unique to a species, and presumably an adaptation for the species's unique way of life, while genes in one species resembling those in other species presumably reveal a legacy from past, common ancestors.

Deducing Polypeptide Sequences

After assembling an ORF from sequence data, a computer program trained in the (nearly) universal genetic code can read the amino acid sequence of the polypeptide as easily as ABC (or ACGT). The technique of reading amino acid sequences from base pair sequences in DNA is not only easy, since it is done by machine, but the substitution of amino acids for triplets of nitrogenous bases promotes samenesses by neutralizing the idiosyncratic "codon bias" of organisms (a preference for one codon above synonyms) and eliminating "neutral substitutions" inherent in redundancy or "wobble" in the genetic code (different codons for the same amino acid or the ability of one transfer RNA [tRNA] to recognize more than one codon).[120] Furthermore, if only theoretically, the deduced amino acid sequence encoded by cDNA sequences offers a distinctive control, since the deduced sequence can be tested by producing the polypeptide *in vitro*, using recombinant cDNA for transcribing mRNA and translating it into the polypeptide itself. In practice, the same logic in reverse has given rise to the widespread practice of using partial amino acid sequences of known proteins as a "handle" to find polypeptide-encoding regions in cDNA libraries.

Comparisons among sequences of amino acid in polypeptides are not unlike the comparisons among sequences of nucleotides in DNA employed in assembling contigs and chromosomes. Databases are searched (using FASTA[121]) to construct a match between gene products. A percentage match score is calculated by dividing the number of matching amino acids by the length of the shorter protein, and entries are retained if they score above some cutoff. A similarity score of 20% might be used to eliminate reasonably independent proteins.

Amino acid sequences deduced from open reading frames of cDNA sequences can also be retrieved from several World Wide Web bioinformatic servers: GenBank, EMBL and SWISS-PROT on the Genmenu computer systems.[122] The opportunity made available through these databanks for switching between sequences of nucleotides and sequences of amino acids is useful as a trick for promoting sameness. One can reduce difference, for example, after aligning amino acids, by converting just the DNA nucleotides in the first two positions of each amino acid codon.[123]

Several computer programs are used to make a first approximation to alignment among polypeptide sequences: GAP, BESTFIT, LINEUP, PROFILE and PROFILEGAP in the UWGCG program package.[124] However, completely objective, "blind," computer-generated systems for multiple, sequence alignment are yet to totally replace "eyeballing" sequences or "manually-assisted" alignment programs.[125] The object of alignment, after all, is not merely to shift sequences around until some maximum similarity falls into place. Structural alignment must serve the larger purpose of complementing functional expectation. For example, GDE2.0 (Genetic Data Environment 2.0) adjusts alignment for the juxtapositioning in coding regions of functionally equivalent amino acid sequences,[126] while other programs provide alignment for phosphorylation sites and ATP-binding sites. Contextualization as well as reiterative reference to higher-order configurations, thus, enter the calculus of alignment (especially wherever phylogenetic relationships are sought).

Typically, initial multiple alignment construction is provided by one program (McCAW obtained from NCBI), whose output (including gaps) serves as input to another program (depending on the type of analysis employed). PREALIGN and ALIGN utilizes progressive pairwise alignment,[127] while CLUSTAL V provides comparisons

Perhaps the major theoretical obstacle to objective alignment is the evaluation of gaps within sequences. Insertions and deletions are common and may not represent functional or structural obstacles for constructing the final protein, and "[m]ost biochemists "tweak" computer-produced alignments to achieve a more pleasing disposition of gaps."[128] If gaps are introduced in the service of alignment without penalty, however, then even totally random sequences could be aligned. A price (gep: gap extensions penalties[129]) is therefore paid for the introduction of gaps. Alignments are scored (assigned normalized alignment score [NAS]) reflecting similarity (the greater the similarity the greater the formal sameness),[130] but only arbitrary decisions determine what score is acceptable.

Evaluating Substitutions in Sequences

Evaluation inevitably requires a degree of subjectivity (which is not necessarily unscientific) and individuality. The problem of "taste" is only problematic when it is exercised without acknowledgment, or, in science, when it is hidden behind jargon. In aligning amino acid sequences, for example, some investigators accept no more than one gap in defining "unambiguously aligned sites" while others tailor their sites even more. The more conversion of data, moreover, the more "fudging." Ultimately taxa are "represented by a composite, tandem pseudo-sequence."[131]

Metricizing is a methods of evaluation preserving more information while setting up a typology based on individual positions. In type 0, all sequences in a position are identical; in types a, b, c, d, etc. (each letter corresponding to a particular sequence), the monomer in a position in one sequence differs from that in the same position of one other sequence; in types x, y, z, etc., the position in two sequences differ from those in the other sequences; and so on for more differences among sequences for monomers in the same position. This classification of positions in the sequences is used to sum the number of differences of each kind (over the length of the sequence) and compute the number of differences between any two sequences (the number of mutations separating them) as a sequence (informational) distance and hence sequence space. The geometrical figure describing this space becomes progressively complex depending on the number of sequences compared and the classes of monomers. Distance categories are then compared through the methods of statistical geometry, requiring the computer to handle all possible combinations of substitutions in every position. In effect, distances (and thus data) are combined (and further reduced) as averages, and conclusions about sameness are drawn from the comparison among averages.

The ostensible point of these comparisons is to discover samenesses and evaluate differences in the sequences, but a much more alluring objective sneaks into the process: explaining substitutions via relatedness. Some substitutions are considered purely coincidental and hence consequences of random mutations ("the ultimate state of blurring of information"[132]). In a macromolecular chain, the substitution of one or another monomer at a position in the polymeric sequence is not necessarily equally likely. For example, changes in DNA from one purine to another or one pyrimidine to another are common compared to transversions which change purines to pyrimidines. In order to estimate complete randomization, therefore, distance traveled in a conversion has to be modified by statistical weighting depending on the type of substitution.

In contrast, other substitutions are considered due to functional constraints. These substitutions thus come under the rubric of adaptation with consequent assumptions of homology and evolution. In the words of Manfred Eigen (b. 1927), physical-chemist and Nobelist,[133]

> For homogeneous mutation rates, the different distance classes change in a quantitatively correlated manner. This quantitative correlation, and deviations from it ... allow the analysis of kinship to be refined and used for comparisons extending over a greater distance.[134]

Gene Families and Classes

The systematic analysis of genomic sequence data relies on computer-driven sequence searches and automated processes to place sequences into categories of polypeptides. With no justification whatsoever, other than the enthusiasm of the investigator for samenesses, polypeptides placed in these categories are said to belong to genetic families and classes.

The most general of genetic families are "open reading frame (ORF) families" or "superfamilies," recognized by any of several sequences said to be characteristic of polypeptide-encoding sequences. An identifying sequence can even be that of a noncoding region containing a promoter (initiation sequences, recognition sequences, binding sequences associated with RNA polymerase, other attachment sites). In yeast, ORF superfamilies (about 60) are recognized by portions of coding sequences containing recognized active sites or other small, diagnostic sequences, sometimes identified as a consensus sequence or "box" (the lipobox at the beginning of an amino acid-encoding sequence which identifies "candidate" lipoproteins in bacteria). Other critical sequences may be quite long, prescribing a hundred amino acids or more, even if identity with specific amino acids of no more than 30%–40% is considered sufficient to make a "match." In any case, most members of ORF families, only reach the specificity of "hypothetical proteins."

Currently, sequence analyzers have refined on nomenclature by recognizing two other (often overlapping) categories of polypeptide-encoding sequences: functional classes and structural families.[135] Functional classes are defined by sequences encoding functionally active proteins, while structural families are defined by sequences encoding small, structurally recognized portions of proteins, whether isolated units or repeated modules. Genomic sequences are scanned for members of these classes or families and "annotated," while lists or "gazetteers" of class or family members are made available for comparison with other sequences. Of course, the categories frequently overlap, but more importantly, at the moment, tests for membership in either category allow many sequences to slip through the cracks. Such sequences are placed into a third category of "Classification not yet clear-cut" or a fourth category of "Unclassified proteins." This awkwardness is, of course, expected to disappear in time.

How ironic that modern sequence categories became snagged on the ancient dichotomy of function versus structure. Of course, one can hope for (if not expect) the eventual merging of functional classes and structural families. At the moment (1998), however, instead of the distance narrowing between them, each category is becoming entrenched in its own sameness. Functional classes are increasingly associated with "[s]equences of common evolutionary origin (homologous *sensu stricto*) reflect[ing] their relatedness by sequence similarities ... into groups ("families")."[136] The idea of homology posits that the similarity of sequences is attributable to their arising from a common ancestor, while their differences are attributable to changes (mutation) accumulated during their separate evolution. Structural families, on the other hand, are associated with "ancient conserved regions" (ACRs)[137] and the shifting of "minigenes"[138] known as "exon shuffling"[139] via recombination and gene copy. Homology and exon shuffling may ultimately find common ground,[140] but the way to that juncture is not yet clear.

ORF Families

The members of ORF families are generally considered as either paralogues or isozymes,[141] appearing within a species, or orthologues or allozymes, appearing in two or more species. With the help of a small "leap of faith," "paralogous genes are coding sequences within a genome that have arisen from duplication events"[142] and paralogous families are groups of genes within a genome that have arisen from duplication events. Paralogues can account for a remarkably large portion of a genomic sequence. In *Bacillus subtilis*, for example, a quarter of the genome corresponds to just a few, greatly expanded paralogous families: 568 gene doublets; 273 triplets; 168 quadruplets; 100 quintuplets. The largest paralogous family contains 77 putative ATP-binding transport proteins.[143] Members of a paralogous family may be clustered, isolated, or both clustered and isolated in different parts of a sequence. Clusters, in turn, may comprise operons with or without members of other functional classes.

Duplication is not thought of as a rare or effete event. "It is believed that the eukaryotic genome has undergone multiple genome duplication events, with the most recent duplication event occurring approximately 300 million years ago, long before the divergence of the lineages leading to the mouse and human genomes."[144]

Regrettably, the question of mechanism is rarely addressed by those recommending duplication as the explanation for similarity among paralogues.[145] Did the alleged duplication occur for a single gene or for an entire genome with subsequent loss of some duplicates? Did the duplication arise from unequal recombination after pairing of nonallelic sequences (recombination followed by gene conversion[146]), through errors catalyzed by repetitive elements or incorrect patterns of cross-over within a cluster having a number of copies of a family?[147] More to the point, instead of gene duplication, could horizontal gene transfer[148] to have taken place, and could the putative duplicate gene be a migrant that entered a genome from outside? This alternative is hardly ever considered as an explanation for familial redundancy.[149] Rather, horizontal gene transfer is invoked to explain rare exceptions, such as "genes with a low codon adaptation index (CAI), re-flect[ing] the optimal codon usage or mutational spectrum of their previous host."[150] Were horizontal gene transfer to be taken seriously as a source of paralogues, all the weight and theoretical baggage dragged along by gene duplication might suddenly be thrown out.

Orthologues do not require a duplication event, although they are not incompatible with one. Assuming that a speciation event has occurred, the "same" gene (or genes, were they previously duplicated) evolving in now "new" species are orthologues. In this case, the "leap of faith" is to function. The polypeptide products of these genes are defined as "proteins of similar function found in different species" and orthologous groups are "groups of genes from different organisms which have the same function." The putative functions of orthologous genes provide the basis of "genome project annotations," which is to say, codification through "similarity searches against sequence databases."[151] Othologues are readily found using BLAST software, and new programs (PROSITE, BLOCKS and PRODOM) are constantly being invented to speed along orthologue analysis. Groups of genes represented in each organism included in an analysis are dubbed a "core set" of common genes, while those genes represented in fewer organisms are demoted to the status of orthologues for those organisms alone.

Despite the automation attached to recognizing paralogues and orthologues, the classes are beset with problems. First, specificity or degree of identity is ambiguous and families are defined at different levels of resolution, for example, as a set of genes that have 10 or more duplicates with a FASTA score greater than 300 in the yeast genome.[152] Second, any one enzyme may perform other functions than those presumably found among members of a functional class. Third, paralogous and orthologous families (sometimes as large as a hundred genes) are identified with putative biochemical and physiological properties but "matched" by sequences despite evidence that function is not necessarily correlated with sequence (primary structure). When it comes to function, higher levels of structures are the most informative, while lower levels of structures are the least.

> The most reliable signature [of a protein domain] will of course be its tertiary structure, since this diverges far more slowly than its primary structure. It is known, for example, that up to 75% of the amino acids of a protein can be changed, with only trivial changes in tertiary structure (less than a 2 Å root mean square divergence in backbone position.... This means that there will be a number of structural motifs or domains in proteins that can be recognized *only* crystallographically. One such example is that of the soybean trypsin-inhibitor protein and the mammalian interleukin-1 beta protein. They are very different at the level of amino-acid sequences but have the same tertiary structure.[153]

Furthermore, conformational similarity is *not necessarily* dependent on primary structure. In effect, the "sequence-to-gene function" approach of the Human Genome Project may not be particularly instructive when it comes to function.[154]

Functional Classes of Polypeptides or Proteins

The functional classes adopted for yeast[155] have been rounded out and adapted for the analysis of other genomes, although their acceptance has not been universal nor have their definitions been settled. In general, a dozen or so (twenty on the outside with some gene categories reserved for particular organisms) functional classes are recognized (plus "unclear classification," "unclassified" and "transposons"). Each class has many subclasses, which are named by the function of most members or, failing in that, by general properties (amino acid transporter, cell division control protein, protein kinase, etc.).

Informational processing is a superclass of enzymes and polypeptide factors involved with some aspect of the cell cycle. The superclass is divided into classes associated primarily with cell growth and division (class 03), replication,[156] repair and recombination, (class 04) with transcription[157] and (class 05) protein synthesis or translation.[158] Histone and histone-like proteins are placed in the replication subclass along with different types and components of topoisomerases, gyrases, and DNA polymerase, replication initiation-specific factors, replication termination factors and partitioning factors. Core RNA polymerase and proteins generally assigned regulatory functions, such as transcription elongation factors exhibiting the helix-turn-helix domain,[159] repressors (some of which resemble aminotransferases), activators, sigma factors, termination factors, and heat-shock response proteins are placed in this class. Other members of the superclass may include products interacting with methylated DNA, endonucleases recognizing newly synthesized strands during mismatch repair and other repair mechanisms. A host of proteins are involved in translation from ribosomal proteins, tRNA synthetases and RNA helicases to RNase, "degradosomes."

A second functional superclass contains enzymes involved in energetics and intermediary metabolism and biosynthetic pathways. The enzymes encoded by genes placed in the (class 02) energy class are concerned with sources of energy or with the generation of reducing power (energy-evolving processes), fermentation, the glycolytic pathway (or Embden–Myerhof–Parnas glycolytic pathway), and the ultimate disposition of the pyruvate it produces, glyconeogenesis, oxidative phosphorylation, the pentose phosphate pathway, substrate-level phosphorylation, the tricarboxylic acid cycle, the utilization of energy reserves (e.g., glycogen), and photosynthesis. Genes encoding products involved in the metabolism of nitrate or other electron acceptors also belong to this class.

The (class 01) metabolism and biosynthetic pathways class is subdivided into several subclasses, each with several subsubclasses including enzymes taking part in the metabolism of amino acids, carbohydrate, cofactors, fatty acid, ionic homeostasis, nitrogen, nucleotides, phosphate, phospholipids/gylcolipids/sterols/sphingolipids, prosthetic groups, sulfur, and vitamins. Proteins in the class of metabolism and biosynthetic pathways are linked to two other class: (class 07) solute transporters and binding proteins and (class 08) intracellular traffic proteins. The former includes enzymes dedicated to the import of organic compounds (the oligopeptide ABC transporters are the most frequent subclass of proteins in *Bacillus subtilis* and *Escherichia coli*) with broad substrate specificity, facilitators, amino acid permeases (antiporters), and sugar transporters. Still other membrane-protein encoding genes prescribe enzymes for linking proteins to membrane phospholipids. The intracellular traffic proteins include transport facilitator proteins in eukaryotes such as ion channel and ion transporter proteins and proteins performing intracellular trafficking between cytoplasm and nucleus, mitochondria, and chloroplasts, vesicles and the golgi network, peroxisomes, vacuoles etc. Other members of this class include genes encoding proteins controlling membrane potential and ATP synthase or translocating-ATPase capable of expelling protons from a cytoplasmic compartment.

Another class (class 09) includes cell-structure proteins responsible for producing and organizing the components of the cell from the inside out: components of the cytosol, the membranous (mitochondria, etc.) and filamentous organelles (elements of the cytoskeleton [microtubules, intermediate and microfilaments in eukaryotes]) plasma membrane, envelopes, walls, etc. Closely related (class 10) contains signal-transduction pathway proteins comprising a large subclass of coupled sensor protein kinases and response regulators (receptor proteins, mediator proteins, G-proteins, second messenger formation proteins, key kinases and phosphatases). This class includes proteins involved in pathways of pheromone responses, morphogenesis, osmosensing and nutritional responses (so-called quorum sensing proteins include aspartate phosphatases, whose products dephosphorylate response regulators, and possibly smaller regulatory peptides).

Another class (class 06) of proteins is involved generally with protein destination and storage, signaling and secretion, including chaperones involved in refolding mature proteins after translocation through membranes, and chaperonins involved in complex-assembly protein complexes, leader peptidases, modification enzymes (glycosylyation, acylation etc.), protein excretion signals (encoded by *sec* genes) of the major secretion pathway, proteolysis, and targeting factors. A subclass of cell-motility and chemotaxis proteins encodes surface materials (bacterial flagella) responsible for cell movement.

Cell rescue/disease/defense is the large class (class 11) of proteins expressed by cells in distress. These include proteins involved in degradation of exogenous nucleic acids, detoxification, DNA repair (base excision, nucleotide excision), and stress responses and rescue. In addition, related proteins involved in aging, apoptosis and lysis or cells in extremis are usually included in this class if only for want of a better place to put them. Finally, the class of proteins prescribing secondary metabolites (class 20), or the things organisms make that render them less attractive, is a hodgepodge of enzymes manufacturing everything from alkaloids and amines to phenolics and tepenoids.

What kind of relationship might one expect among members of any of these functional classes? The classes are convenient, of course, but the connection among their members is not intuitive and their relationship cannot be explained, in general, by gene duplication and deviation. The classes are built by bias, and the prejudice in favor of functional classes is built in before a genome sequence is even completed, for example, when ESTs are mapped, and comparisons are made for genetic resemblance.[160] Similarities between human, mouse, *Drosophila*, *Caenorhabditis elegans*, and yeast genes have blossomed into a minor industry including databases devoted to evaluating functional classes. Samenesses are even sought for genes on human chromosomes of radiation hybrid (RH) panels through the application of FISH, using probes from *Drosophila* clones. As a final touch:

> The regional mapping of the DRES [*Drosophila*-related expressed sequence] clones in the human genome was finally correlated with extensive phenotypic data available in FlyBase by searching Online Mendelian Inheritance in Man (OMIM) for mutant human phenotypes in each region that could plausibly be said to "resemble" the mutant fly phenotype corresponding to the DRES.[161]

Structural Families

Structural families are defined primarily by the structure of small parts within polypeptide sequences. Strictly speaking, structures are motifs in the vicinity of 28 Å,[162] often encoded by single exons,[163] that are mechanically stable, nonbonded, compact conformations, built on electrostatic interactions, hydrogen bonds and van der Waals interactions. When a motif is repeated, it constitutes a module. Using PSEI-BLAST (position-specific iterative BLAST) which combines the GLAST search with a profile analysis, motifs and modules are clustered into several dozen superfamilies, which accommodate several hundred families,[164] and with slightly greater accommodation, motifs and modules are resolved into roughly 1000 families. Among the most widely recognized motifs of structural families are the α-helix, α-helix cage, antiparallel sheet, β-barrel architecture, β-hairpin, β-sheet (strands), calcium-binding, coiled-coil, collagen binding (Von Willebrand type a) compact packaging of α helices (histone H5), homeodomain proteins and other transcriptional factors, ectodermal growth factor (EGF), EGF-like growth factor (laminin), finger-type, four-helix bundle, helix-turn-helix, immunoglobulin repeats, kringle, leucine-leucine interactions (among α-helices), long-linear structure, minikringle, parallel sheet, protease, protease-inhibitor, serine protease-like motifs and domains.

Although motifs are formally defined structurally, ideally, they are also the parts of proteins conveying functional properties. Indeed, the structure–function relations of motifs may be tested, for example, by site-directed mutagenesis or domain replacement strategies. Likewise, a relationship of motifs to sequence is also implied, although it frequently does not pan out in practice. For example, in the case of mammalian immunoglobulin G (antibodies), superposing β-sheet motifs in the variable (V) domains and conserved (C) domains typically demonstrate a common conformation in their central regions surrounded by different structures. This conserved core of 55 amino acids within the 110 amino acids of the V domain and the 100 amino acids of the C domain have only 15% identical amino acids![165]

Members of structural families also have a diachronic dimension as "highly conserved" structures or portion of a sequences which failed to diverge in the course of evolution due to functional constraints. For example, sequence similarities across phyletic lines have been detected using the BLAST program for sequence alignment in database searches. These sequences in presumably related proteins may be gap-free aligned segments or considered gap-free when such segments are separated by unaligned regions. Dubbed "ancient conserved regions" (ACRs), the sequences comprise about 1000 families of "highly conserved protein domains"[166] that freely cross the line from structure to function (from actins to zinc fingers).

"Clusters of orthologous groups" (COGs) are also structural families defined by sequence with one eye toward evolution and another turned toward function. COGs are consistent patterns of sequence similarities across at least three phylogenetic lineages found by computer programs searching for the "best hit" in "all pairwise sequence comparisons" for all the proteins encoded in complete genomic sequences. "Signature" structures, such as the functionally active sites of enzymes, identify large COGs.

For example, a superclass of 863 proteins is defined by "motifs typical of ATPases and GTPases but ... involved in a broad range of processes from DNA replication to metabolite transport."[167] The COG of ATPase ABC transporters and histidine kinase, alone, each include over 100 members.

Structural families defined by domains such as the zinc finger DNA binding domain, the helix-turn-helix motif of transcription factors, and the homeo-box of HOX genes (and other genes, including the MADS-box class of genes in flowering plants[168]) have been lightening rods for speculation on transcription factors and gene-regulatory elements in families of control genes. Likewise, large "families" of polypeptides are defined by their modular construction. The EGF superclass, with the EGF-like module in tandem repeats, includes proteins as functionally diverse as substrate adhesive molecules in extracellular matrix, cell-surface receptors, blood clotting and complement systems, and growth promoting factors (from which it received its name [ectodermal growth factor]). The EGF superclass also crosses species lines. It appears in both human and *Drosophila* laminin [especially in the heparin binding domain of the B1 chain] and in the *lin-12/Notch* gene family of *Drosophila* and *Caenorhabditis elegans*, from *Notch* (36 repeats) to *Delta* (nine tandem repeats), including the *germ-line proliferation-defective-1* genes of *Drosophila*, and the Lin proteins of *Caenorhabditis elegans*. Furthermore, the presence of one motif or module hardly excludes another motif or module in the same protein. For example, the *fat* gene of *Drosophila*, encodes 34 cadherin domains plus four EGF-like repeats,[169] and the extracellular matrix glycoprotein, tenascin, consists almost entirely of repeated EGF and fibronectin domains.[170]

Many proteins are, after all, mosaics, having more than one different domain, and performing more than one function.[171] Thus, motifs, modules, and COGs may be lumped together.

> Proteins may contain two or more distinct regions, each of which belongs to a different conserved [multidomain] family.... [Furthermore,] Some of the COGs may include proteins from different lineages that are paralogs rather than orthologs, primarily because of differential gene loss in the major phylogenetic lineages.[172]

Rather than assigning proteins to separate families for primary, secondary and tertiary structures, for repeats and dominant motifs, for one function rather than another, a compromise would seem appropriate incorporating different concepts of families, including extended families. After all, ACRs may be duplicated, and ORFs may be dominated by single ACRs. Indeed, ACRs blend with ORFs in COGs depending on the criteria for searching databases and the level of allowable noise. Moreover, COGs blend with functional classes, although a COG is a "[m]ore or less arbitrary clustering of genes by similarity."[173]

INTERPRETATION OF GENE FAMILIES

Homology

Homology is hardly a new concept. Darwin himself acknowledged his debt to Geoffroy St. Hillary for the concept of homology (called analogy in the late 18th–early 19th centuries). Homology itself, however, has changed definitions several times. Geoffroy intended to recognize sameness in articulations or relationships of anatomical structures in different species (the scapula, humerus, radius, ulna of the bat, whale, human and cat). These similarities in articulation, now called classical homologies, were considered grounds for testing assertions about natural groupings or links on the Great Chain of Being. Later, when natural groupings were identified with familiar groupings in an evolutionary context, homology fell into the grab-bag of anatomical evidence supporting evolution and was redefined as any alleged similarity inherited from a hypothetical common ancestor resembling existing, "primitive" types. Today, the definition of homology is further eroded, and sequence homologies connote similar molecules allegedly descendant from hypothetical ancestors without any corroborating evidence of a common ancestor.

When macromolecules or, more often, parts of macromolecules, have an average of identical and similar units (frequently in excess of 30% but less than 50%), the macromolecules are said to be homologous. Such an assertion is made without regard to any mechanism underlying the homology, although the attribution of homology is taken to imply (unsupported) sameness due to descent from a common ancestor reinforced by biological inertia and natural selection. Keeping the idea of homology free from assumptions about the source of homology is extremely difficult for biologists imbued with the taxonomist's concept of natural groupings (some degree of relatedness among organisms) and the systematists ideas of Darwinian evolution.[174]

Regrettably genomics is dominated by a devastatingly homogenizing concept of homology, amounting to nothing more than an assertion of sameness puffed up with evolutionary gravity. Homologues are typically defined as "proteins in different organisms that are derived from a common ancestor and have similar functions"[175] without a shred of evidence for such derivation or even the existence of a common ancestor between the species of interest. Ironically, molecular biologists are aware of the inconsistencies in their concept of homology without doing anything about them. Eric Davidson (of the prestigious California Institute of Technology), for example, writes, "No one in the world would conclude that a [sea urchin] spine is homologous with a limb. It just shows that there are these little genetic programs for setting up proximal–distal axes or whatever, that are used over and over and over again."[176] The ideas of conserved and consensus sequences radiate from the same, circular reasoning. In general, molecular sameness is described in terms of frequencies of substitutions, degrees, and ratios in conserved versus variable parts of macromolecules and in identical versus similar sequences of individual monomers in polymers. A conserved region of a macromolecule is one having "greater identity" (Can identity be relative?) among its monomers than a variable region, although a variable region might still have some identical parts (ditto). A pattern (generally short) of frequently identical monomers within a sequence of otherwise similar macromolecules is a consensus sequence,[177] although it may contain monomers at particular sites that are not remotely similar (in which case they are designated "X," meaning any substitution whatsoever).

Constant positions do not necessarily illuminate the ancestral condition, since they are clearly biased by present, functional requirements which may be quite different from the ancestral setting. On the other hand, variable positions may represent functional sequences that would require acute discrimination (in individual tRNAs), and, thus suggest great antiquity. Consensus sequences, which are moderately divergent (practically identical) or highly divergent (alternative substitutions appearing in the same position in half the sequences) are most easily interpreted as having ancient origins, antedating the last common ancestor. Regrettably, molecular biologists rarely consider the horizontal movement of consensus sequences from one phylogeny to another without their having been present in a common ancestor of both phylogenies.

The temptation to assign homology to molecular similarities and to speak of conserved and consensus sequences seems to be overwhelming for some biologists despite the possibilities for dallying in coincidental samenesses and playing with genome juggling. Many molecular biologists seem to be determined to make evolutionary sense of their sequence-recognition paradigms despite the absence of supporting evidence from any sort of fossil footprint:

> The availability of the complete sequence of yeast allows us for the first time to examine the evolution of a eukaryotic species in a truly comprehensive manner. The footprints that indicate the evolutionary path taken by the yeast genome may be recognized by internal similarities between distinct regions of the present-day genome.[178]

If only because small sequences have fewer units to contrast than large sequences, the smaller the sequence, the more likely one is to find similarity, and the greater the similarity, the more likely one is to attribute it to homology. Classical examples are found in signal or recognition sequences, such as the Shine–Dalgarno-type group and the so-called TATA box of eukaryotes. Similarly, the small repeat of bacteria and mtDNA, such as AT-rich regions and GC-rich regions, are widely attributed to homology. The genetic code is a prime example of gross homologizing on a small scale, while attributing homologies to the motifs of polypeptides and entire ORFs is frequently "picky" homologizing on a grand scale.

There is no justification, therefore, for the inconsistency and silliness of some molecular assertions concerning evolution. Moreover, warnings against the practice are numerous. For example, Dan Hartl, admonishes the enthusiast for homology:

> Sequence similarity alone, especially if it is similarity across a small region of a much larger molecule, does not necessarily indicate functional homology, nor does a vague resemblance between a mutant human phenotype and a mutant fly phenotype reveal that homologous genes are at work.[179]

Cladistics: Confusing Molecular Families with Molecular Descent

The formal mystification of sequence homology is done with the tools and tactics of cladistics. Sequences are metricized or converted into metric distances and turned into tree topologies, dendrograms, or cladograms depicting branching models of relationships.[180] Theoretically, nodes, or branching points, represent the points at which new sequences originate; branches, or internodes, represent sequences or character sets; branch length represents numbers of substitutions in sequences. Several methods for reconstructing cladograms alleging phylogenetic relationships from sequence information are available, each with its strengths and weaknesses.

Maximum parsimony analysis[181] looks at actual changes between pairs of sequences and

> attempts to reconstruct the course of events that converted an assumed ancestral sequence into its various descendant lineages, on the assumption that evolution proceeds by some least action principle, i.e., makes the fewest changes possible in going from a given ancestral sequence to any given descendant sequence.[182]

In practice, the analysis depends on selecting phylogenetically informative sites, where two or more residues (nucleotides or amino acids) are represented more than once in a set of sequences, while removing all invariant sites and sites at which the variant residues are found only in a single sequence.

Treeing programs then find the tree topology with the shortest overall length (minimize the total number of substitutions at phylogenetically informative sites). Several computer programs create dendrograms utilizing parsimony algorithms[183] (PAUP [phylogenetic analysis using parsimony], and PHYLIP [phylogeny inference package], FITCHPRO, PROTPARSPRO). In each case, one program feeds another. For example, in the analysis of 16 profilin sequences, an initial alignment served as input to the PHYLIP program SEQBOOT, and the bootstrapped data set served as input to the PHYLIP parsimony program PROTPARS.[184] The cladogram was finished when *Tetrahymena* profilin, TRYR, was used as the outgroup to root the otherwise unrooted tree.

Parsimony methods, which assume that the "best" tree has the shortest overall distance, attempt to minimize the total number of substitutions. These methods may find the correct branch topology in simulation experiments and predict actual branch lengths correctly for all branches, but, because parsimony analysis looks at actual changes, results may suffer from distortions due to multiple changes (and back changes) in distantly related lineages. Parsimony analysis also soaks up vast amounts of computer time and becomes inefficient when the number of sequences being compared exceeds five. Furthermore, the methods often find more than one minimal topology. Additional assumptions such as concordance with the results of other treeing methods, are then introduced to select the best "best" tree.

The maximum likelihood method (DNAML likelihood program in the PHYLIP 3.4 program package) adopts a model of nucleotide substitution in order to estimate the likelihood of a particular topology. Because the method

Domains of Life[a]

Bacteria	Archaea	Eucarya
Former names and current vernacular names		
Prokaryota (prokaryotes)		**Eukaryota (eukaryotes)**
Eubacteria	Archaeobacteria	Eukaryota Protoctista: Algae [brown, green, red lines]; Ciliates; Diplomonads; Flagellates; Microsporidia; Oomycetes; slime molds Animals Fungi Plants
Kingdoms (or other high taxa within domains)		
Aquifex-Hydrogenobacter, Aquificales (Hyperthermophilic)	Crenarchaeota (Hyperthermophiles): surface [S–] layers, mostly glycoprotein	Amitochondrial protists Type I (internal mitotic spindle): Archamoebae (single basal body associated with flagella) & diplomonad flagellates (*Giardia*)
Thermotoga. Thermotogales (Hyperthermophilic)	Euryarchaeota (Methanogens, etc.): rigid cell wall polymers	Amitochondrial protists Type II (paired basal bodies): retortomonad–diplomonad; microsporidia (lack flagella); trichomonad-hypermastigid lineages
Flavobacteria		Primitive (discoidal) mitochondrial protists: euglenoids and kinetoplastids
Gram-positive bacteria & acetogens		Alveolates: ciliates, dinoflagellates, apicomplexans
Gram-negative bacteria		I. Tubulocristate lineages: (1) Stramenopiles or Chromista (less cryptomonads but including chromophyte algae) (bi- or tripartite flagellar hairs; chloroplast endoplasmic reticulum [CER] in most photosynthetic members) (bicosoecids, brown algae, chrysophytes, diatoms, labrinthulids, oomycetes, opalinids, thrausto chytrids, xanthophytes [according to Sogin, 1994]) (2) Dinoflagellate–Ciliate–Apicomplexan lineage (cortical alveolation; mainly closed mitosis with internal spindle) II: Platycristate lineage: red algae, chytrids, choanoflagellates, cryptomonads, prasinomonads (green flagellates) Green algae
Green sliding bacteria		Animals
Green sulfur bacteria		Fungi
Purple sulfur bacteria		Plants
Cyanobacteria		

[a]Domains defined by Woese et al., 1990. Members of lineages from various sources: Sogin, 1994; Taylor, 1994.

utilizes complete nucleotide sequence information in determining phylogenetic trees, it soaks up computer time. but computer time may be reduced by clustering sequences (of fifty or less) into groups in advance and calculating their overall log-likelihood from their weighted mean. The method provides an estimate of the probability associated with given branches or distances, and may also help evaluate the relative likelihood of two topologies.

Distance matrix analysis looks only at the number of changes between pairs of sequences, or the distance between two aligned sequences (the extent of difference among the differing positions) without taking account of the actual changes (DRAWTREE and distance matrix methods in the PHYLIP package of programs[185]). Each pairwise comparison for all sets of data is reduced to a single figure which generates a distance matrix used to estimate evolutionary relationships. Depending on the data, distance matrix analysis of various types select (from a matrix of pairwise comparisons) the branching order and branch lengths that best accommodates the measured distances. UPGM (unweighted pair-group means) is one of the simplest, although it assumes equal rates of change down different branches of phylogenetic trees. Because distance matrix analysis looks only at numbers, it can correct statistically for underestimation of observed versus actual changes separating two sequences.

The extremely fast and efficient **neighbor-joining method**[186] works sequentially to find the nearest pairs of neighboring sequences providing the shortest overall length of the tree. It is, in effect, a cluster method using a Kimura parameter or LogDet transformation for calculating branch lengths proportional to number of inferred substitutions. For sequences with 20–30% similarity, neighbor-joining is performed with NEIGHBOR or some version of the PAM matrix, available in the PRODIST program.

Some methods specifically confront idiosyncrasies of distance matrix analysis, biological processes and computing. **Evolutionary (maximum) parsimony analysis**, corrects for distortions due to unequal rate effects and artifacts caused by rapidly evolving lineages of distantly related sequences.[187] **Transversion parsimony methods**[188] correct for differences in the likelihood of one sort of substitution over another in nucleotide sequences. **Bootstrapping** measures the reliability of nodes by determining how many times they emerge in the "best" tree following their deletion and replacement at random. PROTRARS bootstrap values and CLUSTAL V running 100 to 1000 replications evaluate the level of bootstrap support for branches in neighbor-joining trees. Similarly, consensus trees, generated by CONSENSE have values for the number of times pairs appear.

Ultimately, comparisons are comparisons and never rise above the hypothesis of sameness. The idea of functional classes, structural families, and

ORF families, whether paralogues or orthologous, rests on the assumption of relatedness in the first place. Rather than testing relatedness as an hypothesis, molecular biologists bend every effort to trace molecular evolution. The very premise of branching, that all variation of a sequence is traceable to the same, original template, is problematic. Certainly the possibility of horizontal gene transfer looms as a very viable alternative. However well-suited one hypothesis is to a given set of data, trees neither detect every false homology, or homoplasy, resulting from convergent similarities nor eliminate all the noise inherent in sequence comparisons.[189] The analysis of complexity in genomic sequence data must not be allowed to proceed by folding sequences back upon themselves, allowing them to become patterns of substitutions. Evolutionary scenarios for molecules will not be elucidated while molecular biologists continue to be enthralled to linear concepts of change, to repetition and to a dialectical bias favoring sameness.

SUMMING UP: CATEGORIES OF SAMENESS

This chapter, has explored many facets of the same problem: Molecular biology's persistent reliance on sameness, both to define questions and to find answers. Sameness is determinative, from the individuation of DNA fragments to complete genomic sequences, from the initial alignment of sequences (whether by "eyeballing" or by disinterested computer programs) to the final disposition of functional and structural genetic families, from linear sequences to dichotomous phylogenetic trees.

Sameness, or more of the same, is what biologists have to sell. It is the product that biologists take to the market place and the commodity the public is eager to purchase. The central conjecture of contemporary biology is that DNA is reducible to categories (classes or families) of similar sequences, the members of which category behave in much the same way and have evolved through many of the same mechanisms. This is the conjecture that has so far sustained the biotechnology industry and its spin-offs in medicine, agriculture, hygiene and environmental (pollution) control. In retrospect, three types of formal sameness seem to underlie the biologist's central conjecture: (1) samenesses defined by material units, (2) samenesses defined by processes and (3) samenesses defined by the organization of classes.

The first category of samenesses (epitomized by biology's myopia or habit of viewing life in too narrow a field) draws upon material similarities to define material units and proceeds to draw upon material units to define material similarities. Formal samenesses in this category are among the oldest in biology and classically encompass things as large as whole organisms. More often, and more recently, samenesses based on direct, material comparisons concern smaller things (from organs down to ions) and frequently the monotonously repeating units of larger things (the monomers of polymers). Formal sameness has been attributed to pigments, cell wall materials and cell storage products widely distributed among a variety of living things. For example, chlorophyll c_2 found in dinoflagellates (distantly related to the chromophytes) is equated with chlorophyll c_2 found in cryptomonads (not related).[190] Likewise, cellulose, a conspicuous component of the cell wall of green algae and higher plants, is also identified in the walls of dinoflagellates and tunicates. Calcium carbonate (produced by dinoflagellates, red and green algae, mollusks, etc.), silica (in dinoflagellates, radiolarians, diatoms, chrysomonads), chitin or sporopollenin (in arthropods, dinoflagellates, green algae, higher plants) are all formally the same wall materials occurring in broadly different biological groups. Starch and glycogen, and other storage polymers of glucose, are also identified in a variety of organisms.

The tradition of identifying sameness in material units reaches its apotheosis in identifying genes and their products (RNA and, if once removed, polypeptides) by the sequence of nitrogenous bases in DNA. The steps in making this identification are technically overdetermined and, since so many of these steps are performed by machines and depend on computers, rise above criticism to the rarefied atmosphere of (inaccessible) scientific objectivity. The problem is that machine logic is inevitably circular, and samenesses among defined material units are mere tautologies. For example, identifying a gene in one organism by a comparison with the sequence of a putative gene in another organism has the appearance of certainty while telling one precious little about the genes or the organisms beyond the "already known."

The second category of samenesses draws upon similarities in products to assign sameness to processes, and proceeds to draw upon sameness in processes to assign similarities in products. The processes include information processing (replication, transcription and translation), intermediary metabolism (through which biologically important chemicals are made or utilized), energy metabolism (through which cells are supplied with utilizable forms of energy), sensory functions, and transport into and out of cells, etc. The logic of these samenesses is often unvarnished pragmatism: products (RNA, phosphoglyceraldehyde, ATP, receptors, and permeases, etc.) are ascertainable, whereas processes are sometimes only vaguely known and, often, only in theory.

Frequently, sameness along the product–process axis is assigned on a provisional basis, with apologies, while efforts are made to discover more precisely the characteristics of a process (enzymes, intermediaries, kinetics, energetics). But "filling the gaps," cannot rectify the situation, because the problem is more fundamental: The problem is that in living things the choice of endpoint or even intermediate product, is arbitrary and arbitrarily influences the assignment of sameness to a process. There are, after all, "many ways to skin a cat." For example, similar polypeptides[191] may result from the translation of different messenger RNAs (mRNAs), that is, messages having different codons[192] transcribed from different genes. Even translation, considered the most universal biological information-processing system,[193] is performed by different types of ribosomes in prokaryotes and eukaryotes.

As a corollary problem, sameness is sometimes assigned to products on the basis of similarities in processes or steps within a process. For example, coenzymes that catalyze a large class of reactions in biosynthesis and metabolism[194] are sometimes relegated to a narrow slot of single and simple activities.[195] Even more problematic, enzymes that catalyze different reactions are generally identified strictly with one reaction.[196] Convenience and time constraints may justify the practice but not the resulting confusion (or the confusion that permits the practice in the first place).

The third category of formal samenesses draws upon similarities in things to attribute samenesses to the organization of biological classes or families and vice versa. Frequently, these similarities are not well delineated but constitute a *Gestalt* or overall sense of the things placed within a class. Assignments in this category are most conspicuously, culturally hide-bound. Presently, the most familiar representative of these samenesses is homology, the alleged existence of rudimentary structures (or molecules) in hypothetical ancestors used to explain the presence of similar organs (tissues, cells, etc.) in fossils or extant organisms.[197]

One cannot be critical enough of cultural forces compelling *Gestalt*-sameness. They are the source of the worse excesses of racism, sexism and nationalism. But why are the same forces tolerated in science? Rudimentary, Popperian science (science by negation) should be a sufficient corrective, since it requires the scientist to ask the question: "How much difference is necessary to overwhelm a *Gestalt*?" The current literature does suggest that molecular biologists are neither asking that question nor are they willing to repudiate the organizing sameness of naive evolutionary scenarios.

Formal samenesses in science are reflections of formal samenesses in society. They are derived from culture. They are not panspecific but heavily encrusted with cultural values. The prevailing habit of observing and comparing is not attributable to a deep structure or to genetic determinism, and biologists relying on formal samenesses are not simply "doing what comes naturally," while using the best data and tools available. Rather, economic and social contexts determine value and thereby decide what data are accumulated, while political and historical forces shape what comparisons are warranted.

Value begins in the market place and proceeds backward into the laboratories and classrooms of biologists and their minions. This is not to say that biologists are not part of the market place themselves, and, indeed, great corporate slugfests have been fought over products, such as PCR, consumed by biologists themselves. The value of biologists as consumers, however, pales in comparison to the value of biologists as producers of products and propagators of paradigms. As long as formal samenesses make work for biologists, biologists will continue to grind them out. At the same time, biologists in their role as teachers will continue to propagandize on behalf of sameness.

CHAPTER THREE

Biology's Shifting Paradigms: Contending with Difference

It is thus in the nature of genera to remain the same in themselves while becoming other in the differences which divide them.

Gilles Deleuze (Difference & Repetition, *p. 31*)

Thomas Kuhn[1] did not suggest that the passing of the old guard was well mourned following a successful "scientific revolution." The new guardians of the gate dispensed easily with people. Their problem, was what to do with old paradigms following a shift to new ones.

Old paradigms define and determine the legitimate fields of scientific inquiry. They are thus profoundly problematic for the new guardians, since, if their revolution is to solidify, the legitimacy of the old must pass onto the new. What is more, the ability of new paradigms to spell out answers to old problems is precisely what makes the new paradigms a salable commodity in the academic market place.

Some old paradigms were easily brought under the aegis of the new, following biology's latest "revolution" led by Watson and Crick under the banner of DNA's double helix. Genetics, for one (from genes and mutations to heredity and protein synthesis), was readily brought into conformity with the new paradigm, and converts flocked in droves to the new molecular biology. Indeed, the virtue of genes as physical instruments of heredity was that they were easily reconfigured into the virtual physical-chemistry of complementarity among nitrogenous bases. The hold of the revolutionaries was thereby strengthened, and the base of counterrevolutionaries eroded.[2]

Other old paradigms were not that easily dispatched. The case in point, is systematics. Systematics was built on the belief that biological classes (taxa) can be organized through the construction of hierarchies of difference. In the past, the systematists have used the same differences which taxonomists apply while naming organisms, or, recently, which evolutionary biologists employ as criteria for relatedness among organisms. The problem faced by postrevolutionary biology was to bring differences and hierarchical systematics into line with the rubric of molecular biology. The "fix" did not work out. Instead, systematics has had its evolutionary rationale eroded and its taxonomic prop kicked out from under it.

In the wake of the Human Genome Project and the advent of cladistics as the dominant method of molecular systematists, molecular sequences became the preeminent data for systematic biology and the defining differences among biological classes. Today, small molecular differences completely dominate thinking on large biological relationships, and tracking differences in molecules has become the standard for deciphering evolution even in the face of incongruity and contradictions.[3] Here is a classic case of "the tail wagging the dog": the systematics of sequence differences no longer concerns the systematics of organisms; sequences change without regard to the organisms containing these sequences. Molecular biology has captured systematics but has not offered any superior explanations for difference and its organization into hierarchies.

These problems of postrevolutionary biology are confronted in this chapter, which attempts to lay out the best case for the methods of cladistics and received evolutionary arguments, reserving critical judgment for the end. The chapter begins with a brief history of biology's treatment of difference. Cladistic methods (extending the discussion from Chapter 2) used in the construction of hierarchies are described, and molecular systematics's most notable contribution, its organization of life into three domains, based upon molecular differences, is fleshed out. The chapter moves on to examine each domain and similarities in pairs of domains. The final critique draws on the problems of differences that reach across domains and do not define hierarchical relationships. These differences suggest that biological origins may be broader than points in life's universal phylogenetic tree.

A BRIEF HISTORY OF DIFFERENCES IN BIOLOGY

The habit of placing living things into classes defined by difference has a long and eventful history. Classically, classes served various organizational/philosophical/religious functions, such as determining what could and could not be eaten. Animals were supposed to differ from plants by way of sensation, while plants (vegetation) differed from animals by not requiring nourishment. Sensation, which human beings shared with nonhuman animals, justified the greater value placed on animal life compared to plant life, a value which reached its apotheosis in the sanctification of animal life and prescriptions against slaughtering animals. In contrast, plants could be massacred and their bodies burned without compunction. Differences such as these between plants and animals were useful and conflicts did not necessarily require resolution.

By the 18th century, however, biology was immersed in conflict. Biologists were caught up in the rush to divide living things according to a right of ownership. At the time (and since), vast collections of things (removed from their place of origin [and original owner]), accrued to victors, to those who could afford to buy the collections and maintain them in the new circumstances, and to those empowered to study them. Collections were then catalogued and organized according to the new owner's value systems. As for living things, a premium was placed on complexity and stability (parallel to the adult, European [Anglo-Saxon, etc., as the case may be] male, or Man), with everything else (and everyone else, including women) falling in line as "higher" and "lower" depending on their resemblance to life at the pinnacle.[4]

The scientific formalism in the 18th century accommodated readily to emerging bourgeois values of higher and lower. The qualities allowed men to propose natural laws and organize nature systematically, and the struggle to organize all the collected booty on scientific principles was waged in tooth and claw. A few victories were scored in the wars of classification, the most far reaching of which was the solidification of taxonomic units, especially the species.

Taxonomy, the branch of biology concerned with formal systems of nomenclature, was the passion of the 18th century, and none was more passionate than Carolus Linnaeus (Carl von Linné; 1707–1778). Like others before him, notably John Ray (1627–1705), and contemporaries, Bernard Jussieu (c. 1699–1777), Linnaeus used differences to define groups[5] and to place all living things into a nested system of hierarchies (beginning with species within a genus). Zoological taxonomy began with the publication of Linnaeus's *Systema Naturae*, the 10th edition of which, published in 1758, became the standard for zoological nomenclature. It was, nevertheless, considerably altered by, among others, Peter Simon Pallas (1741–1811) at the low end of taxa and Georges Léopold Chrétien Frédéric Dagobert, Baron Cuvier (1769–1832), at the high end. The publication of Linnaeus's *Species Plantarum* in 1753, with important additions from Bernard Jussieu's nephew, Antoine Laurent de Jussieu (1748–1836), among others, became the basis for plant taxonomy.

Natural laws brook no exceptions, and the rules established by Linnaeus for naming living things are nothing if not rigid. Even today,

> Scientific names are words of the Latin language ... [and are] subject to the rules of its grammar.... The name of a species consists of the name of the genus to which it belongs followed by one epithet, ordinarily an adjective, occasionally a noun in apposition or in the genitive.... Names of groups higher than genera are in the plural. Some are proper nouns; the remainder are adjectives used as proper nouns, agreeing in gender with the names of the kingdoms in which they are included; either expressing characters of the groups which they designate, or consisting of generic names modified by terminations signifying "resembling" or "of the group of."[6]

The hallmark of 18th century formalism was the Great Chain of Being. It epitomized law, constancy and inflexibility ("Chain" never seemed quite the right metaphor), having no gaps and no room for change,[7] but it was not entirely above criticism. George-Louis Leclerc, Comte de Buffon (1707–1788), and later, Etienne Geoffroy Saint-Hilaire (1772–1844) drove a time axis through the Chain, and Jean-Baptiste Pierre Antoine de Monet, Chavalier de Lamarck (1744–1829) attempted to explain apparent breaks in the Chain with an evolutionary principle. The German nature philosophers tugged the Chain into a recapitulation of life itself, and Johannes Peter Müller (1801–1858) turned its links into echelons on a ladder of life, each step representing the completion of an act of development.

By mid-18th century, even the changeless Chain had chinks, and, under the impact of evolutionary theory, it cracked. Ernst Heinrich Philipp August Haeckel (1834–1919), germinated branches from the Chain and turned it into a great Tree of Life.[8] The idea of a "Tree of Life" is, of course, ancient but it flowered in the 19th century under Haeckel's able hands. The point of departure for biology was Haeckel's portraying the branches of life's tree as bearers of related organisms and the branching points as indicative of more remote relationships. Formerly, nothing whatsoever about the grouping of similar organisms into classes separated by difference implied relatedness in the sense of a linear family. Even the hierarchical groupings of Linnaeus and the later concept of natural groupings had not betrayed anything approaching a familial relationship. Darwin and his allies, of course, are given credit for having derived an arborized view of life, but Haeckel's stature as a biologist and his enormous success as an author of popular books on biology turned the concept of related groups into a theory.

Haeckel permitted himself a great deal of latitude and drew the tree of life as he saw fit.[9] He created a Kingdom Protista (including bacteria) and positioned it between the kingdoms of Metazoa (animals) and Metaphyta (plants), and he imagined an ancient Kingdom Protophyta (of bacteria, in general, also known as Moneres or Monera, and Schizophyta) at the root leading to the Kingdom Protozoa, forming the base of life's "trunk."[10] In the early 20th century, microbiologists bifurcated the trunk and squeezed all living things into just two incommensurable branches, which Edouard Chatton (1883–1947)[11] called prokaryotes (i.e., microbes), consisting of any sort of unicellular (uniglobular) organism seen with the microscope (e.g., yeast and *Paramecium* as well as bacteria) and eukaryotes, consisting of everything else. The branches were not entirely self-sufficient, and soon criteria were refined. Eukaryotes were redefined positively as organisms whose cells had nuclei, while prokaryotes were redefined negatively as organisms whose cells lacked nuclei.

Soon thereafter, Herbert Faulkner Copland (1902–1968)[12] proceeded to distinguish four kingdoms of living things: a Kingdom of prokaryotes called Mychota, containing bacteria and bluegreen algae (since called bluegreen bacteria) and three kingdoms of eukaryotes: Protoctista containing protozoa, the red and brown algae, and fungi; Plantae with walled cells containing chloroplasts having chlorophyll *a*, *b*, carotene and xanthophyll, which produce sucrose, starch and cellulose; Animalia consisting of wall-less cells and attaining a high degree of complexity by passing through embryonic stages of development.

The growth of physiological-ecology as a discipline then inspired Robert Harding Whittaker (1920–1980)[13] to shuffle Protoctista back to Protista (without Haeckel's Monera) while extracting Fungi and adding green algae. He viewed life as divided among producers (consisting of plants and algae), consumers (animals), and reducers (fungi and bacteria). Later still in the century, the "emerging emphasis on similarities, made clear by ultrastructure, rather than differences, ... encouraged a holistic view of unicellular eukaryotes"[14] and chipped away at the botanical/zoological barrier, for example, by breaking off euglenoids from plant-like organisms and placing them with protozoans bearing kinetosomes. A compromise was then worked out with Five Kingdoms: Bacteria (also known as Monera or Prokaryota), Protoctista (or Protista including red, brown and green lines of algae and protozoa), Fungi, Plantae and Animalia.[15] The Five Kingdoms did not last either. Thomas Cavalier-Smith recognized only four divisions of life (fashioned around cell types), but he placed them at levels within yet another hierarchy. Bacteria (comprising all prokaryotes) and Eukaryota were assigned to the level of Empires, the highest taxonomic category alluded to by Linnaeus. In addition to the presence versus absence of a nucleus, Eukaryota was distinguished from Bacteria by four fundamental departures:[16]

1. Internal cytoskeleton and internal system of membranes (endomembranous system); compartmentalization and transfer of materials; surface modification especially in unicellular forms and extracellular materials capable of interactions during development of multicellular forms.

2. Novel mode of nutrition: phagocytosis and internal digestion by lysosomes; uptake of cellular endosymbionts (leading to their conversion to permanent cell organelles or plastids.[17]

3. Loss of bacterial cell wall in favor of more flexible plasma membrane and thin glycoprotein surface coat; fusion of sex cells enabled complete genome merger (allopolyploidy).

4. Changes in genomic organization, split genes and chromosomes "packaged" in nucleosomes, with large amounts of repeat DNA attached to nuclear matrix. In addition, Eukaryota uses three RNA polymerases rather than one used in Bacteria.

Bacteria broke down unambiguously into kingdoms Eubacteria and Archaebacteria, while Eukaryota were much more difficult to differentiate. The eukaryotes separated somewhat ambiguously into two superkingdoms of Archezoa and Metakaryota. The Archezoa, containing one kingdom (also called Archezoa), with three phyla, consisted of organisms whose cells lacked mitochondria, peroxisomes and golgi apparatus (GA or dictyosome [although they might possess membranes of the smooth endoplasmic reticulum resembling golgi transport elements, and mitochondrial-like hydrogenosomes]).[18] The Metakaryota, with five kingdoms, consisted of organisms (with the exception of the proteromonads) whose cells had the full complement of endomembranes and cellular organelles. The five kingdoms of Metakaryota were: Protozoa, Chromista (consisting of algae with chloroplasts [plastids] not resembling those of plants), Fungi, Plantae and Animalia,[19] each with the exception of Fungi, divided into two or more subkingdoms. The big loser in the new classes was the Protists (or Protoctista) which lost major groups to Archezoa and Chromista.

Plants were not entirely forgotten in the battle of names. Adolf Engler (1844–1930), leader of the Berlin school of plant taxonomy, wrote *Syllabus der Pflanzenfamilien*, followed by the American, Arthur Cronquist (1919–1992), who developed a phylogeny of flowering plants at and above the level of family. Modified by George Ledyard Stebbins, the "Cronquist system" of plant classification became standard.

It was microbiology, however, that made the least progress, and many lost faith that microbes would ever yield to phylogenetic classification.[20] Cornelis Bernardus Kees van Niel of the Hopkins Marine Station of Stanford University, who pioneered the study of photosynthesis in bacteria, and Roger Yate Stainier, of Berkeley and the Pasteur Institute, who studied tryptophan metabolism and bacteriochlorophyll, and author of *The Microbial World*, despaired of ever finding a satisfactory taxonomy of microbes.

Meanwhile, Carl R. Woese (b. 1928), at the University of Illinois, was "looking" at the small subunit of bacterial ribosomal rRNA. He had enzymatically broken down the subunit into oligonucleotide fragments of six to 20 nucleotides, dispersed in two dimensions on film where their presence could be detected as "fuzzy spots." Woese subsequently identified the nucleotides present in spots (their sequence or profile) chemically, and showed that the patterns of oligonucleotides on film were unique to each species he studied and that the matches, similarities and differences he observed in the sequences from different species was evidence of their phylogenetic relationship.

After a decade of "mind-numbing tedium," Woese had sequences for the small subunit rRNA (actually the ribonuclease [RNase] T1 oligonucleotide content of 16S rRNAs) of about 60 species of bacteria. Following a suggestion by his colleague, Ralph Wolfe, Woese discovered that the rRNA profile of microbes thought to be methane-producing bacteria (methanogens) was not that of bacteria at all.[21] Soon thereafter, Otto Kandler and Wolfram Zillig supported the separation of Bacteria from several methanogens on grounds of membrane lipids and modified nucleotides of tRNAs, and added salt-loving halophiles, heat-loving thermoacidophiles, and sulfur-metabolizing methanogens to the new group, Archaebacteria.[22] The 16S rRNA sequences led to other gratifying results, including vindication for claims of bacterial origins for cell organelles (chloroplasts[23] and mitochondria[24]).

At this point, the argument on behalf of rRNA's use in taxonomy seemed quite straightforward: molecules are part of a phenotype and, especially for very small organisms, provide genuinely useful criteria for evaluating relatedness. But Woese and colleagues[25] did not stop there. They continued their program of organizing life around molecular data, especially the wealth of new sequence data becoming available. The method of choice for organizing molecular data was the new cladistic methods of geometrical representation that relied on information theory. According to Woese and others, molecular divergence measured evolutionary separation while reflecting evolutionary time. Ribosomal RNA would henceforth be the ultimate molecular chronometer.[26]

THE NEW LOOK OF DIFFERENCE

Richard W. Hamming introduced biology to information theory as a device for evaluating difference in the form of changes (substitutions) in otherwise similar data. The question was, "What is the probability of going from one (character) state to another (character) state?" Answers were sought using the methods of statistical geometry. The probability of a substitution is, of course, nothing more than a guess, but, inevitably, a guess comes to be thought of as a likelihood, and its graphic representation, or cladogram, merges with paths of change, hence, an evolutionary scenario. The "Hamming metric" was first applied to problems of evolution by Ingo Rechenberg,[27] and biology has not been the same since.

Cladistics Methods

Cladistics is currently the chief player in the game of defining large, evolutionary differences among living things using small, molecular differences in their DNA sequences. Cladistics is an adept player, since it offers the possibility of juggling many variables at the same time, but its main asset is that it easily incorporates sequence information from the vast databases of molecular biology, and draws explicit conclusions about the probability of substitutions with the help of computer programs.

Today, DNA sequencing is routinely performed in service laboratories at the behest of scientists studying evolutionary relationships. DNA is the material of choice for sequencing for several reasons. Above all, DNA sequence technology is far advanced over other sequencing technologies. In addition, DNA is easily obtained, and specific sequences are available in virtually unlimited amounts through cloning and the polymerization chain reaction (PCR) with specific primers. Moreover, "nuisance" introns are readily eliminated from DNA sequences. Thus, while one may be interested in ribosomal RNA (rRNA), sequencing is performed on rDNA,[28] the DNA encoding rRNA, and while one may be interested in polypeptides, sequencing is performed on the encoding DNA regions. The DNA sequences are converted by computer to polypeptide sequences which ignore introns and "smooth out" idiosyncrasies due to codon bias and other misdemeanors of genomic DNA.

After obtaining the DNA from the species of interest, or more likely, a clone or primer sequence from a commercial source, the investigator routinely performs PCR and obtains ample quantities of DNA from the desired gene. Following complete sequencing, if not already available, a consensus sequence of the entire gene is obtained with programmed contig alignment (DNASIS). The resulting sequence is then used for queries in BLAST searches against sequences in databases such as GenBank.[29] Sequences with high similarities are thus located for evolutionary analysis.

Alignment of sequences in general, as always, relies on "eyeball" assisted computer programs. Some version of CLUSTAL (V or W) is employed and sequences are "adjusted manually" or excluded when they cannot be aligned even with the help of gaps. When alignments cannot be made across an entire sequence length or a comparison is desired for a closed sequence, so-called conserved regions are identified by dot-plot analysis (SeqApp) set with a given window and a minimum number of matches for identification. These regions may subsequently be realigned by CLUSTAL or a data editor (PAUP). Gaps may be accompanied by penalties or may be treated as missing data. Other "adjustments" sometimes smooth out problems at the price of reduced reliability. For example, after aligning by amino acid sequences, ambiguous nucleotides may be discarded by back-translating to nucleotides, and bias arising from the sequence used initially for alignment (reference sequence) may be minimized by realigning by subunits.[30]

Different treeing methods are commonly employed, but, in general, trees are preferred which are constructed with the fewest number of steps. In trees built by maximum parsimony methods, the total length of branches separating species is proportional to the number of inferred substitutions, and the branching pattern is based on the assumption that sequences are connected with the fewest number of changes (parsimony). The trees show actual changes between pairs of sequences and reflect the greatest statistical likelihood (log-likelihood and bootstrap probability) of observed sequences. Maximum parsimony methods are liable to be confounded, however, by widely varying rates of substitution (changes or mutations) among species, since observed changes in sequences tend to underestimate the number of actual changes separating two species, especially when they are distantly related.

David Swofford's PAUP (phylogenetic analysis using parsimony)[31] is a popular treeing program. PAUP uses the "branch-and-bound" method (a heuristic search) in which sequences are reshuffled exhaustively in a different order of entry in order to find the tree that minimizes the total number of substitutions necessary to link all sequences. PAUP is easy to use, and performs well by way of giving good estimates of correct trees and branch lengths in simulations with synthesized data sets containing generally divergent evolutionary patterns. Various criteria may be introduced through minimum evolutionary criteria (HKY85 distances), log determinant distances (LogDet) or weighted parsimony (which assigns greater cost to transversions than to transitions in DNA sequences). The problem is that PAUP searches use very large amounts of computational time, and are frequently performed, therefore, with limited numbers of sequences or minimal-information parts of sequences.

Wayne Madison's and David Madison's MacClade is a highly flexible treeing program which allows one to freely alternate the distribution of species on parsimonious trees, although it is suitable only for character state (four nucleotides; 20 amino acids) data. The program permits the detection of possible horizontal gene transfer through the use of differential weighting biases for sequence gain or loss (by ACCTRAN and DELTRAN optimization[32]). Constraint trees (permitting strictly vertical transmission of sequences [forced monophyly]) may also be used in alternate topologies to test "fine branching" hypotheses.

Ideally, one finds the most parsimonious tree or optimal tree with the fewest branches and shortest overall path length connecting all sequences under consideration. In practice, several equally parsimonious trees are usually derived from the same data, inferring different phylogenetic relationships with equal parity or distance among the molecules and their presumed ancestor. Additional data may not tilt the balance toward any one or even a subset of such trees. In addition, polytomies, or origins of two of more "descendants" from a single node (multiple, simultaneous branching events from a polytomous node) are common and often cannot be resolved even with additional data.[33]

In trees built by distance matrix methods, branch length measures the extent (distance) to which a fraction of positions differ in pairs of aligned sequences and genetic distance is assumed to be proportional to phylogenetic distance. For polypeptides, a tree is produced from a matrix of pairwise distances (Margaret Dayhoff's PAM 250 scale for data more than 50% identical; the Gonnet–Cohen–Benner [GCB] scale; BLOSUM tables for data less than 50% identical). Distance matrixes may be calculated and pairings evaluated with PRODIST for sequences with as little as 20–30% similarity. SCORE makes pairwise comparisons and calculates distance values. Because distance matrix methods look only at the number of changes between pairs of sequences, they do not depend on the number of changes in a sequence. Statistical corrections, thus, are possible for the weakness of parsimony methods. On the other hand, distance matrix methods are subject to large sampling errors as well as to systematic biases.

Trees are frequently constructed from DNA sequences using programs (DNAML) in Joseph Felsenstein's PHYLIP (phylogeny inference package[34]). Several variables must be estimated even if their effects are not well-defined: a transition/transversion ratio; the relative probabilities for change in the first, second, and third bases of codons; the purine/pyrimidine ratio in extant sequences used to estimate the purine/pyrimidine pool.[35] Data converted to sequences of amino acid residues in polypeptides (proteins) may use maximum likelihood methods to examine accumulated results simultaneously, allowing the synthesis of tree topologies for different proteins, and making it possible to estimate and statistically test (by the Kishino–Hasegawa test) the overall reliability of a particular tree topology.[36]

Each treeing program has problems of statistical consistency, and none is guaranteed to find the correct tree as the number of characters grows. These problems are reduced by assuming a common mechanism for changes in all characters, but such an assumption requires new transformations to be consistent with tree selection methods.

In simulation experiments, where the "right" tree is built in, simple parsimony methods and other methods, uncorrected for multiple changes to character states (amino acids or nucleotides), find the wrong tree when branches are evolving at different rates or parallel changes occur on different lineages. Corrections applied for distance (Jukes–Cantor, Kimura 2- and 3-parameter models), can be used with maximum likelihood and parsimony methods to enhance convergence toward the right tree. The problem becomes making biologically sound corrections, and, "the more we know about the mechanisms of evolution, the better the corrections we can make, the higher the chance of getting the correct tree."[37]

Drawing Evolutionary Inferences from Molecular Differences

Cladistic methods utilizes differences in sequence data to draw phylogenetic inferences. The methods shuffle through "character states" (alternative possibilities for sequence pairs) until they fall into place among the nodes and clades of a cladogram. In all cladograms, species occupy the distal end of branches connected proximally through paths of nodes and internodes. Only the order of splitting of character states is inferred for species, but cladograms are routinely interpreted as phylogenetic maps or scenarios. Further distortion arises when branch length is imagined to reflect evolutionary time instead of diversity.

Correlations between rates of evolution and amino acid or nucleotide substitutions were sought since the advent of molecular sequences (Kimura's and Ohta's principles of molecular evolution). As one might expect, given the sensitivity of parsimony methods to problems of substitution rates, correlations between evolutionary time and molecular diversity must be sought judiciously. Regrettably, this is not usually the case.

Molecular biologists do not attempt to hide the exchange of evolutionary time for molecular diversity. The dean of the new molecular taxonomists, Carl Woese, for example, puts the confusion clearly:

> Since the genetic sequences with which we deal exist only in one state (their present state), past states cannot be known directly. However, two different versions of the same molecule (from two different organisms) have at some past time shared a common ancestor. Therefore, knowing the differences between the two is almost as good as knowing the difference between the present state of either one and the earlier, ancestral state: in other words, the difference between two extant (homologous) sequences is a (relative) measure of evolutionary time, the time since the two sequences shared a common ancestor. (This difference is, of course, akin to distance, not time. Thus, any time interpretation given it requires knowledge or assumptions regarding relative rates of change in both lineages).[38]

This is not to say that all biologists were comfortable with the exchange of time for diversity. Cavalier-Smith, for one, warned, "we are certain to be wrong if we treat branch lengths on rRNA trees as completely accurate indicators of evolutionary time."[39] A few biologists held out for a balanced approach[40] and some believed that "the hypothesis that the behavior of a single gene, whether an RNA or a protein gene, faithfully represents the evolutionary history of an entire organism is rapidly losing support; just in time ... for the tidal wave of sequences originating from genomic sequencing projects and the resulting possibility to compare and analyze simultaneously many homologous gene collections."[41] One might have imagined that some sort of rationalization might have accompanied the glut of molecular data (such as a "majority rule"), but nothing of the sort was forthcoming. The seduction of reading phylogenetic scenarios into molecular cladograms simply proved irresistible; its consequences are ongoing.

Cladistic theory does not depend on concrete ancestors existing at nodes. Nevertheless, "An organism with the expected evolutionary features and present in the expected time range is generally nominated for the role of ancestor to subsequent species."[42] As a consequence, several terms take on dual meanings which, while not necessarily contradictory, are not necessarily synonymous. Symplesiomorphies or just plesiomorphies (most forms) are defining characteristics, the set of characters shared by all species in the class but sometimes redefined as primitive features "in relatively unmodified states."[43] Synapomorphies are shared characters present in a clade of sibling (sister) groups but sometimes equated to "derivative states evolved from the primitive features within members of the clade."[44] Autopomorphies are the seen characters or the distinguishing characteristics of species but sometimes interpreted as advanced features.

The methods of cladistics are not to blame for the distortions made of them. MacClade, PAUP, Treetool and other treeing programs (see above) produce parsimonious unrooted trees, having no point of departure around which an evolutionary scenario can be written (imagine looking at a tree from above). Rooting (linearizing) a tree is easily accomplished, however, in any of several ways. In the simplest, theoretical possibility, rooting is accomplished by default: symplesiomorphies point to one member bereft of any positive synapomorphy. This member becomes the sibling (sister) group of all other groups in the clade. Other efforts to root trees depend on choosing an outgroup (lacking one or another symplesiomorphy) to linearize the ingroup. The internode where the groups converge is considered the point of origin of the ingroup. Still other rooting devices require assumptions about nested (hierarchialized) relationships among characters considered homologous.[45] When the homologues are in the genome of a single species (clustered or diffuse), they are said to be paralogues and may be isozymes (alternate versions of enzymes often produced by cells under different physiological functions) or members of gene families. When the homologues are present (whether singly or multiply) in different species, they are said to be orthologues or allozymes, and also considered members of gene families. The point where these homologous sequences collide is considered the root of the clade.

NONCELLULAR LIFE: VIRUSES AND RETROTRANSPOSONS

Those studying noncellular life have traditionally relegated many differences to "boxes" of sameness. Noncellular life forms seem to have fallen victim to a conspiracy to conflate them into as few varieties as possible. Their small size may be the excuse, but the myth of viral sameness should have been exploded by the results of sampling techniques, such as fingerprinting, that resolve allelic heterogeneity in populations.[46] Still, noncellular life is said to contain few varieties of biomolecules and to have these in sparse amounts. The strategy for noncellular life is described in textbooks as one of compacting and compressing. Viruses are apotheosized for using the same viral genes (virogenes) in several capacities and having functionally overlapping virogenes comprising almost half the genomes of DNA and RNA viruses. Polypeptides encoded in several genomic regions are also put together through alternative splicing or differential sequence expression.

Noncellular life exists in a great variety of forms, nevertheless. Some virions are naked (plant viruses, bacteriophage[47]), showing only their proteinic capsule (capsid); others have capsid proteins decorated with glycoproteins (adenovirus); still others are enveloped in membrane (influenzavirus). Their capsid may contain capsomeres of one or different polypeptides and capsid form varies from iscosahedral and isometric to pleomorphic, lemon-shaped, rod-shaped, and helical; infectivity may be accomplished through an intromissive device or the unadorned capsid; genomes may encode a few polypeptides or a variety of enzymes functioning in DNA synthesis, nucleic acid metabolism and the regulation of protein processing.

Viral classification has traditionally depended on host selectivity, particle size and structure, and serological (antigenic) specificity and cross-reactivity, to which are added genomic size and type (namely, the type of nucleic acid strand or strands contained in a virion), as suggested by David Baltimore. Not every criterion is useful "locally" within groups and across the whole spectrum of viruses and retrotransposons. Genomic size, for example, varies enormously, from the smallest single-stranded RNA genomes of approximately 3.5 kilobases (kb) in bacteriophages such as MS2 and Qβ and 3.0–3.3 kb in the sense strand of hepadnaviruses to the largest double-stranded DNA genomes of up to 280 kilobase pairs (kbp) in cytomegaloviruses. The problem is that genomic size overlaps considerably among viruses grouped together by other criteria and differs significantly within these groups.[48]

The type of nucleic acid present is a more consistent and hence useful taxonomic criterion, since the virions of similar viruses generally contain only one form of nucleic acid, and that, generally, in only one configuration. The nucleic acid may be single-stranded RNA (ssRNA), in which case the RNA may serve as a positive (+)[49] mRNA strand (picornaviruses, togaviruses, flaviviruses, and coronaviruses), or as a negative (–) template for mRNA synthesis (bunyaviruses, arenaviruses, orthomyxoviruses, paramyxoviruses, and rhabdoviruses). In addition, viruses may switch from RNA to DNA, using ssRNA as a template for DNA synthesis (retroviruses), in the presence of the virus's own reverse transcriptase, or DNA as a template for RNA synthesis (hepadnaviruses and caulimoviruses). The viral genome may also be composed of double-stranded RNA (dsRNA: reoviruses), double-stranded DNA (dsDNA: papovavirus, adenovirus, herpesvirus, poxvirus) or single-stranded DNA (ssDNA: parvovirus).

Sequence data also challenge the simplistic view of noncellular life, and, if these data are allowed to stand on their own (and not shoved into a box of samenesses), noncellular life may come into its own. Defined generally, but not exclusively by sequence data, noncellular life encompasses RNA and DNA viruses and a superfamily of "switching" retroviruses, pararetroviruses and retrotransposons. By analogy with the domains of cellular life (see below), noncellular life would then constitute a superdomain, and the groups of viruses and retroviral-like elements would be consigned to domains.

RNA Viruses

The genome of all (nondefective) RNA viruses contain a gene encoding an RNA directed (dependent) RNA polymerase.[50] These genes share six to eight short sequence motifs in clusters, and the phylogenetic tree inferred from those sequences agrees with supergroups "merely based on the strandness, on the structure and organization of the respective genomes, and on additional conserved genes."[51] All virogenes may be present on each piece of viral nucleic acid in a virion, hence nonsegmented, or different genes may be present on separate nucleic acids in the viral genome, hence segmented.[52]

RNA viruses do not perform recombination *sensu stricto*, joining pieces from different molecules without replication. Many RNA viruses (Picornaviridae and Coronaviridae) may, however, undergo copy-choice, or template-switching by the replicase during synthesis of a new strand, with the same effect as recombination. Segmented viruses may achieve a similar effect when coinfection results in new combinations of genomes within the new virions.[53]

(+)ssRNA Viruses

The genomes of positive (+)single strand RNA viruses play the role of mRNA and are capable of infecting their host cell in the absence of viral proteins, although their normal capsid greatly enhances the efficiency of infection. Viruses of this type (picorna-like, alpha- [Sindbis- or toga-] like, flavi-like, carmo-like, sobemo- [luteo-] like, and corona-like), are the most abundant virus class, comprising close to 80% of the RNA virus families or groups, but they do not resemble each other closely (with the exception of subsequences or domains, especially of nonstructural proteins) and are frequently broken down into positive-stranded nonsegmented (monopartite) viruses (picornaviruses, caliciviruses, poty/bymoviruses), bisegmented como/nepoviruses and polysegmented viruses (bromovirus).[54]

Picorna- and alpha-like (previously Group A arboviruses) supergroups of (+)ssRNA viruses infect both plants and animals. The picorna-like or "simple-set" viruses include families of animal viruses (Picornaviridae and Caliciviridae[55]) and several groups of plant viruses (Potyviridae, Comoviridae). Picornaviruses (entero-, rhino-, hepato-, cardio- and aphthovirus) have genomes between 7.2 kb (human rhinoviruses) and 8.5 kb (foot-and-mouth disease virus). Structural and nonstructural (enzymatic and regulatory) proteins are translated as a polyprotein (2100 to 2400 amino acid residues long) from a central portion of the genome. The 5′ end, covalently linked to a small viral protein gene (encoding VPg), contains an untranslated region (600 to 1200 bases) involved in translation, virulence and possibly encapsidation. The 3′ end has an untranslated region (50 to 100 bases) involved in replication of the negative (–) strand. While these features are "viral-like," the 3′ end resembles eukaryotic mRNA by way of enzymatically acquiring a polyadenylic acid (3′ poly[A]) tail.

Togaviruses (Western equine encephalomyelitis virus) have genomes of about 11.7 kb. The resemblance to eukaryotic mRNA extends to both ends, with a 3′ poly(A) tail and a 5′ methylated cap.[56] The genome also illustrates "differential expression," producing nonstructural proteins in a first round of translation and structural proteins in a second round.

Some (+)ssRNA supergroups infect only plants. Sobemo-like virus has a 5′ VPg gene, but it lacks a 3′poly(A) tail. Carmo-like virus likewise has a 5′ methylated cap and lacks a 3′ poly(A) tail but produces separate mRNAs.

Other (+)ssRNA supergroups infect only animals. Flaviviruses (previously known as Group B arboviruses), transmitted by mosquitoes and ticks, have genomes of about 10.5 kb which are "eukaryote-like" at the 5′ end, having a methylated cap, but lack a 3′ poly(A) tail. Like picornaviruses, flaviviruses encode their structural and nonstructural proteins in one polyprotein. Coronaviruses have the largest genomes, 27 to 30 kb, among the (+)ssRNA viruses. Like togaviruses, coronaviruses have both a 3′ poly(A) tail and a 5′ methylated cap, but unlike other ssRNA viruses, coronaviruses translate a viral RNA-polymerase (or replicase) from the 5′ end of the genome. The plus strand then spins off the complete negative (–) strand of RNA that acts as the template for the production of a variety of mRNAs by differential transcription. These mRNAs have different lengths, but each begins with a 5′ nontranslated leader sequence containing a control region (they are monocistronic) and their own 3′ poly(A) tail.

(–)ssRNA and dsRNA Viruses

The genomes of negative (–) single strand RNA viruses play the role of template for mRNA (+) synthesis. In similar double-stranded RNA (dsRNA) viruses, the genome is ambisense, or ambistranded, having both positive and negative strands. All these viruses have nucleoprotein particles surrounded by a lipid envelope containing one or more glycoproteins. Unlike the naked genome of positive-strand RNA viruses, which are infectious on their own, the naked genome of negative-strand RNA viruses is not infectious. The explanation for this lack of infectivity seems to be that RNA-dependent RNA polymerase responsible for producing the mRNA strand is not normally present in the host cell but introduced by the virus upon infection.

A rather homogeneous group of Filoviridae, Paramyxoviridae, and Rhabdoviridae, constitute the Mononegrivales, an Order or superfamily[57] of monopartite (nonsegmented: with a physically continuous genome), negative-stranded RNA viruses. They have a moderately small genome ranging from about 11 kb in rhabdoviruses to 17–20 kb in paramyxoviruses (Sendai virus). Negative-strand RNA genomes are never polyadenylated or capped as such, but the genomes of paramyxovirus and rhabdovirus are eukaryote-like in encoding a polyadenylation signal at the end of each gene. The genomes are also eukaryote-like by way of their linearly arranged genes separated by an intergenic sequences (guanine–adenine–adenine), and each gene's beginning with a translational start signal.

In other (–)ssRNA viruses (Arenaviridae, Bunyaviridae, Orthomyxoviridae) and dsRNA viruses (Birnaviridae, Reoviridae), the RNA genome is segmented and physically divided into two (bipartite) or more (polypartite) linear helices within the same virion. Influenza C virus (Orthomyxoviridae) has a seven segment genome, and influenza A and B viruses have an eight segment genome. The segments are not equal (the three negative sense strands of a bunyavirus are 8.5 kb, 5.7 kb and 0.9 kb, while the two strands of an arenavirus are 2.8 kb and 5.7 kb). In single-component viruses, each virion carries the entire complement of segments (the ssRNA *Reovirus* with 10 to 12 segments ranging from 680 to 4500 bases). In multicomponent viruses, virions do not tend to carry the full complement of segments, and coinfection of the same host cell seems to be necessary for a successful infection (Tobravirus, Cucumovirus, Bromovirus).

DNA Viruses

ssDNA Viruses

Single strand (ss) DNA viruses are represented by parvovirus (canine parvovirus). Its genome consists of 5 kb of ssDNA which is packaged primarily as antisense strands although some sense strands will also occur in the same virion. Like (+)ssRNA viruses, ssDNA viruses undergo replication before transcription. The small genome contains a *rep* gene, encoding polypeptides involved in transcription, and a *cap* gene, encoding capsid proteins. Variety is created through alternative splicing. In some parvoviruses, replication is dependent on helper-adenoviruses, herpesviruses or treatment with an "inducing" agent such as ultraviolet light, cycloheximide, or some carcinogens.

Plant bipartite DNA viruses (geminivirus) are unusual in segregating their genomes among virions. Two or more genome segments are actually present in different virions, but these can "get it together" where horizontal movement or inoculation by sap-sucking insects can result in multiple infections of a single cell. Both strands of the DNA code for overlapping reading frames.

dsDNA Viruses

Double-stranded (ds) DNA viruses are grouped into "large," a "smaller large" (or intermediate), and "small" varieties based on the size of their genomes. A striking characteristic, well documented in the case of the "large" mammalian herpesviruses, is that sequences of nonstructural (enzymatic and control) viral proteins are similar to those of nonviral host proteins, suggesting that these viruses devolved from the host species (see regressive evolution in Chapter 4). Sequences in other herpesviruses, such as catfish herpesviruses, do not resemble those of mammalian herpesvirus, however, and no major virion structural protein, in the shell, core, or surface membrane, has an unambiguous cellular homologue.[58] Host genes in the dsDNA viral genome would, therefore, seem to have gotten there by recombination.

The "large" variety of dsDNA genomes are linear and more than 100 kbp. For example, in herpes simplex virus, a member of the Herpesviridae, the genome consists of 152 kbp containing 75 genes. The genome encodes a variety of enzymes, inverted repeats and multiple repeated sequences. The genes are frequently structured and organized the same way but show little sequence similarity even across the family. Although overlapping, each gene is expressed from its own promoter.

A "smaller large" (or intermediate) DNA genome occurs in the Adenoviridae family where linear dsDNA is 20–38 kbp and contains 30–40 genes. Promoters are shared by clusters of genes, and, via alternative splicing, several mRNAs are synthesized encoding different polypeptides. Sequence similarity is great among members of six adenovirus groups, but much less between groups.

Polyomaviruses (members of Papovaviridae) package a "small" variety of dsDNA approximately 5 kbp, but it is circular, rather than linear, with supercoiling within the virion, and associated with eukaryote-like histones (H2A, H2B, H3 and H4). Apart from the noncoding regions surrounding the origin of replication and controlling transcription, six genes are encoded within the two strands of DNA, of which one has a unique open reading frame, while five share overlapping sequences.

Another "small" variety of dsDNA genomes occurs in bacteriophage. Size varies among bacteriophage genomes, and their genomes are expanded in mutant varieties used as cloning vectors (see Chapter 2). Normally, single genomes occur in λ phage (49 kbp [46–54 kbp]); genomes lengthened by the terminal redundancy of genes occur in T4 phage (160 kbp). In the filamentous bacteriophage M13, the genome is expandable with extra sequences of nonessential intergenic elements. In a newly infected bacterium, a phage-encoded endonuclease cleaves the phage DNA at a *cos* site, leaving "sticky ends" capable of reannealing. With the help of DNA ligase, the DNA heals into a circular molecule which may then undergo vegetative replication or, by cleavage at an *att* site, become integrated into the bacterium's own chromosome.

DNA/RNA and RNA/DNA Switching Viruses (Pararetroviruses and Retroviruses) and Retrotransposons

Retroviruses and pararetroviruses, with DNA-containing virions, and "related," nuclear-bound retrotransposons, are characterized primarily by encoded reverse transcriptase. Because sequence similarity is low (only 25% of the amino acid residues are similar in the relatively good match between the reverse transcriptases of murine leukemia retrovirus and human immunodeficiency retrovirus [HIV-1]), alignment is based on "conserved" amino acids (42 positions containing largely invariant or chemically similar residues in 88% of the reverse transcriptases). These residues define seven, short domains (of 6 to 16 amino acids) encoded in all reverse transcriptase sequences. Most of the remaining amino acid residues, bearing no sign of similarity, occur in unique "expansion regions" between the "conserved" domains.

The X-ray crystal structure of reverse transcriptase from HIV-1 (actually the p66 subunit containing the active catalytic site for polymerization) resembles a right hand. The "palm" subdomain and all but the first 40 amino acids at the amine-terminal of the "finger" subdomains show significant amino acid sequence similarities when compared to all other reverse transcriptases, while the "thumb" subdomain shows no sequence similarities between reverse transcriptases encoded in retrotransposons. Expansion regions extend from the active site, suggesting that their amino acids would not have major affects on enzyme activity, and, hence, their evolution would not be functionally constrained.

Similarity among reverse transcriptases is weighted so heavily by some virologists that noncellular forms producing them have been linked in a retroviral superfamily,[59] which would seem more nearly equivalent to a domain in the superdomain of noncellularity. The relationships among pararetroviruses, retroviruses and retrotransposons are, of course, steeped in the uncertainties of sequencing.

Pararetroviruses

In the **pararetroviruses**, hepadnaviruses (primarily of mammalian and avian liver cells) and caulimoviruses (represented by cauliflower mosaic virus [CaMV]), integration into a host genome is not necessary for completion of the life cycle, and DNA is packaged in virions following the action of reverse transcriptase. Hepatitis B virus, the prototypic hepadnaviruses, has a "gapped," partially double-stranded, DNA genome, consisting of a variable antisense strand (1.7–2.8 kb) and a sense strand (3.0–3.3 kb) encoding four genes, including *gag* and *pol* with reverse transcriptase and RNase I domains, similar to retroviral genes. Upon infection and presumably after entering the host-cell's nucleus, the "gap" is filled, and the DNA closes into a circular form, and three strands of mRNA are transcribed. Virions also contain a reverse transcriptase (an RNA-dependent DNA polymerase) which converts the RNA transcripts to DNA, providing a core for cytoplasmic virions.

Cauliflower mosaic virus, the prototypic caulimoviruses, has a "gapped," circular DNA genome of about 8 kbp, the α strand of which has a single gap, while the complementary strand contains two gaps. Eight genes are encoded. After migration to the host-cell's nucleus and repair of the gaps, transcription yields two mRNA transcripts that acquire polyadenylate tails.

The shorter transcript is translated into proteins that become incorporated into large inclusion bodies, and, on these "replication complexes," the longer transcript is "replicated" by reverse transcriptase into DNA and packaged into virions.

Retroviruses

A temporal form of genomic apportionment occurs in **retroviruses** (avian reticuloendotheliosis virus; Friend virus, spleen focus-forming viruses). Retroviruses are dimeric, containing two strands of genomic RNA with identical sets of genes but not necessarily identical genes. Genes may differ from each other as a result of notoriously error-prone replication, and they may differ from their parental strand as a result of recombination and copy-choice. Retroviruses also contain cellular tRNA used as a primer during DNA synthesis, and a proteinic reverse transcriptase. The retroviral genome encodes some minor, downstream genes and three primary, upstream genes: *gag*, *pol*, and *env*, controlled by the same promoter (although *gag* and *pol* are not necessarily translated with the same reading frame), and flanked by terminal repeats (long terminal repeats [LTR]) of several hundred base pairs. *gag* encodes a polyprotein that is processed into low molecular weight structural proteins, including one or two nucleic acid-binding domains. *pol* encodes a polyprotein which is processed into catalytic components of the mature virions, including a DNA polymerase (capable of functioning both as RNA directed and DNA directed), a ribonuclease (RNase H), an integrase and an aspartate proteinase responsible for processing *gag*'s and *pol*'s polyproteins. *env* encodes a protein that interacts with the host-cell's receptors, thus mediating viral entry, and another protein that probably mediates fusion of the virus's envelope and the host-cell's membrane. The gene's mRNA is removed from the full-length retroviral RNA transcript before translation.

Overlapping virogenes are also a feature of the retroviral genome. For example, in the lentiviruses, including human immunodeficiency viruses (HIV 1 and 2), 3′-terminal coding exons of regulatory genes *rev* and *tat* overlap with each other and with the *env* gene, and in HIV 1, an extra *vpu* gene also overlaps with *env*, while in HIV 2, an extra *vpu* gene overlaps with *vif* and *vpr* genes.

Following infection, reverse transcriptase uses the RNA templates to transcribe a DNA which turns out slightly longer than the RNA due to duplication of the LTRs (resembling the replication of retrotransposons; see below and Chapter 4). The resulting dsDNA (probably a linear rather than a circular form), in combination with viral proteins, is transported into the host nucleus and incorporated into a host's chromatin under the influence of integrase polypeptides encoded by the virus's *pol* gene. The integrated provirus thereafter acts as a normal set of genes in the host cell's nucleus. Without any mechanism for excision, the provirus (like any retrotransposon element) might become "fossilized" in the host cell's genome or integrated by modification. The full-length products of transcription less the LTRs, meanwhile, enter the cytoplasm and are packaged in pairs and released as new retroviral virions.

Retrotransposons

Retrotransposons encode an RNA directed DNA polymerase (reverse transcriptase) which utilizes its RNA as a template for a DNA copy, known as a retroelement,[60] that is subsequently inserted in nuclear DNA. They are abundant and broadly distributed throughout the eukaryotes (animals, plants, fungi, protozoa). Similar patterns of insertion may be found in plasmids (the Mauriceville and Varkud plasmid of *Neurospora crassa* mitochondria), group II introns (of yeast [*cox1,* al1 & al2] and plant mitochondria and algal plastids), RNA viruses, DNA viruses, "switch" viruses (retroviruses [mammals, birds], hepadnaviruses [mammals, birds] and caulimoviruses [plants]) and retroelements from organellar genes (RTL [reverse transcriptase-like] in *Chlamydomonas reinhardtii* mitochondria) and multicopy single-stranded DNAs- (msDNAs-) associated reverse transcriptases of prokaryotes (Purple bacteria; *Myxococcus xanthus*, *Escherichia coli*) that potentially jockey between RNA and DNA forms on their way to and from genomes.

The identity and location of various structural and enzymatic domains in the genomes of retrotransposons are usually inferred by sequence similarity with retroviral elements. The gene organization of retrotransposons is similar to that of retroviruses, but the retrotransposon's genome is more variable than that of retroviruses, and while LTR-retrotransposons have retrovirus-like LTRs, nonLTR retrotransposons lack them (although they may have a TAA repeat in *Drosophila melanogaster* or a polyadenylic acid tail). Similarity is extensive between LTR-retrotransposons of the gypsy variety (Gypsy-Ty3 subgroup, in insects, sea urchins, plants and fungi) but limited for LTR-retrotransposons of the copia variety (Copia-Ty1 in *Drosophila melanogaster, Saccharomyces cerevisiae*, the true slime mold, *Physarum polycephalum*, and probably vertebrates). Similarity is limited in the latter to a nucleic acid-binding motif of *gag* and four enzymatic domains of *pol* of retroviruses (which are also similar to those elements in gypsy-like retrotransposons). The order of enzymatic domains is different as well. Other LTR-retrotransposons in the cellular slime mold *Dictyostelium discoideum*, and the silkworm moth, *Bombyx mori* and nematodes, do not fit into either of these varieties. Copia and Ty1 elements of yeast (*Saccharomyces cerevisiae*) seem tantalizingly viral-like, however, when they make particles within cells, even if "these particles do not appear to be infectious."[61]

NonLTR retrotransposons (I factor in *Drosophila melanogaster* and R2 of *Bombyx mori*; L1 of mammals; cin4 in *Zea mays*; ingi of *Trypanosoma brucei*) generally show characteristics of *gag* and *pol* genes of retroviruses which may be translated out of frame or separated by a termination codon, although some nonLTR retrotransposons lack *gag*-like genes. The reverse transcriptase-encoding region of nonLTR retrotransposons resembles a portion of the retrovirus *pol* gene, generally in the center of the retroelement's genes, but aspartate proteinase domains are absent and RNase H and integrase domains may be absent or upstream as in copia-like elements and retroviruses. Like retroviruses, I and L1 elements would seem to transpose by means of an RNA intermediate, since they excise introns inserted into them at the same time they undergo transposition in the host genome.

A high level of sequence diversity is found among the reverse transcriptases of retrotransposons, and phylogenetic analysis of these sequences suggests, however tentatively, ancient divergence. Retroviruses, interspersed among retrotransposons on the phylogenetic tree, would seem to have independent origins, while the calimo-variety of pararetroviruses may (or may not) be linked to the Gypsy-Ty3 variety of LTR-transposons. In general, the high degree of divergence suggests that little confidence should be placed in the tree's specific branching pattern.[62]

THE NEW LOOK AT CELLULAR LIFE

The introduction of sequence data and cladistic methods revolutionized prokaryotic and eukaryotic systematics. Carl Woese and colleagues[63] affirmed molecules as life's be-all and end-all, declaring that "systematics in the future will be based primarily upon the sequences, structure, and relationships of molecules."[64] Marching off to battle the old systematists, the armies of molecular biologists proclaimed a sanctuary in the field of viruses, ignoring them simply because they did not have rRNA, so convenient for constructing branching trees, and other "right" kinds of molecules.[65] The sanctuary will, no doubt, become a no-man's land some day, and the battle will move to viruses, but, in the meantime, the molecular biologists seem content to confine their conflict to the otherwise-"universal" life forms.

Reinforced with (1) "group invariant" structures in small regions of ribosomal ribonucleic acid (rRNA) and (2) the number of and subunit patterns in RNA polymerase, Woese and colleagues grafted all living things together into a "Universal Tree of Life" and cleaved it into three "domains"[66] (or urkingdoms): Bacteria (originally Eubacteria), Archaea (originally Archaebacteria or Metabacteria[67]) and Eucarya (or Eukaryota; "eukaryotes" is retained as an acceptable common synonym).[68]

Finding the Root of Difference

The process of determining the point (common ancestor) from which lineages diverge in the interior of a network of branches is called rooting a tree. Pinning the root on life's "Universal tree" would seem impossible, since "there are no outgroups for contemporary life forms,"[69] defined as forms having rRNA and thus perched on the "universal" phylogenetic tree. Rooting life's "universal" tree did not require imagining qualities of a nonliving outgroup or including viruses in life's orbit but merely tracing pairs of alleged paralogues back to a consensus origin. In effect, "Paralogous genes [proved to be] extremely useful in rooting evolutionary trees, since one set of sequences can be used as an outgroup for the other."[70]

One root (the protein root), pursued by Gogarten et al., and Iwabe et al., in 1989[71] traced "ancient genes" to "aboriginally duplicated" genes (paralogous genes) present in the common ancestral condition prior to the branching of the three primary lineages. Based upon unambiguously aligned, "conserved" coding regions (EF-1 and EF-2 of about 130 amino acids) for elongation factors EF-Tu and EF-G, and the F1-α and F1-β subunits of the proton-pump in F1-ATPase, the root separated the Bacteria on one side from a common branch of Archaea and Eucarya.[72] Supporting this divide were molecular similarities found between Archaea and Eucarya, in their ribosomal proteins, RNA polymerases, and the presence of histone-like protein in Eucarya and, at least, in one thermoacidophilic archaean.[73]

As an alternative, aligning the EF-1 and EF-2 sequences separately and combining the conserved regions of each into a tandem composite of about 600 amino acids (the hypothetical condition of a duplicated gene), furnishes a single tree that not only supports the root on the bacterial branch but the relationship of Archaea to Eucarya. The archaean, *Sulfolobus*, would then seem to be the closest living prokaryotic relative of the Eucarya.[74] Still, one cannot help but wonder where the root will be placed when other "ancient genes" (such as thymidylate synthetase or ribonucleotide reductases) are followed to their hypothetical origins in a duplication event.

A second root (the rRNA root) based on similarities in rRNA, however, separates a Eucarya branch on one side from an Archaea–Bacteria branch on the other side. The implied ancestral populations are dubbed the proto-eucayotic and proto-bacterial lineages. Supporting this divide is the similarity in genomic organization among prokaryotes (Archaea and Bacteria) in which a single chromosome (or gonophore) carries the entire genomic organization, and the transcription of the nucleic acid message is closely linked to the translation of the message to protein. In Eucarya, on the other hand, transcription is uncoupled from translation, first by the requirement for a "splicing mechanism" joining the operating portions (exons) of so-called "split genes," and, second, by the necessity to move messenger RNA (mRNA) from the nucleus to the cytoplasm where, alone, protein synthesis takes place.[75]

Recently, the universal phylogenetic tree has been tested as a consequence of the publication of the complete genome for a member of the Archaea, *Methanococcus jannaschii*, from sediment at a hydrothermal chimney about 2600 m beneath the Pacific Ocean.[76] As it turns out, 56% or 971 of *M. jannaschii*'s 1738 genes are unique (i.e., unlike those in either the Bacteria or Eucarya), vindicating the separate "domain" status of Archaea. Of the remaining genes, however, some are similar to genes of Bacteria and some to genes in Eucarya. One of two conclusions are possible: either the three domains are not monophyletic (Archaea being paraphyletic at best), or the very idea of monophyletic groups must be abandoned at the level of domains.[77]

The universal phylogenetic tree with Archaea and Eucarya as sibling (sister) domains on one branch and Bacteria on the second branch suggested that the last common ancestor for all cellular living things was something of a cross between a bacterium and a mixed archaeal-eukaryan. Mitchell Sogin,[78] for one, validated this concept by computing structural similarities in the 16S-like rRNA sequence. Once again, the sequences converged toward three major assemblages, corresponding to Bacteria, Archaea and Eucarya (or, in Woese's old names, Eubacteria, Archaebacteria and Eukaryota).

Clearly, the renaissance of systematics has not run its course, if only because so much DNA and RNA remains to be sequenced. The small subunit rRNA that got the whole thing started, after all, is not the last word in sequencing data. The evolutionist Rudolf Raff, for example, urges, "At the very least, phylogenies should be drawn from more than a single gene, and those genes must yield concordant results."[79] Those sequencing DNA are not short of advice:

> Ribosomal RNA has been used for phylogenies at all taxonomic levels, from Ur-Kingdom ... to subspecies. However, at the highest taxonomic levels, the burgeoning availability of DNA sequence for all domains of life offers opportunities to confirm or challenge the rRNA "Tree of Life." ATPase gene families..., central intermediary metabolism phylogenies ... and information-processing systems ... have already been explored with differing results.[80]

In the present day of electronic communication, the longevity of any "Tree of Life," is hardly at issue. Today, the worldwide web is a magnet for data on species and speciation, nucleotide and amino acid residue sequences, and web sites are burgeoning as new pages are added almost daily. The Maddison brothers, David and Wayne, authors of *MacClade: Analysis of Phylogeny and Character Evolution*,[81] now offer the "Tree of Life" at http://phylogeny.arizona.edu/tree/phylogeny.html, an incredible, up-to-date and yet easily accessible database on taxa, while the "very much under construction" "TreeBase" at http://phylogeny.harvard.edu/treebase promises even more current and complete taxonomic data, and comparative genomic sequences are overflowing at The Institute for Genomic Research.

Putting an Age on Difference

Ever since the late 1970s, molecular sequences and assumptions about the rate of nucleotide substitution provided a basis for speculation on the origin of life's taxa. The problem was that nucleotide substitution, like the radioactive decay of different isotopes, could run at different rates and, unlike radioactive decay, could run at inconstant rates over periods of time. For example, the genes for tRNA, 5S RNA and 16S-like rRNA in several nuclear genomes seem to have evolved slowly compared to many polypeptide-encoding genes.[82] Moreover, "biotic radiation,"[83] or changes in species, did not necessarily reflect "molecular radiation," or changes in their molecules. Nevertheless, substitutions were frequently advanced as "ticks" of "molecular clocks" or chronometers of evolutionary time.

Calibrating these clocks based on sequence data is not straightforward. Of course, the ubiquitous problem of aligning sequences remains, but it can be set aside by the assignment of "reasonable" gaps. More problematic is setting aside substitutions possibly arising from horizontal gene transfer. The idea of molecular clocks cannot be tested as such, since it is basically tautological, but clocks can be legitimized as theory and presumably evaluated for consistency later on. Calibration must also ignore uncertainties in geochronological dating and ambiguity in the fossil record for particular taxa. But, aside from all this uncertainty, the rate of substitution for organisms (the slope of substitutions over time) belonging to selected taxa can be calculated, and the point of divergence of all other organisms with similar molecules extrapolated from the obtained rate.

How good is the agreement among clocks? Based on sequence data from 5S RNA, the assumption that the "molecular clock" runs at a constant rate per year, and a chronometer calibrated by the separation of plants and animals, the "cenancestor," or last common ancestor of prokaryotes and eukaryotes, is said to have diverged about 1.5 billion years (Ga) ago.[84] Similar assumptions for protein clocks place the bifurcation at 1.3, 2.0, 2.5, 2.6, 2.8 and 3.5 Ga ago.

In 1996, Russell Doolittle et al., published what was to be a highly controversial protein clock based explicitly on the assumption that larger data sets should cancel anomalies and smooth the rate of change. With sequence data from 57 families of enzymes, comprising 531 different sequences from 15 principal groups of organisms, with "tests for self-consistency among the data themselves, [and] adjustments for observed changes in rate along different lineages,"[85] with a chronometer calibrated for the separation of echinoderms and vertebrates (570 million years [Ma] ago), and weighting scaled values and rate adjustments with the plant–animal divergence (set at 1000 Ma ago), the last common ancestor of all cellular living things was found to have diverged about 2000 Ma ago.

A large part of this estimate depended on continued variability among the fraction of the sequences that were variable at any one time (the fraction of covariation). If, however, amino acid positions, or sites, in a protein sequence that varied once did not remain variable, a correction would be required (the data, originally corrected for a Poisson distribution, would have to be fitted with a gamma distribution), and, had such sites spent considerable periods of time in the invariable category, the cenancestor might have diverged as far back as 3500 Ma.[86] Furthermore, were horizontal gene transfers to have entered the data, the "short circuit" would have artificially reduced the overall distances among taxa and led to an underestimate of divergence time.[87] Likewise, were some paralogues to have entered the data in the guise of orthologues, the overall distances would have been artificially expanded and led to an overestimate of divergence time.[88] Using precisely the same data, but taking account of the rate of variation among sites and patterns of amino acid substitution, Xun Gu[89] placed the end of the cenancestor at 2500 ± 200 Ma ago, excluding estimates of 3500 Ma at the high end (1% significance limit) and less than 2000 Ma at the low end (5% significance limit).

This remarkable estimate (with an estimated error of only 8%) is probably most interesting for suggesting that life spent nearly half of its existence (from ~3.8 Ga to 2.0 Ga = ~1.8 Ga) as "a common ancestor." What was life doing during this common ancestral phase? Some ideas may be gathered from fossils, but most guesses are based on sequences of DNA and polypeptides from extant organisms belonging to one or another of life's domains. Possibly life spent its time mixing. Separation into individuals would then have come in life's later half.

DOMAINS IN LIFE'S "UNIVERSAL" PHYLOGENETIC TREE

Judging from the history of taxonomy, one should not expect either the domains of life or their subdivisions (kingdoms) to remain the same for very long. Today's "universal" phylogenetic tree,[90] drawn originally from small subunit 16S-like rRNA sequences, "is consistent with findings concerning DNA structure..., the ribosomal A protein..., the presence of introns in tRNA genes of *Sulfolobus* ... and in the halobacteria, although the introns are less eukaryote-like in the latter."[91] The prokaryotes are broken into Bacteria and Archaea while the eukaryotes are maintained intact. The Archaea are a new group; Bacteria are considerably revamped from what they had been, and several eukaryotic groups find themselves in new relationships if not new groups.

Bacteria (Also Called Eubacteria)

Bacteria comprise all typical bacteria. Cells are prokaryotic, generally noncomposite (unicellular) with conspicuous exceptions among the cyanobacteria and myxobacteria, with a structurally stable circular DNA genome with low G + C content. The genome size and content are reduced by the compacting of genes and minimization of regulatory elements. tRNAs contain the modified sequence TψCG,[92] and ribosomes contain a bacterial type of rRNA. A murein cell wall (rigid sacculus) renders cells highly resistant to mechanical and osmotic stress. Cytoplasmic membranes contain predominantly acyl ester and fatty acid lipids (shared with eukaryotes).

Eleven main bacterial groupings (= kingdoms) are inferred from small subunit rRNA sequences by the maximum likelihood method of phylogenetic analysis.[93] According to Cavalier-Smith, "four major phyla contain photoautotrophs (green bacteria, cyanobacteria, purple bacteria, heliobacteria) and in each of these some (or all) can photosynthesize under anaerobic conditions."[94] Phylogenetic analysis of 16S rRNA finds five groups containing photosynthetic species, each with a different type of photosynthetic apparatus. Water functions as a reductant only among the cyanobacteria, associated prochlorophytes and chloroplasts [plastids], although the water-dependent photosystem-II RC is functionally similar to the quinone-based RC of purple sulfur, non-sulfur bacteria and the Chloroflexaceae.

1. Green Nonsulfur Bacteria

Green nonsulfur bacteria (*Herpetosiphon*, *Thermomicrobium*) and relatives (*Chloroflexus*) fall out (next to *Thermotoga*) on the deepest branch of the Bacteria. Predominantly photoheterotrophic bacteria (*Chloroflexus aurantiacus*), some strains grow autotrophically, using a simple (unique) cyclic pathway for CO_2 fixation involving carboxylation of acetyl-CoA to form 3-hydroxypropionate (an unusual intermediate) followed by reduction and carboxylation to produce succinate.

Woese and Pace argue, that, "Unless there has been lateral transfer of the system, photosynthesis evolved at an early stage in the evolution of Bacteria.... [T]he deeply branching phototrophic genus *Chloroflexus* ... [is] consistent with an aboriginal autotrophy."[95]

2. Gram-Positive Bacteria

Gram-positive bacteria, including some formerly Gram-negative bacteria (*Planctomyces*), and mycoplasm (*Mycoplasma*) encompass a second deep (ancient) subdivision of Bacteria. Their cells have a peptidoglycan (murein) wall (exoskeleton and osmotic barrier).[96] A periplasmic gel between two lipid bilayers[97] is present in the formerly Gram-negative members, and no wall is found in mycoplasm.[98] Many Gram-positive bacteria are

Archaeal Enzymes with Similarities to Viral, Bacterial and Eukaryal Enzymes[a]

Enzymes in Archaea	Similar enzymes in viruses	Similar enzymes in Bacteria	Similar enzymes in Eucarya
DNA gyrase			
DNA gyrase subunit B	DNA topoisomerase large subunit	DNA gyrase subunit B	DNA topoisomerase II
Acidic ribosomal protein			
Ribosomal protein L20/L12		Protein homologue PO	Acidic ribosomal protein PO
DNA-directed RNA, polymerase A			
DNA-directed RNA, polymerase A	DNA-directed RNA, polymerase	DNA-directed RNA, polymerase	DNA-directed RNA, polymerase
DNA-directed RNA, polymerase B			
DNA-directed RNA, polymerase B	DNA-directed RNA, polymerase	DNA-directed RNA, polymerase	DNA-directed RNA, polymerase
DNA-directed RNA, polymerase C			
DNA directed RNA, polymerases C	DNA directed RNA, polymerases	DNA directed RNA, polymerases	DNA directed RNA, polymerases
Elongation factors			
Elongation factors[b]		Elongation factors	Elongation factors

[a]Data condensed from Benner et al., 1993.

[b]For elongation factors A, E, and B: a matching of Archaeal elongation factor sequences against the databases shows matches that are likely to be not significant at a level that excludes some nonarchaeal elongation factors.

especially sensitive to antibiotics that interfere with peptidoglycan synthesis. Sequence analysis divides the Gram-positive bacteria into two groups as a function of their guanine and cytosine (G +C) content.

The low G + C subdivision includes *Clostridium*, *Staphylococcus*, *Streptococcus* (*faecalis*), Mycoplasma (*Mycoplasma genitalium* "thought to contain the smallest genome for a self-replicating organism [508-kbp])"[99]; Bacillus/Lactobacillus: (*Bacillus* [*subtilis*]) are photosynthetic species. Other members of the group (*Heliobacterium*) are photoheterotrophic, lacking autotrophy.

The high G + C subdivision arises from a shallow branch of the phylogenetic tree, suggesting recent origins. Members of this group (*Actinomyces*,

Streptomyces, Actinoplanes, Arthrobacter, Micrococcus, Bifidobacterium, Frankia, Mycobacterium, Corynebacterium) are acetogenic, H_2/CO_2 chemolithoautotrophs. They generate energy and fix CO_2 through mechanisms resembling those of Archaean methanogens but produce acetic acid instead of methane as a "waste" product. Other members of this subdivision include the "true clostridia," formerly Gram-negative, walled organisms with a second lipoprotein membrane (*Megasphaera, Sporomusa*).

3. Cyanobacteria: Blue-Green Bacteria and Chloroplasts

Chlorophyll-based phototrophs share the Calvin–Benson cycle for CO_2 fixation. Cyanobacteria (*Oscillatoria, Nostoc, Synecoccus, Prochloron, Anabaena, Anaystis, Calothrix*) may be related at a deep level to Gram-positive bacteria (chlorophyll *a* is similar in structure to heliobacterial chlorophyll *g* in the bacteriochlorophyll [Bchl] containing reaction center [RC]). The diversity in light-harvesting systems in microbial mats is due primarily to the presence of Chl *a* and phycobilin pigments in cyanobacteria and Bchls *a*, *b*, *c*, *d*, and *e* in anoxygenic phototrophs (Bchl *g* is not detected in mats). The Cyanobacteria are broken down into anoxygenic, oxygenic and chloroplast categories on the basis of physiology and locale.

The anoxygenic cyanobacteria comprise two groups. Bchl-*c*-containing, filamentous autotrophs are found in marine hypersaline as well as in hot-spring mats.[100] Bchl-*a*-containing organisms are found in temperate environments.

The oxygenic cyanobacteria include picoplanktic prochloralean unicells. They contain Bchl *c* or *d*,[101] possess phycoerythrin, and are adjusted to low nutrient levels in open-ocean species, while the coastal species have only phycocyanin and thrive on elevated nutrient levels. The prochloralean combination of chlorophyll *a* and chlorophyll *b*, found in the photosynthetic apparatus is found in several branches within the cyanobacterial cluster, in green algae, mosses, and vascular plants. Cyanobacteria also run chlorophyll *a* in combination with phycobilins, similar to that of red algae and cryptomonads.

Chloroplasts are thought to be derived from prochloralean unicells. The most compelling evidence comes from sequence analysis: The *GapAB* genes of Metaphyta and algae chloroplasts (plastids), including those of red algae and *Euglena*, resemble *gapA* genes of free-living cyanobacteria (*Anabaena* and *Synechocystis*). All GapAB proteins are chloroplast-specific. *GapA* genes appear to have been lost in the lineage leading to trypanosomes after divergence of *Euglena*.[102]

*4. Purple Bacteria and Mitochondria**

Many purple bacteria and formerly Gram-negative bacteria attached to the purple bacteria by sequence analysis are primitively photosynthetic. Some have the ability to fix molecular nitrogen form intracellular gas vesicles, and many are intracellular symbionts of eukaryotic cells. Presumably one class of these bacteria became a mitochondrial subdivision. The remaining purple bacteria fall into four subdivisions on the basis of physiology and structure. Possibly, a fifth subdivision contains myxobacteria (*Bdellovibrios* + *Wolinella* and *Camphlobacters*).

α purple bacteria comprise the purple nonsulfur bacteria. These include photoheterotrophic bacteria, autotrophs using Calvin–Benson Cycle for CO_2 fixation (*Rhodobacter*, *Rhodopseudomonas*), rhizobacteria, agrobacteria, and rickettsiae (*Nitrobacter*, *Thiobacillus* [some species], *Azospirrillum*, *Caulobacter*), β purple bacteria comprise *Rhodocyclus* (some species), *Thiobacillus* (some species), *Alcaligenes*, *Bordetella*, *Spirillum*, *Nitrosovibrio*, *Neisseria*.

γ purple bacteria are enterics (*Acinetobacter*, *Erwinia*, *Escherichia*, *Klebsiella*, *Salmonella*, *Serratia*, *Shigella*, *Yersinia*), vibrios, fluorescent pseudomonads (*Pseudomonas aeruginosa*, *P. putida*), *Legionella* (some species), *Azobacter*, *Beggiatoa*, *Thiobacillus* (some species), *Photobacterium*, *Nanthomonas*. Like other former Gram-negative bacteria, members of this subdivision have two membranes separated by peptidylglycan forming a complex collectively referred to as the cell envelope. The outer membrane is specialized, having an outer monolayer composed of lipopolysaccharides instead of the usual phospholipids,[103] including "porins" comprising integral membrane channel structures.[104]

δ purple bacteria are sulfur and sulfate reducers (*Desulfovibrio*).

The most compelling evidence for a purple-bacterial origin of mitochondria is the sequence of integral mitochondrial proteins and purple bacteria proteins: The *gapC* gene of *E. coli* (a purple bacterium) is more similar to *GapC* genes of plants, animals and fungi than to *gapC* of cyanobacteria. The gene presumably moved from a purple bacterial protomitochondrial endosymbiont to the ancestral eukaryote's chromatin (lateral gene transfer) prior to the origin of Archaezoa, since the gene is found in the nucleus of several mitochondria-less archaezoans. Differences with the sequence in the *GapC* gene of kinetoplastids and *Euglena* are sufficient to suggest an independent source for the gene in these organisms. The monophyletic origin of mitochondria, thus, remains debatable.[105]

---

*A mitochondrial/purple bacteria relationship is supported by recent sequence data reported. See Andersson *et al.* (1998).

5. Spirochetes

Spirochetes comprise two groups. The first consist of treponemes and borrelias. The second, Leptospira group includes the *Leptospira* and *Leptonema*.

6. Flavobacteria

Flavobacteria (yellow bacteria) are nonphotosynthetic, obligate autotrophs. An aerobic subdivision includes *Flavobacterium, Cytophaga, Saprospira, Flexibacterer.* An anaerobic, Bacteroides group, includes *Bacteroides* and *Fusobacterium.*

7. Green Sulfur Bacteria and Green Bacteria

Green sulfur bacteria and a photosynthetic sibling (sister) group, Green bacteria (*Chlorobium, Chloroherpton*) lack enzymes of Calvin–Benson cycle. Instead, these bacteria use the reductive citric-acid cycle for CO_2 incorporation.

8. Planctomyces

The Planctomyces (*Planctomyces, Pasteuria*) and thermophilic *Isocystis pallida*, have an unusual type cell wall. Phylogenetic analysis of their 16S rRNA shows them branching deeply (below the Thermotogales) near the base bacteria.

9. Chlamydiae

These radio-resistant micrococci and relatives (*Chlamydia psittaci, C. trachomatis*) also branch deeply from the base of bacteria, possibly in a sibling (sister) relationship to planctomyces. Chlamydiae also have an unusual type of cell wall.

10. Deinococci

Dinococcus radiodurans comprise the Deinococci.

11. Thermotogales

Thermotogales also branch near the base of Bacteria. They possess a sheath-like structure surrounding cells (*Thermotoga maritima* and *T. Neopolitana*; both hyperthermophiles) and their "overballoon" at ends, containing porins, "is most likely homologous to the outer membrane of Gram-negative bacteria."[106]

Archaea (Also Called Archaebacteria)

The Archaea is a new kingdom whose members are identified by sequencing data and molecular evidence such as the ether-linked lipids in cell walls. In particular, archaean "16S rRNAs are readily identified by the unique structure they show in the region between positions 180 and 197 or that between positions 405 and 498."[107] "Archaea" may be a misnomer for this kingdom, however, since "the speciation of the domain Bacteria presumably occurred earlier [than the domain Archaea]."[108]

Cells are prokaryotic, noncomposite (unicellular), with structurally stable circular DNA, and a stable genome. tRNAs lack the modified sequence TψCG, and ribosomes contain an archael type of rRNA. Cell walls lack peptidoglycan, and cell membrane lipids are predominantly prenoid glycerol diethers or diglycerol tetraethers (distinctive biphytanyl isoprenoid ether lipids). According to Cavalier-Smith,

> The isoprenoidal ether lipids [of Archaea] are more stable to heat or acid than are acyl ester lipids [of Bacteria].... The most extreme thermophiles have lipid monolayers with biphytanyl lipids while mesophiles have lipid bilayers with monophytanyl lipids. For mesophilic bacteria it may not matter much whether they have a bilayer of isoprenoidal ethers or acyl ester. But for extreme thermoacidophiles, a monolayer of biphytanyl isoprenoids is clearly an advantage.[109]

Based on transversion distance among small subunit rRNA sequences, Woese and Pace divided the Archaea into two kingdoms (but see discussion on paraphyletic view of Archaea).[110] One kingdom, Crenarchaeota, contains only sulfur-dependent thermophilic organisms; the other kingdom, Euryarchaeota, contains organisms metabolizing a range of oxidized compounds as electron acceptors and has representatives of four phenotypes: methanogens, extreme halophiles and alcaliphiles, sulfate-reducing archaea, and extreme thermophiles.

Euryarchaeota (Common Name: Euryotes or Euryarchaeotes)

Euryotes comprise methanogens and species exhibiting similar sequence data but of diverse phenotypes. They display varied patterns of metabolism, a broad range of niches (hence "eury-") and rigid cell walls that mimic (converge with) the murein cell walls of bacteria while containing a glycoprotein S-layer, protein sheath, methanochondroitin, glycocalyx, protein S-layer, pseudomurein, sulfated heteropolysaccharide, and glucosaminoglycan. Their ribosomes contain a euryarchaeal type of rRNA. Arguably, euryotes arose above the root indicated when bacterial sequences are considered an outgroup.

Methanogens

Methanogens are anaerobic and exhibit mesophilic growth. They are present in virtually all terrestrial and aquatic anaerobic habitats on Earth. Their methanogenic metabolism is based on reduction of carbon dioxide to methane (unrelated to methanogenesis in bacteria). The methanogen's 16S rRNAs are comparable in size to bacterial 16S rRNAs but not in sequence or pattern of base modification. They are divided into three (or four) distinct groups (I–III [or IV otherwise grouped with III]) in addition to a very deeply branching, thermophilic lineage (*Methanopyrus*).

I. Methanococcus (Methanococcales) group: *Methanococcus (Mc.). jannaschii*; *Mc. thermolithtrophicus*, *Mc. vannielii*, *Mc. voltae.*

II. Methanobacter (Methanobacteriales) group: *Methanothermus (Mt.) fervidus*, *Methanobacterium (M.) thermoautotrophicum*, *M. formicicum*, *M. arboriphilus*, *M. ruminantium*, strain M-1, *M.* sp., Cariaco-isolate JR-1; *Methanobrevibacter*, *Methanosphaera stadmaniae.*

III. "Methanosarcina" group (Methanomicrobiales lineage): *Methanosarcina (Msa.) bakeri*, *Methanococcoides methylutens*, *Methanothrix (Mth.) soehngenii*; (IV) Methanospirillum group: *Methanospirillum (Ms.) hungatei*, *Methanogenium*, *Methanoplanus (Mp.) limicola.*

The complete DNA sequencing of *Methanococcus jannaschii* held several surprises. First, "cell division in *M. jannaschii* might occur by a mechanism specific for the Archaea."[111] Second, "Despite the availability for comparison of two complete bacterial genomes and several hundred megabase pairs of eukaryotic sequence data, the majority of genes in *M. jannaschii* cannot be identified on the basis of sequence similarity."[112]

Extreme Halophiles and Alcaliphiles

These euryotes are present in continental salt and soda lakes. They comprise a sibling (sister) group to Methanomicrobiales with relatively superficial branching (*Halobacterium volcanii*, *Halococcus morrhuae*).

Fully aerobic and mesophilic, they exhibit photo-organotrophic growth (on ATP formed photically) but have no photosynthetic CO_2 assimilation. Although Archaea have no chlorophylls and no exclusive phototrophs, retinal or "bacteriorhodopsin-based photosynthesis characteristic of the halobacterial group [occurs] within the archaebacteria,"[113] and "*Halobacterium* admittedly can use sunlight for energy, but it cannot exist purely autotrophically and is also an obligate aerobe."[114]

Sulfate-Reducing Archaea

A new phenotype of H_2/SO_4^{2-} chemolithoautotrophs has been identified from sequencing data: *Archaeoglobus fulgidus*.

Thermophiles (Hyperthermophiles)

The deepest branch of euryotes is the thermophilic *Thermococcus-Pyrococcus* group and *Thermoplasm* (possibly a sibling group to the Methanomicrobiales). Hyperthermophilic thermophilies include *Thermococcales celer*, *Methanopyrus kandleri*, moderately thermophilic (mesophilic) thermophiles include the microaerophilic heterotroph *Thermoplasm celer*, and kryophilic species, forming a component of marine microplankton.

Crenarchaeota

(common name: crenarcheotes or crenotes; also comprising most of the thermoacidophiles, sulfur-dependent archaebacteria, extreme thermophiles and eocytes)

Hyperthermophilic crenotes (some growing optimally at temperatures above 100°C) are sulfur-dependent (metabolize sulfur and sulfur compounds) while exhibiting little metabolic and ecological diversification. They thrive in volcanic areas. All extant representatives are H_2/S^0 and methanogenic H_2/CO_2 chemolithoautotrophs, although hyperthermophilic sulfolobales exhibit H_2/O_2 chemolithoautotrophy under microaerophilic conditions. They have elaborate cell walls and envelopes mostly of glycoprotein.

Their circular DNA has a high GC content, and their ribosomes contain a crenarchaeal type of rRNA. The organization of their rRNA genes resembles that in eukaryotes more than in euryarchaeotes: (1) relatively high levels of modified bases in rRNA and tRNA; (2) ribosomes having relatively high protein/RNA ratios; (3) lack of a tRNA gene in the spacer region of the rRNA operon; (4) no 5 S rRNA gene linked to the rRNA operon at its 3′ end. They also resemble eukaryotes by way of the subunit structure and sequence(s) of DNA-dependent RNA polymerase; transcription signals; common sensitivity to diphtheria toxin; presence of true eukaryotic histone.

Deep branching crenotes arguably arose below the root established by use of bacterial sequence for an outgroup. Lineages are short (reflecting rRNA compositional disparity rather than true phylogeny). Crenotes separate into two groups: (1) The *Thermoproteus-Pyrodictium* cluster (also *Sulfolobus*-like): *Desulfurococcus mobilis*, *Pyrodictium occultum*, *Sulfolobus solfataricus*; (2) *Thermofilum pendens*, *Thermoproteus tenax*.

Eucarya

Eucaryans generally have lavish genomes (microsporidians being an exception) contained within a nucleus. Their genes are at least ten times larger than those of prokaryotes, containing nuclear introns and other non-coding sequences, and are packaged with histones or other basic proteins and nucleoproteins as chromatin or chromosomes during mitotic division. Genomic organization is, likewise, expansive and genes are not generally bunched in functional units (cistrons) as they are in prokaryotic genomes, although members of some of the abundant gene families may be organized in clusters. In addition, three DNA-dependent RNA polymerases operate in eukaryotes in contrast to one in bacteria and archaeans. The eukaryotic genome contains regulatory sites such as distal promoter elements and upstream activation sites (UAS) or enhancers (such as TGACTCA) that bind gene-specific activators (such as steroid receptor and products of oncogenes) as well as promoters of genes containing a TATA box or a CCAAT box that bind basic transcriptional factors (TFs) and cooperate in the initiation of transcription by RNA polymerase II. Frequently, multiple gene products are required to activate transcription at particular sites, and heteromeric regulatory complexes are a feature of many eukaryotic transcriptional activators.

The most conspicuous difference between the eucaryan genome and prokaryotic genomes is not the eukaryan genome as such but the presence within eukaryotic cells of additional genomes resembling those of bacteria. These are the genomes of chloroplasts, present in plants and algae, and mitochondria, present with few exceptions throughout the Eucarya. These organelles have an "extended genome," since some of their genes are ensconced within the organelles while other genes are installed in the eukaryotic cell's nucleus.[115]

Extended Chloroplast Genomes

Chloroplast DNA (cpDNA) exists in multiple copies present as multimeric complexes within structures known as nucleoids. Chloroplast genes recombine only within an individual organelle and not between biparentally inherited chloroplasts.[116] cpDNA is generally circular, consisting of 120 to 220 kb in higher plants. Single-copy regions of genes are separated by segments of a characteristic large inverted repeat (IR), typically 20–25 kb long, present in chromophytic algae (with chlorophyll *a* and *c*) and chlorophytic algae (with chlorophyll *a* and *b*) except *Euglena*, but absent in rhodophytic algae (with chlorophyll *a* and phycobilins).[117] Ribosomal genes and other genes included within the IRs are thus doubled. *Euglena gracilis*'s chloroplast DNA contains three tandem repeats, each containing an rRNA gene cluster. The genome is otherwise rather compact, even containing some overlapping genes (*psbD* overlaps *psbC* by 50 bases pairs in the cpDNA of land plants), although pseudogenes may be present and introns are common (group I in *Chlamydomonas* cpDNA; group II in land plants, and group III in *Euglena*).[118] An occasional cpDNA intron may contain internal elements encoding a reverse transcriptase and act as a mobile genetic element or retrotransposon.[119]

The chloroplast is the site in plants and algae where light fuels photosynthesis and phosphorylation via electron transport. cpDNA contains 80–100 open reading frames (ORFs; presumptive polypeptide-encoding genes) generally transcribed from a bacterial-type promoter (as a polycistronic precursor). These genes encode some of the enzymes of photosynthesis and the enzymes and polypeptides of photophosphorylation and electron transport: photosystem II genes *psbA*, *psbF*, and *psbD* and the ATPase subunit *atpH*. Other cpDNA genes encode about 20 of the chloroplasts 60 ribosomal proteins, about 30 tRNAs (all lacking a 3′-CCA end) and four rRNAs (small subunit 16S rRNA and large subunit 23S, 4.5S and 5S rRNAs). The 4.5S rRNA resembles the 3′ end of the 23S rRNA of prokaryotes, and the 5S rRNA is "typically" bacterial.[120] Noncoding regions of the cpDNA tend to accumulate additions or suffer deletions, and much of the variation in cpDNA is attributable to the IR.

Other chloroplast genes are encoded in the host-cell's nuclear DNA (nDNA).[121] To begin with, the nDNA encodes the chloroplast version of the more or less "universal" enzymes, glyceraldehyde-3-phosphate dehydrogenase and phosphoglucose isomerase. Other genes encode products specific to the peculiarities of translation in the chloroplast. These include genes encoding two thirds of the chloroplast's ribosomal proteins (*rpl22*), RNA-binding proteins (28–33 kd CS-RBD-type), RNA polymerase and elongation factor (*tufA*). Still other genes are more idiosyncratic and fluctuate with species, especially the massive number of genes involved in photosynthesis. For example, in land plants, these include chlorophyll genes of the photosystem II core complex operating in water oxidation, its extrinsic complex (lumenal side of the thylakoid membrane) involved in docking (*psaD*, *psaE*, *psaF*, *psaG*, *psaH*, *psaK*, *psaL*) and stabilizing (*psbO*, *psbP*, *psbQ*, *psbR*, *psbS*); several genes (nuclear *cab*) components; genes belonging to chlorophyll *a/b* families encoding CAB proteins (antennal proteins) of the light harvesting chlorophyll complex I (LHC I) and LHC II and CP apoproteins; plastocyanin (transfers electrons from cytochrome b_6/f complex to photosystem I; ferredoxin, ferredoxin $NADP^+$ oxidoreductase, thioredoxins and ferredoxin-thioredoxin oxidoreductase, transferring electrons to the chloroplast's matrix and ATP synthase. In addition, the nucleus contains genes encoding proteins regulating the expression of several chloroplasts genes and responsible for splicing, stabilizing and accumulating their transcripts as well as translation.[122]

Extended Mitochondrial Genomes

Mitochondrial DNA (mtDNA) is usually a single sequence present in multiple copies (10^2–10^4). It is usually circular (with exceptions: the linear, small genome of *Chlamydomonas reinhardtii* and some medusazoan cnidarians, and both circular and linear mtDNA in plants), although some mtDNA may be incomplete and may contain catanes, closed circular molecules topologically linked to open circular molecules (maxicircles). mtDNA consists of about 14 to 39 kbp in animals, 17.6 to 115 kbp in ascomycetes (yeast), 2040 kbp in the maxicircle DNA of the kinetoplastic (trypanosomid) protozoa, 50 kbp in *Achlya ambisexualis* (oomycete) and 120 to 2400 kbp in angiosperm plants. It encodes 75–100 genes, frequently similar to coding regions in bacteria. For example, "A protein domain of the peripheral benzodiazepine receptor from rat mitochondria has excellent sequence similarity to a domain of the CrtK protein of *Rhodobacter capsulatus*, a photosynthetic purple bacterium."[123]

Mitochondria are the homes for enzymes of the citric acid cycle and the proteins and polypeptides of the electron transport chain driving chemiosmosis and oxidative phosphorylation. Mitochondria also contribute to numerous biosynthetic pathways, including those leading to amino acids and heme, cholesterol derivatives, folate coenzymes, nucleotides, pyrimidines, phospholipids, and urea. A variety of genes encoding the enzymes and polypeptides operating in all these biosynthetic pathways are located in the mitochondrion's own mtDNA, but many more are sequestered in the cell's nucleus, frequently as slightly different duplicated genes (coding minor isozymes), and, particularly in yeast, have been identified through the isolation of petite (*pet*) mutations as well as sequencing data. mtDNA is distinguished by its rapid rate of change, notably in vertebrates (mammals), Hawaiian *Drosophila*, and *Tetrahymena*, compared to its nuclear counterpart, but sea urchin mtDNA appears to diverge with an unremarkable rate, and change is extremely slow in plant mtDNA.[124] The mtDNA genome encodes up to three rRNA genes (large subunit and small subunit) utilized in intramitochondrial translation and about 30 different tRNA genes including those involved in decoding unusual mitochondrial codons (exceptions being ciliates, kinetoplastid flagellates, flowering plants, green alga, and, once again, Cnidaria among the Metazoa). Another set of genes specify components of complexes in the electron transport chain, and unique ORFs encode respiratory polypeptides: subunits of cytochrome *c* oxidase (*cos 1*), the (apo)cytochrome *b* component of ubiquinol cytochrome *c* reductase (*cob*), portions of ATPase, NADH dehydrogenase subunits, some of which may overlap.[125] Gene order may be related (human mtDNA, sea urchin, *Drosophila*) or show little relationship (clustered tRNA in echinoderms in contrast to uniform distribution in vertebrates, insects, *Ascaris suum* [a nematode]).

Introns are absent in animal mtDNA except for two group I introns in the cnidarian, *Metridium senile*, but group I and II introns are present in other mtDNA, conspicuously fungal mtDNA, along with direct (recombination) repeats of 1 to 10 kb accompanied by coding regions for maturases involved in intron removal or transposition. "Junk" sequences, such as pseudogenes, are found in plant mtDNA. Mitochondrial genomes in some fungi have a large inverted repeat carrying ribosomal RNA genes and some other genes, and metazoan mtDNA carries a "control region" of varied length (the displaced- [D-] loop in mammals) containing the origin of replication and the promoters for transcription and sometimes tandem repeats. Mitochondria fuse and those in yeast (but not *Paramecium* and animals) may share their mtDNA. Plant and *Chlamydomonas* mitochondrial genomes recombine repeatedly and indiscriminately during the cell cycle and gametogenesis.[126]

The part of mitochondria's apparatus encoded by nDNA includes genes encoding cytochromes (*CYC1*, *COX6* [subunit VI of cytochrome *c* oxidase], the F_1 ATPase) and other proteins directly involved in electron transport and oxidative phosphorylation. nDNA genes also encode products required for the expression of the mitochondrial genome: mitochondrial RNA polymerase, regulators required for mitochondrial translation of particular genes, and maturases required for splicing type II introns. Nuclear genes may also play parts in such bizarre mitochondrial behavior as adding uridine bases to transcripts (trypanosomids). These genes have not lost their "primitive touch." The mitochondrion's RNA polymerase gene, located in the cell's nucleus, for example, resembles bacteriophage T3 and T7 RNA polymerase, and is also represented in the chloroplast genome and the nuclear genome of some unicellular eukaryotes.

The mitochondrion is intimately in contact with its environment (cytosol and nucleus) and many signals, feedback controls and regulatory functions pass between them. For example, nuclear encoded RNA is only imported by mitochondrial membrane as part of a cytoplasmic ribonucleoprotein complex,[127] and targeting of cytoplasmic proteins is generally the task of mitochondrial signal sequences of dedicated presequence or terminal amino acids. Mitochondrial functions and protein synthesis are repressed by cytoplasmic glucose and catabolic repression pathways, while, drug-induced respiratory distress in mitochondria is communicated to the nucleus, promoting cytochrome *c* expression. Heme in the presence of iron, emanating from mitochondria, seems to regulate the transcription of the *CYC1* gene and the *COX* genes and the accumulation of the nuclear-encoded subunits of cytochrome oxidase.[128] Mitochondrial release of heme also mediates some features of the cells's response to oxygen. Trafficking between mitochondria and nuclear gene products is not entirely without hazard. In addition to "mitochondrial diseases" in vertebrates (mitochondrial cytopathy syndromes) and in maize, genes for the cytoplasmic male sterile phenotype are located in mitochondria.

Uniquely Eukaryan Genomes

Many modern systematists rely on structures accessible to molecular analysis for differential criteria of systematic importance. In the case of eukaryotes, these structures are frequently parts of the cell's cytoskeleton and endomembranous system. For example, specific intermediate filaments (considered synonyms for specific types of differentiation) set Eucarya apart: the actomysin system which powers phagocytosis, cytokinesis,

Precambrian Geological Timetable: Geological and Paleobiological Subdivisions of the Precambrian[a]

Age in Ga (billion years)	Eon	Era	Period or cycle	Geological and biological events
~4.50 to .53 (duration: ~4 Ga)	Prephanerozoic[b] (= Proterozoic + Archean)			
2.50–0.53 (duration: ~2 Ga)	Proterozoic[c]			Continental crust breakup and dispersal; continent-derived sedimentary rock; solar luminosity increases; strong tectonic activity
0.90 (1.0)–0.53 (0.60–0.55)[b]		Neoproterozoic; Late Proterozoic (III); Late Precambrian	Terminal Proterozoic Ediacaran System[d]	Coalescing of eastern and western Gondwana; little new crust (Arabian-Nubian shield); global glaciation; reduced erosion associated with Pan-African uplift; Acitarch (large cells or cysts) extinctions; increase in oxygen; evolution macroscopic size multicellular organisms
(0.70) 0.60–0.57			Vendian	Vendian biota
~0.65–0.61			Early Vendian	Acritarch assemblages rare and of low diversity
0.65–0.54			Tommotian	
			Cryogenian; Varanger Ice Ages	Earliest recorded extinction event; Increase in seawater $^{87}Sr/^{86}Sr$; reduced hydrothermal flux; separation West Gondwana from North America; formation juvenile crust; reduced atmospheric CO_2 induces iceages
~0.59				Decrease in hydrothermal iron flux and increase in oxygen concentrations remove major phosphorus sink; more phosphorus available for recycling; increase in fertility of oceans; coeval biological events

Age in Ga (billion years)	Eon	Era	Period or cycle	Geological and biological events
~1.0–0.65 1.0–0.60			"Upper (late) Riphean"	Rich and diverse biota; cyanobacterial and eukaryotic protists; decline of stromatolites in abundance and diversity; possible radiation of eukaryae
0.80 (0.70)–(0.68) 0.60			Tonian	Iron formations oxidized; oxygen concentrations relatively low; rapid subsiding extensional basins; opening Iapetus and other ocean basins; high hydrothermal flux
~0.85				Major episode of continental extension and juvenile crust formation begins accompanied by high fluxes of reduced materials into oceans; increased proportion of organic carbon buried
1.20–0.60			Nubian Cycle	Post-Grenville amalgamation of continental blocks; rifting & mountain formation
1.2–1.0				Solitary spheroid and colonial acritarchs
1.60 (1.25)–0.90 (1.00–0.75) ("post-1.00 Ga")		Middle Proterozoic (II) or Mesoproterozoic	Lower Riphean	Precambian crust almost complete; CO_2 levels decline; eucarya fossilation; major clades higher eukaryotes differentiated
1.2–1.0			Stenian	Protistan radiation; decline in stromatolite diversity; possible fecal pellets
			Ectasian	
			Calymmian	
~1.4 and 1.7				Solitary spheroid and colonial acritarchs
2.00 2.50 (2.10)– 1.60 (1.4)		Early Proterozoic (I) or Palaeoproterozoic Age of Bacteria: Stromatolite-builders	Siderian; Colorodo Cycle	Worldwide red bed formation; atmosphere oxidizing; deep ocean reducing; continental crust; variety of bacterial and cyanobacterial microbial-mat microfossils (extremely rare acritarch cysts, macroscopic ribbons & sterane biomarkers)

Age in Ga (billion years)	Eon	Era	Period or cycle	Geological and biological events
1.69			Straherian	Membrane lipids of archaea, bacteria & possible eucarya
1.70			Orosirian	Bacteria with unusual lipids
1.85			Rhyacian	Atmosphere & oceans highly oxygenated; deep oceans oxic; depleted in iron
~2.1				Appearance eucaryae (megascopic alga, *Gypania*)
4.5–2.5 (duration: ~2 Ga)	Archean[e]			Oceanic lithosphere; volcanic islands and microcontinental blocks form principal land areas
2.7–2.1		Age of Methanotrophs		
(3.2) 2.9–2.5 (2.6)		Late Archean	Superior Cycle	Enormous accretionary complexes; vast crustal growth (>50–60% Precambrian crust); reducing atmosphere devoid of free oxygen
≥2.7				Onset of methane cycling (methanogenesis); production and consumption of both O_2 and CH_4 in surface environments
~2.8				Archean microbiota: stromatolithic cherty carbonates
(3.3) 3.5–3.2 (2.9)		Middle Archean		CO_2 atmosphere; 30–50°C

Age in Ga (billion years)	Eon	Era	Period or cycle	Geological and biological events
3.9–3.3		Early Archean	Barberton Cycle	Early Archean microbiota: stromatolites, microfossils in cherts
(3.55) ~3.5 – ~3.3				Dissolved ferrous iron available; earliest banded iron-formations
Pre-3.5			Isua Cycle & Pilbara cherts	
3.8				Falloff of impactors; sedimentary rock (mainly volcaniclastic); bicarbonate rich ocean; pH ~ 6
3.7–4.0				Origins life on ocean surface (?)
4.2–4.0		(Hadean)		Origins of life on midocean ridges (?); bulk of ocean accumulated; high CO_2 & H_2O pressure; 80–100°C; abiotic organic synthesis
4.5–3.9				
4.5				Accretion; magma; atmosphere >1200°C
Duration: 0.01–0.10				

[a]Data summarized and harmonized from Chang, 1994; Conway Morris, 1994b; Hayes et al., 1992; Kasting & Chang, 1942; Knoll, 1994; Lowe, 1992, 1994; Schopf, 1992; Summons, 1992; Towe, 1994.

[b]The Precambrian portion of the Earth's history is composed of the Archean and Proterozoic eons. Its borders are not yet defined by the International Commission on Stratigraphy.

[c]The Proterozoic Eon is commonly identified with the period between 2.50 and 0.57 Ga ago.

[d]The Ediacarian System (Fauna) of South Australia; last Precambrian rocks containing megascopic fossils; equivalent to Vendian and/or Eocambrian.

[e]The Archean Eon extends from the Hadean (4.50 to 3.90 Ga) to the Proterozoic Eon (2.50 Ga).

amoeboid motion and contractility; microtubules and the kinesin/dynein/dynamin-ATPases which power the segregation of chromosomes during karyokinesis, directed movement of transport vesicles, and the 9 + 2 assembly of sliding microtubules enclosed in a ciliary membrane.[129] Among other molecular systematists, however, the identifying feature of eukaryotes is their "group-invariant rRNA characteristics," especially that found in the region in the small subunit rRNA between nucleotides 585 and 655 (using the numbering adopted for *Escherichia coli*) where members of both prokaryotic groups (Bacteria and Archaea) exhibit a common characteristic structure which "is never seen in eukaryotes."[130] Likewise, rRNA figures prominently in configuring intra-eukaryotic classes, but reconciling branching patterns for sequence data with more familiar groupings of eukaryotes can be a daunting task.

Mitchell Sogin[131] began the monumental task of unraveling the evolutionary history of eukaryotes by identifying similarities from comparisons of 1020 sites (on the way to 1600) that could be "unambiguously" aligned in full-length, small subunit rRNA (16S-like rRNA [alias 18S rRNA]) from 195 eukaryan and prokaryotic taxa. He constructed a phylogenetic tree based on maximum likelihood. Molecular distances were based on parsimony, statistical measurements around branch points, and congruence with patterns of ultrastructure organization. As a result, he discovered that, contrary to the traditional view of relatively recent (1–2 Ga ago) origins for eukaryotes, divergence among the eukaryotes translated to extremely long branches. "The extraordinary depths of branching in the eukaryotic subtree eclipse those seen within the entire prokaryotic world."[132] Along with animals, plants, and fungi, at the "crown" of the eukaryan branches, were two more groups and a tail of archezoan branches.

The First Branches

The first branches of the eukaryan lineage (closest to the prokaryotic lineages), include several members of Cavalier-Smith's Archezoa (see above), while the "crown" of the eukaryan stem is filled by branches of Cavalier-Smith's Metakaryota with some additional archezoa.[133] Sequences for the small subunit rRNA genes,[134] large subunit rRNA,[135] RNA polymerase subunits,[136] isoleucyl-tRNA synthetase,[137] and elongation factors EF-1a and EF-2[138] demonstrated deep branching for Metamonads, Parabasalia and Microsporidia[139] (although "the order of the three groups remains contentious in rRNA phylogeny"[140]), although Archaeamoeba might branch much higher from the eukaryan stem (although "other molecular trees disagree with rRNA on the order of these deep branching taxa"[141]). Many members of these archezoan groups are primarily or exclusively intracellular parasites and, along with heterotrophs and extracellular parasites, lack mitochondria[142] (and cytochromes), relying on ATP produced by oxidative phosphorylation in host cells or glycolysis for energy-rich compounds. These groups also lack peroxisomes (or microbodies), in most cases, a golgi apparatus, and in some instances, flagella, and they have ribosomes resembling in size those of prokaryotes.

The Metamonada, represented by the human gut parasite, *Giardia lamblia*, branches first from the eukaryan stem of the rRNA phylogenetic tree.[143] This inference is drawn from the parsimony and distance matrix analyses of the "unambiguously" aligned sites of rRNA, but the appearance is reinforced by several features shared with prokaryotes: the retention of prokaryotic rRNA structural features, and the presence of the prokaryote-type Shine–Dalgarno mRNA binding site in the nuclear rRNA gene, a plasmid, and sensitivity to antibacterial drugs, especially nitroimidazoles.[144] Corroborating their early divergence is the absence of rough endoplasmic reticulum (ER), golgi apparatus (GA), and their lysosomes's lack of glycosidases, while all their hydrolytic enzymes are bound to the lysosomal membrane. They also lack a sexual life-cycle, have only a few cytoskeletal proteins, and 5′ regions in genes of tubulin, surface protein, and a heat shock protein are more characteristic of prokaryotes than eukaryotes.

The Parabasalia, represented by trichomonads, *Trichomonas vaginalis* and *T. foetus*, occupy the next branch on the eukaryan tree inferred from 16S-like rRNA, although Cavalier-Smith removed them from the Archezoa. Unlike typical Archezoa, trichomonads have a GA and an extranuclear mitotic spindle (resembling that of dinoflagellates). They also contain hydrogenosomes (also present in ciliates and chytrid fungi lacking mitochondria), a double membrane-bound organelle that converts pyruvate or malate to acetate, carbon dioxide and hydrogen gas.

Microsporidia, which Cavalier-Smith included in the Archezoa, represented by *Vairimorpha necatrix*, are the next to branch from the eukaryan stem of the 16S-like rRNA tree, although, "Microsporidia [may have] actually evolved recently from so-called 'crown' eukaryotes (the twigs, such as animals, plants, and fungi) and may share a close ancestry with fungi."[145] Microsporidia lack cilia and centrioles and undergo mitosis within the nucleus (like yeast). Their most unusual feature is a differentiated polar tube ("a hybrid between a harpoon and a hypodermic needle"[146]) present in spores which is everted in the presence of a susceptible host and allows the spore's contents to move directly into a host cell. The microsporidian's 70S ribosome resembles bacteria's and their sequences of 16S and 23S rRNA are divergent from those of other eukaryotes. Microsporidia also share some characteristics with fungi and animals, however, such as an insertion in the elongation factor EF-1a[147] and, with fungi but not animals, such as similarities in alpha- and beta-tubulins.[148]

A heterogeneous group of organisms, formerly classified as protoctistans (or protistans) are next to leave the eukaryan stem. Bracketed by the plasmodial slime mold, *Physarum physarum*, at the lower end and the cellular slime mold, *Dictyostelium discoideum*, at the upper end, the group includes the Euglenoids and kinetoplastids, heterolobosean, *Tetramitus*, *Vahlkampfia* and the soil amoebae, *Naegleria gruberi*, the Amoebamastigote and *Entamoeba histolytica* and *E. invadens* (Archamoeba). Red algae branch off next in trees based on 16S-like small subunit rRNA, although phylogenetic trees based on parsimony of sequences in 5S and large subunit rRNA place red algae nearer the base of the eukaryan stem.[149]

Crown Clusters

Crown clusters branch off the eukaryan stem in loose phylogenetic groupings of organisms, during a "radiative period." The alveolates (dinoflagellates, apicomplexans, ciliated protozoans[150] and haplosporidians [obligate intracellular parasites, predominantly of marine invertebrates][151]) constitute a cluster of more deeply branching organisms, characteristically having a small number of nuclei (oligonuclear), sometimes of different sizes, within a shared cytoplasm and cyto- (plasma-) membrane. Stramenopiles (comprising brown algae, labyrinthulids, chrysophytes, xanthophytes, diatoms and oomycetes) contain mitochondria with tubular cristae, and many have tripartite mastigonemes not found in any other eukaryotes. The remaining eukaryotes comprise the kingdoms of plants, fungi, animals. Higher plants (Metaphyta) would seem to share a recent common ancestry with chlorophytes (exclusive of euglenoids[152]). Chytrids, considered by some to be flagellated protists, seem to branch from an early-diverging higher fungal lineage. Choanoflagellates branch from the base of the animal radiation (Metazoa), followed by a polyphyletic assortment of sponges and the radial cnidarian/placozoan clade.[153] Oddly, in a remarkable example of phylogenetic devolution, the parasitic, oligonuclear myxozoans appear to have branched off the cnidarians.[154] The branches of all bilateral tridermic (triploblastic) metazoans converge upon a common ancestor(s) and thus appear to be monophyletic, although those lacking a coelom (noncoelomates [acoelomates and pseudocoelomates]) are easily distinguished from coelomates (both protostomes [schizocoelous] and deuterostomes [enterocoelous]).[155]

Phylogenetic trees based on sequences from large ribosomal subunit rRNA also show a distinction between radial metazoans (various sponges, Placozoa, Cnidaria, Ctenophora) and bilateral metazoans. Again, the bilateral metazoans are monophyletic, but they emerge from a deep branching point unlike the situation in the trees drawn from small subunit rRNA data. Moreover, the 28S rRNA sequence-tree shows rapid radiation of deuterostomous phyla, closely pursued by the protostomous phyla. Closely spaced radiations give rise to fungi, plants, diploblastic metazoans, triploblastic metazoans, and several protoctistans (*Chlorogonium elongatum* and *Pyramimonas parkeae* [two chlorophytes], *Cryptomonas ovata* and *Chilomonas paramecium* [two cryptophytes], *Acathocystis longiseta* [a heliozoan], and *Porphyridium purpureum* [a rhodophyte]).[156]

Mitchell Sogin's plea that "The branching patterns in the eukaryotic subtree demands a reconsideration of what kingdom level phylogenetic boundaries are meant to represent"[157] has not gone unheard. Rudolf Raff has probably struggled hardest among molecular biologists to bring molecular sequence data into line with traditional, systematic data for "crown" eukaryotes. He argues that "[p]reliminary inferences drawn from [a consensus of maximum parsimony and distance methods using] full-length 18S rDNA sequences ... approximates traditional views,"[158] although trees inferred from partial 28S rRNA sequences[159] show poor agreement with those drawn from 18S rRNA sequences. Rates of change in sequences may be the culprit, since for example,

> [s]imple inspection of intergroup distances makes it evident that the sequences from fungi have been changing faster than those of plants and animals.... These observations are in full accord with recent reports that suggest that animals and fungi are more recently related than animals and plants.[160]

Raff also finds comfort in discovering that

> The deepest animal group, the Cnidaria (jelly fish and their relatives), serve as the outgroup for all bilaterally symmetric metazoans. The platyhelminthes (flatworms) are the sister group of all coelomate metazoans (Eubilateria), which include the major animal phyla characterized by possession of a true coelomic cavity. Protostome and deuterostome superphyla are resolved. Taxa in the protostome coelomates, which are the sister group of the arthropods, are unresolved and shown as a polychotomy.[161]

R. F. Doolittle et al., can even place a time estimate on crucial events: "Our best estimate of the deuterostome–protostome divergence is 670 Ma, with the schizocoelomate–pseudocoelomate divergence occurring 50 to 100 Ma before that."[162]

WHERE DOMAINS MEET

A variety of genes are shared, or are, at least, similar in members of the three domains, Archaea, Bacteria and Eucarya. Typically, sharing is attributed to the presence of genes in common ancestors, and the more widespread the genes, the more profoundly they are thought to be required by life.

Shared Across the Three Domains

Given life's reliance on replication, transcription and translation, one is not surprised that the single DNA-dependent polymerase (a member of the B family of polymerases) encoded by the archaean *Methanococcus jannaschii*, shares sequence similarity and motifs with several other archaeal polymerases, with eukaryotic α, γ, and ε polymerases, and bacterial polymerase II (but not I). Furthermore, most (but not all) proteins associated with ribosomal subunits, especially the small subunit, are common to Archaea, Eucarya and Bacteria. Moreover, "The structural and functional conservation of the signal peptide of secreted proteins in Archaea, Bacteria, and Eukaryotes suggests that the basic mechanisms of membrane targeting and translocation may be similar among all three domains of life."[163]

The elongation factors EF-Tu and EF-1α also provide classic examples of cross-domain sharing. Considered members of the guanosine triphosphatase supergene family, these elongation factors are similar to *ras*, and thus ubiquitous in prokaryotes, mitochondria, chloroplast and eukaryotes.[164] Indeed, representatives of both kinds of elongation factors are present in the three domains. One elongation factor, EF-1 (EF-Tu or EF1α) assists in binding of activated tRNAs to ribosomes; another, EF-2 (EF-G), is involved in the transfer of amino acids from activated tRNAs to the nascent protein. "These two kinds of molecules are homologous (paralogous within organisms), [presumably] having been produced by a gene duplication in the line leading to the common ancestor of modern life."[165]

Shared Across Two Domains

Other genes are shared in members of two domains. Possibly, pairs of shared genes can be attributed to the loss of genes in one or another domain. Alternatively, a gene originating in a common ancestor of two domains might be represented in extant members of those domains and absent in members of the third domain.

Similarities in Archaea and Bacteria

Archaea and Bacteria were formerly considered members of the same class and even today are grouped as "prokaryotes." Both groups exhibit streamlined genomes: small numbers of genes are frequently arranged in clusters of dissimilar unique genes (rRNA genes being an exception) sometimes organized in operons of functionally related genes. Several similarities are found in organisms "exhibiting putative archaic traits" such as hyperthermophilic Crenarchaeota (Archaea) and Thermotogales (Bacteria). For example, porphyrins and related compounds functioning in electron transport are similar, except for those functioning in photochemistry. Likewise, the composition and order of genes in clusters associated with methanogenesis, polysaccharide biosynthetic enzymes, as well as all the genes associated with the ability to fix nitrogen are similar in both Archaea and Bacteria, but the "neighborhood" of these clusters are not especially comparable. "The assembly of genes and gene clusters into coherent circular genomes might therefore have happened twice independently, once each in Bacteria and Archaea."[166]

Other similarities are more difficult to dismiss as convergence, although they do not constitute unambiguous evidence on behalf of a common ancestor for Archaea and Bacteria. These similarities include sequences in a variety of archaean and bacterial metabolic proteins such as more than 20 putative polysaccharide biosynthetic enzymes and proteins associated with the transport of small inorganic ions into cells. Furthermore, tandem repeats of short segments in both Bacteria (*Escherichia coli*) and Archaea (*Methanococcus jannaschii*) are "hypothesized to participate in chromosomal partitioning during cell division."[167] Similarities in the holoenzyme ribulose bisphosphate carboxylase/oxidase (RUBISCO) of spinach chloroplast and an archaeal halobacteria are usually dismissed as a consequence of a horizontal gene transfer rather than a genuine step of RUBISCO evolution within the Archaea.[168]

Still other similarities are attributed to ecological constraints. For example, several genome sequences in the archaean *Archaeoglobus fulgidus* resemble sequences in bacterial genes. Lateral gene transfer among protobacteria and protoarchaea living in ecological continuity or primitive endosymbiosis coupled to lateral gene transfer might accounted for these similarities.[169]

Similarities in Archaea and Eucarya

Similarities among archaeans and eukaryotes are abundant and frequently welcomed as evidence of common ancestry. Sequences are frequently suggestive of an origin in a common ancestor for Archaea and Eucarya, although similarities may be easily stretched back to an origin in a common ancestor for the three domains. For example, the 7S RNA component of the archaeal signal recognition particle (SRP) contains "a highly conserved structural domain shared by other Archaea, Bacteria, and Eukaryotes.... However, the predicted secondary structure of the 7S RNA SRP component of Archaea is more like that found in Eukaryotes than in Bacteria."[170].

The sequences with greatest similarity among Archaea and Eucarya include the large subunit DNA-dependent RNA polymerase, several ribosomal proteins, and subunits of ATPase.[171] Furthermore, "the *Methanococcus* translation elongation factors EF-1α (EF-Tu in Bacteria) and EF-2 (EF-G in Bacteria) are most similar to their eukaryotic counterparts."[172] Likewise, "Aminoacyl-tRNA synthetases of *M. jannaschii* and other Archaea resemble eukaryotic synthetases more closely than they resemble bacterial forms," and "[a]ll of the subunits [of DNA-dependent RNA polymerase (11 genes)] in *M. jannaschii* show greater similarity to their eukaryotic counterparts than to the bacterial homologs." At the same time, "The archaeal transcription initiation system is essentially the same as that found in Eukaryotes and is radically different from the bacterial version [and] ... the rfc [replication factor complex] sequence for *M. jannaschii* shares the characteristic eukaryotic signature in this domain [sequence of domain VI]."[173] Finally, "[w]ith respect to the related cellular processes of replication initiation and cell division, the *M. jannaschii* genome contains two genes that are putative homologs of Cdc54, a yeast protein that belongs to a family of putative DNA replication initiation proteins.... A third potential regulator of cell division in *M. jannaschii* is 55% similar at the amino acid level to *pelota*, a *Drosophila* protein involved in the regulation of the early phase of meiotic and mitotic cell division."[174]

Other similarities are a little more esoteric. For example, an archaean heat-shock protein, thought to function as a chaperone, assisting protein folding, is similar to polypeptide-1 of yeast and to a polypeptide encoded in the mouse t- [tailless-] complex expressed in developing sperm.[175] Likewise, archaean DNA contains introns in both rRNA and tRNA genes, resembling eukaryan introns.[176]

Carl Woese is persuaded that, "for functions common to all three domains, the archaeal version is usually closer in sequence to its eucaryal homolog than either is to their bacterial homolog."[177] For example, an archaeal histone (from *Methanotherma fervidus*) is not only "specifically related to the eucaryal versions but its sequence is closer to the (consensus) sequences of the *four* eucaryal histones (H2A, H2B, H3, and H4) than these four are to one another."[178] Cavalier-Smith[179] also finds abundant similarities between Archaea and Eucarya which he attributes either to their immediate common ancestor or to a bacterial line (Thermotogales) from which they were derived. The similarities occur in N-linked glycoproteins, in short introns in tRNA spliced by proteins using a 3′-phosphate 5′-OH mechanism, in certain detailed properties of DNA polymerases, and in RNA polymerases and ribosomal proteins.

Similarities in Bacteria and Eucarya

Several "bacterial" genes are present in eukaryotes. A triosphosphate isomerase gene, possibly of alpha-protobacterial origin[180] and several mitochondrial-like genes even occur in amitochondrial eukaryotes (Archaezoa). Most "bacterial" genes in eukaryotes, however, belong to the "extended" genomes of mitochondria and chloroplasts (see above and Chapter 4), some within the organellar genomes (cpDNA and mtDNA) proper, and others in the eukaryan nucleus (nDNA).

CRITIQUE

Incongruity and contradictions are nothing new to systematics. They are the "bread and butter" of the trade and have reemerged over the centuries despite, if not because of, numerous scientific revolutions ("Everything changes so that everything remains the same"). New paradigms merely draw out old problems; paradigms justify themselves; the new guardians of the gate offer new techniques, data and terms to replace old ones, but difference is repetition. The changes which even staid systematics has undergone since the "revolution" launched by Watson and Crick have not solved these problems.

"Been There; Done It"

Thus, an 18th century naturalist could easily read the Great Chain of Being in today's "universal" phylogenetic tree. Of course, a great many more organisms discovered in the meantime have been placed in the new "Chain," but there, as in the original, the links or branches are still recognizable as entirely separate groups of organisms with each group organized hierarchically through an evaluation of difference. The chain or tree, and each of its links or internodes, is still hierarchical, and organized by differences, even if these differences are now to be circumscribed by symplesiomorphies, synapomorphies and autopomorphies and explained by evolutionary homologies, gene duplications, paralogues and orthologues.

What the 18th century naturalist would be missing is only that the "Chain" is no longer a scalar, but the quality of direction, that natural historians had only begun to slip into the early chain, is now a dominant feature. The "evolutionary revolution" of the 19th century legitimized direction in the chain by using it to provide a better explanation than previously available for biological hierarchies. Indeed, the current hegemony of evolutionary theory requires the modern systematists to draw out the chain and every point on it along a vector, originating in an ancestor and stretching to extinct and extant species.

Then as now, the chief characteristic of all biological hierarchies is that points of branching always represent the origin of the branch and never its terminus, or point of merging with another branch. A great deal of biological folklore supports the notion that species can separate but never merge. Furthermore, in high (secondary) schools, colleges (tertiary) and universities, students are assiduously indoctrinated in evolutionary theory: that different species spring from a common ancestor. The geometry of evolution is purely linear: a lineage is a line that originates at a point and goes to a point in the present or to extinction.

The alternative is never stated: "origins" may not be points but blobs spread out over segments of curves or across curves. Thus, one might speak of "common ancestors" at the origin of species instead of "a common ancestor," and acknowledge that even evolution by natural selection requires populations, and populations of living things are inevitably heterogeneous. The tradition of linearity in systematics dictated that, when systematists began looking at molecular data, sequences were turned into just another set of bifurcations and passive carriers of biological hierarchies. Data were bent where inconsistency and incongruities cropped up, but few systematists questioned premises. Perhaps that important failure will be corrected soon.

Lineages Do Not Necessarily Come and Go from Points

The distribution of genes shared by members of two domains is problematic for hierarchical systematics. The paleobotanist Otto Kandler recognized the problem of sorting out life's properties to the three domains when considering characters shared in pairs of domains:

> "The distribution of such features found in only two domains is not unequivocally compatible with the branching order at the basis of the phylogenetic tree.... For instance, glycerol fatty-acid esters, catabolic mode of glycolysis, etc., occur only in Eucarya and Bacteria; circular genome structures, Shine Dalgarno sequences, etc., are common characteristics of Bacteria and Archaea; while V-ATPases, glycoproteins, etc., are only found in Eucarya and Archaea."[181]

The problem is thus that each pair of domains would seem to have had a common ancestor which is one too many common ancestors. If nodes can only give rise to branches, and only three domains have arisen, the separation of one domain from a common stem can leave only one common ancestor for the remaining two domains. The question is, which pair of domains had this last common ancestor?

Answers are ambiguous, since phylogenetic trees drawn for different sequence data are incongruous. Beyond problems with rRNA data (see above), sequences for ATPases and translation factors show Bacteria branching off prior to the separation of Archaea and Eucarya, while sequences for archaeal malate dehydrogenase show the Archaea branching off prior to the separation of Bacteria and Eucarya. Similarly, while sequences from eukaryotic DNA-dependent RNA polymerases 2 and 3 (pol2 and pol3) branch off in the vicinity of the corresponding archaeal components, polymerase 1 (pol1) branches from the connection between archaeal and bacterial polymerases, significantly remote from pol2 and 3.[182]

One way out of the dilemma is to challenge assumptions about linearity and premises about group dynamics. Can lineages fuse? Could once separated groups return to their source in common ancestors or even merge with other separate groups?[183] Eucarya, for example, is broadly acknowledged to have had chimeric origins by way of endosymbiosis, but did it also have "biphyletic" origins involving the fusion of bacteria and archaeans or possibly "oligophyletic" origins involving multiple and evolving fusion events.[184]

Another way out of the dilemma is to broaden the concept of origins from that of points to something larger, something spread over a range of living things. Imagine a time when life was not segregated into individuals (the predominant way it is today), but consisted of transient systems that existed mainly as a consequence of the benefits of cooperation. Then imagine that other or additional benefits might sometimes have arisen through new combinations within or between systems; parts that cooperated in other or new ways.

How might today's individuals have arisen in such a paradise? Answers are suggested by the possibilities of competition (the subject of so much of today's theory). The "collectives" of life's forms might have narrowed into today's preponderant, individual life forms. What began as so many blobs would have collapsed to many points by the replacement of cooperation with competition and the breakdown of partnerships into life's individualistic entrepreneurs.

CHAPTER FOUR

Devolution: Reconciling Sameness and Difference

Repetitions repeat themselves, while the differenciator differenciates itself. The task of life is to make all these repetitions coexist in a space in which difference is distributed.

Gilles Deleuze (Difference & Repetition, *p. xix*)

Mechanisms are the scientist's answer to the question, "What's happening?" They are at the heart of "revolutionary" science and "ordinary" science, of causal thinking, reductionism, scientific enterprise and criticism. In theory, when a mechanism is understood, one can make a desired product or avoid a dreaded consequence. In practice, the efficiency (bottom line) of one or another mechanism (for producing or destroying commodities) determines its usefulness in the market place. What, then, are the mechanisms for producing or removing the samenesses and differences of life?

Undoubtedly, many scientists alive today would answer this question with a reference to deoxyribonucleic acid (DNA), alluding to the massive sequences of nitrogenous bases currently being stored in databases. Undoubtedly, Watson's and Crick's successful unraveling of DNA's secondary structure and the advent of the computer in the epoch following World War II were singular events shaping modern biology's attitude toward sameness and difference. Watson and Crick can be credited with mobilizing biologists for that "mad pursuit"[1] and pointing biology in the direction of sequencing, while computers allowed biologists to move in the desired direction culminating in the Human Genome Project. Not only are computers built into and essential for the operation of biotechnology, but computers determine how data are obtained, managed, stored and rendered useful in virtually all biological applications.

Molecular biologists gravitated toward DNA because it was, after all, "the gene," and because it was remarkably easy to purify, compared to proteins. Thus, DNA chemistry progressed rapidly; restriction fragments of DNA became available in virtually unlimited quantity through cloning and the polymerization chain reaction (PCR); reverse transcriptase made possible the replacement of ribonucleic acid (RNA) with DNA; sequences for coding regions were translated to the amino acid residues of polypeptides and vice versa; techniques for sequencing DNA crept up on their goal of cheap, "high-throughput" output; the goal, reading the structure of entire genomes, quickly moved from the realm of possibilities to mordant reality.

Computers were also irresistible to molecular biologists. Computers solved the problem of what to do with all the data amassing through molecular analysis, especially as analysis became increasingly automated. The German entomologist, Willi Hennig, was the first to work out algorithms that could transform data on primary sequences in macromolecules into bifurcational trees. Soon, with the addition of techniques for data-management borrowed from other fields, molecular cladistics was born "to find the true historical pattern of descent and to base a classification scheme upon it."[2]

Chapters 2 and 3 have documented the vast superstructure of molecular evolution built up and around sequence data and computer programs. In essence, genes took on a life of their own, with their own evolutionary histories (not limited to those of the organisms in which the genes resided). These chapters also demonstrated that the solutions posed to the question, "What's happening?" were not always salubrious. Instead of mechanisms for producing or reducing sameness and difference, biologists adopted rationalizations and wrenched data through formulas such as:

> There are [enter some number] apparent violations of the clade's phylogeny on the gene's inferred tree, and, although these incongruities could be due to old horizontal transfers [or enter some other nonconformist mechanism] ... it is more likely that different active lineages of the gene have been propagated in various members of the taxon.[3]

Other molecular biologists adopted gross generalizations, naive interpretations of evolution and crude distortions of biology's history to wrench data into conformity with models of life. Still other molecular biologists, claiming that evolution is, after all, nothing but metaphysics anyway, disclaimed any way of knowing what actually happened in the course of life's genesis.

Chapter 4 discounts these rationalizations, claims and counterclaims while making no apologies for the silliness and contradictions of biologists. Instead, the chapter attempts to reclaim data, to work out of them and, in a deconstructionist sense, take flight to other plateaus.[4] From this vantage point, the chapter brings into focus, if only at a distance, a concept of devolution, a mechanism of history that escapes sameness and difference. Devolution,[5] is the descent (collapse) of qualities upon successors through assimilation into chimeras and parsing out into fragments. In its failure to "differentiate" and "equate," devolution exchanges differences for samenesses through exclusion and samenesses for difference through repetition. In contrast to difference and sameness, life's history is drawn through redundancy and reassortment, both locally and globally, spatially and temporally, over virtually limitless time and instantaneous reactions. Devolution is sameness's and difference's Möbius band, their fusion and plane of turning, their ever becoming one another.

The chapter begins with the origins of life and posits the impossibility of life's singularity as opposed to multiple origins collapsing (devolving) into life. The chapter continues with an examination of devolution along a broad spectrum of subjects (from the macromolecular to the organismic) for which suitable data are compiled from biology's literature: RNA-based genomes, the RNA world, early introns hypotheses and genetic code(s), DNA-based genomes, genes, and chromosomes, gene superfamilies, the parsing out of life's units, viruses and cells, Archaea, Bacteria and the Eucarya mix. The choices of subjects is not intended to be exhaustive but sufficient to move devolution beyond the sameness/difference dialectic and over the threshold to Deleuze's coexistence space.[6]

EARLY EARTHLIVES

The ambiguity that inevitably surrounds early life on Earth is a source of inspiration for study and should not be abridged with premature certitude. The insistence of systematists on monophyletic solutions to branching problems, which dictates a single origin of life, is both regrettable and unnecessary.[7] At the moment (1998), one cannot even say with confidence that life originated on Earth. It may very well have arrived here from abroad in some form or other. Indeed, "the earliest ETs were like to have been simple organic molecules!"[8]

Life on Earth, therefore, is called Earthlife here, to distinguish it from whatever other forms of life may be "out there." Moreover, since Earthlife may not have been confined to a single origin, it might better be called Earthlives, at least, until (if ever) the mechanism of its unification is revealed.

Multiple early Earthlives is congenial to the idea of devolution which departs under any set of assumptions from a monophyletic view of life at the beginning. Devolution virtually replaces a beginning of life with the assimilation and fragmentation of whatever and how many early Earthlives there were or might have been. This is not to say that monophylety is not an alternative; only that monophylety is one among several alternatives.

A Geochronometric View of the Early Earth

Remarkably, those who study the beginning of Earth are broadly in agreement about what happened (at least after the first 500 million years of the Earth's history), to the degree one can trust those who write, "It is generally believed that."[9] Agreement does not mean that the literature is easy to read or follow, however, and disagreement abounds when the discussion of geology turns toward bioneogenesis.

Tracing the physical evidence for the origins of life has occupied the attention of mineralogists, metamorphic petrologists, economic geologists, and sedimentologists as well as paleobiologists. The methods of such a diverse group of scientists are so different that one can scarcely be surprised (albeit frustrated) when they employ different nomenclature (if not languages) for what may yet turn out to be the same thing. To be fair, data accumulated by practitioners of different scientific disciplines do not necessarily match, or, at least, not *ab initio*. Furthermore, these scientists have different priorities, which may not include correlating their data with those from other fields: geochemists are interested in dating rocks from their content of isotopes; sedimentologists in explaining anomalous inclusions; paleobiologists in identifying fossils of one kind or another (molecular, micro and macro). But why can't they come to terms on terms? The geochemists identifies changes in Precambrian continental crust formation by "cycles" (Isua, Barberton, Superior, Nubian Cycles); the paleobiologist is more likely to identify the same periods as Middle and Early Proterozoic Eras and the Archean Eon. Fortunately, in most cases, the names assigned to the history of the Earth are synonyms. More mind-stretching is necessary where the names do not overlap completely (e.g., Eocambrian versus the Ediacaran System and Vendian). Even something as trivial as the notation for "years" defies agreement: the geochemist prefers "a" for *annus*; the paleontologist prefers "y" for years. Thus the literature is burdened with Ga and Gyr for a billion years and My as well as Ma for a million years.[10]

Nevertheless, many geochemists[11] seem to agree that the age of the solar system is approximately 4.6 Ga, and, beginning about 4.5 Ga ago and continuing for some 0.3 Ga, Earth enlarged by accretion; its interior heated to several thousand degrees Celsius; its core and mantle formed, and seas of molten magma percolated beneath a superheated, nonreducing atmosphere (possibly a runaway "greenhouse" effect if water was present) derived by outgassing of volatile components of the mantle. Accretionary pressure declined between 4.4 Ga and 3.9 Ga ago; temperature dropped below 300°C, and liquid water became stable by 4.2 Ga ago. By about 3.8 Ga ago, temperature declined to an average of 80 to 100°C, although very large impacts could have raised temperature transiently by 100°C. High vapor pressure of water in the atmosphere dampened boiling, and a bicarbonate-rich ocean (pH ~ 6) covered the world except for land masses of volcanic islands and microcontinental blocks formed from accretionary complexes. The conditions were thus established in the Hadean world for the accumulation of organic compounds and the origin(s) of life, or something like it.

Earthlife is assumed to have been present somewhere in the vicinity of 4.0 ± 0.2 Ga ago, and isotopic signals ($^{13}C/^{12}C$ ratios) from carbon occluded in apatite grains in metamorphosed sediments of Greenland's Itsaq Gneiss Complex suggest that oceanic life existed some 3.85 Gyr ago.[12] By 3.5–3.2 Ga ago, temperature reached 30–50°C, supported by the faint, young Sun and reduced green house gases (e.g., carbon dioxide declined to levels merely 100 to 1000 times present levels), and Earthlife was, presumably, evolving rapidly.

Origins

Biologists have enough trouble understanding the origins of extant forms of Earthlife, so one is not surprised that mystery and guesses surround the origins of early Earthlife. Speculation falls into two opposing camps (members of which may consort on occasion) depending on point of origin: Earthly origins versus extraterrestrial origins.

Earthly Origins

The benchmark theory for Earthly organic "origins" is Alexander Ivanovich Oparin's (1894–1980), 1924 proposal that life originated in a strongly reducing atmosphere (rich in hydrogen, ammonia, methane, and saturated and unsaturated hydrocarbons). Published in Russian, Oparin's idea was ignored in the West,[13] but the English language translation of his book, *Origin of Life*,[14] in 1938, attracted some attention, including that of Harold Urey who developed a quantitative model and testable hypotheses for an early reducing atmosphere.[15] Stanley Miller's celebrated experiment[16] followed, successfully testing the efficacy of a reducing atmosphere for manufacturing biological materials. Miller's results showed that his hypothetical reducing atmosphere (with the addition of electrostatic discharge and heat) could produce abundant amino acids, an index molecule for Earthlife.

The primitive atmosphere, formed of volatile compounds from degassing the Earth's interior, would have been nonreducing, however. What is more, a predominantly oxidizing atmosphere, consisting of water vapor (instead of hydrogen), nitrogen (instead of ammonia) and carbon dioxide (instead of methane and carbon monoxide) would have been vastly less efficient at abiotic synthesis than a predominantly reducing atmosphere. Possibly, "way down, beneath the oceans," dissolved ferrous iron would have provided a reducing environment. The Earth might yet have been the source of organic compounds, formed by geochemical synthesis.

Extraterrestrial Origins

The idea of extraterrestrial, or panspermic, origins of Earthlife are as flowing with speculation and controversy as ideas of terrestrial origins. Exotic as some of these speculations are, testable concepts of panspermia are quickly reduced to two mundane possibilities: provisioning Earth with raw materials (i.e., providing material of low information content); introducing some sort of hereditary material (i.e., high-information macromolecule) or process (template-copying) that either introduced life directly or led to the origin of life via an evolutionary process.

Provisioning Earth with Low-Information-Containing (Raw) Material

Extraterrestrial material is certainly added to Earth and may very well have represented a significant source of chemically derived (abiotic), organic compounds. The material provided may or may not have been unique in the sense that nothing like it was synthesized abiotically on Earth, and the extraterrestrial material may or may not have played a role different from that of indigenous and indigenously produced Earth materials. Analysis of amino acids found in the Murchison meteorite (recovered near Murchison, Victoria, Australia September 28, 1969) in the late 1980s illustrates the dilemma of choosing between these terrestrial and extraterrestrial sources of ultimately biological building blocks: the meteorite contained a racemic mixture of 74 amino acids whereas amino acids produced by Earthlife are generally optically left-handed; eight of the Murchison amino acids are present in Earthlife proteins, and eleven play other biological roles, but 55 are unknown in Earthlife.[17]

Extraterrestrial space is not a trivial source of primitive, raw materials for life, especially given the abundance of hydrogen, carbon, cyanide (CN), other simple organic species in the solar upper atmosphere and the presence of a variety of organic molecules and radicals, phosphate, ammonia, water, and possibly polymers of hydrogen cyanide, adenine and other purines in comets.[18] A comet on a grazing trajectory or shattering and landing on the Earth's surface intact and melting slowly might well have supplied several organic-rich ponds. Indeed, the amount of material delivered during the Hadean Era of Earth-building would have been massive, given the dimension of accretion (as much as one third the Earth's mass). In the 10^8 years of this period, interplanetary dust particles (IDP), meteoritic and cometary fallout could have added 10^{16} to 10^{18} kg of organic carbon to Earth, more than the total estimated for the present biomass![19] Impacts might also have provided energy for prebiotic organic synthesis.[20] Given that the purine, adenine, amino acids and other organic compounds have been synthesized from the condensation of HCN and NH_3, "cometary collisions may have provided the primitive Earth with an important source of volatiles. ... [and] probably the starting point for the nonbiological synthesis of biochemical molecules that preceded the first organisms."[21]

Some (relatively) low-information containing materials reaching Earth from galactic space might also have been (relatively) complex. Carbonaceous meteorites might very well have been the source of lipid "amphiphiles," the organic lollipops with a polarized "head" and a neutral "tail" present in the lipid bilayers of eukaryal membranes.[22] The evidence for extraterrestrial source of amphiphiles is "wishful thinking," since, on theoretical grounds, the jump from short chains and cyclic aliphatic hydrocarbons to long chain hydrocarbons and fatty acids would have been difficult for strictly terrestrial sources operating under geochemical constraints. On the other hand, the Murchison meteorite contains a complex mixture of oxidized hydrocarbons including nonanoic (C9) monocarboxylic acid, which is sufficiently amphiphilic to form lipid bilayers.[23] The origins of Earth's protocells may thus have depended on nonanoic "manna."

Space As a Source of High-Information-Containing Material

Space may also have injected high-information containing or "semantophoretic" molecules (conveyers of meaning following translation on the order of hereditary material) onto planet Earth. The name most associated with the notion of information-rich seeding is that of the eclectic molecular biologist, Francis Crick, although, earlier in this century, the Swedish chemist, Svante August Arrhenius (1859–1927), came up with the idea of spores or microorganisms escaping from stellar–planetary systems and traveling under the pressure of starlight to the nearest similar system. The British physicist, William Thomson Kelvin, also played with the idea of life in interstellar space. Upon reaching Earth, the ubiquitous spores were to have colonized the planet, fertilizing it with life (hence "panspermia"). Crick, in collaboration with, the biochemist, Leslie Orgel, added the possibility of an intentional (or could it have been accidental?) seeding of Earth by extraterrestrial, intelligent beings who delivered the spores via spacecraft.[24] Crick dubbed this version of the idea, "directed panspermia."[25]

Michael Behe, today's most outspoken advocate for life's basis in intelligent design, can now be added to the ranks of enthusiasts for directed panspermia. In Behe's words, "a civilization capable of sending rocket ships to other planets is also likely to be capable of designing life.... [T]here is no logical barrier to thinking that an advanced civilization on another world might design an artificial cell from scratch."[26]

The astronomer, Fred Hoyle has also proposed a galactic biosphere ("the whole of our galaxy as a biological system"[27]) and an extraterrestrial intervention in the creation of Earthlife. Based mainly on astronomical data (interstellar extinction curves [the reddening law], light scatter [albedo], polarization, and spectroscopic properties [emission bands, signature absorption spectra and infrared radiation]) on density and distribution of elements in space, Hoyle and Chandra Wickramasinghe[28] first suggested that a high proportion of available matter in the galaxy is present in clouds of interstellar dust, and dust particles contain a large fraction of galactic carbon, nitrogen and oxygen. The idea of organic polymers in interstellar dust followed from radioastronomical data and agreement of flux curves for galactic spectra (the galactic center IRS 7, the Trapezium nebula and the OH maser source, OH 26.5 + 0.6) in the infrared band (3 to about 3.6 μm) resonated with the calibrated infrared spectrum of polyformaldehyde (or polyoxymethylene, a linear polymer of formaldehyde [H_2CO]), dry polysaccharides and even dry bacteria.

Hoyle's next step was the intellectual jump from interstellar dust to microbial grains resembling bacteria. This step to cosmic biology was a long

Origins of Life

"Making a living"	Catabolic pathway	Anabolic pathway
"RNA World"		
Solar energy supports RNA replication		Early association of RNA genomes, ribosomes and membranes
"Obcell"		
Photophosphorylation; heterotroph utilized prebiotic "soup"		Efficient DNA replication, transcription, protein synthesis, membrane growth and division on outside of liposome-like vesicle
"Progenote"		
Spontaneous pyrite formation as energy source	Primordial chemolitho-autotrophs	
	Primitive chemo-lithoautotrophs	Reductive citric acid cycle
	Various redox energy sources based on H_2 oxidation	Primitive hydrogenase activity

and perilous one but one that could not be avoided. To flesh out, so to speak, the idea of extraterrestrial life, Hoyle developed the "microbial grain model" to include the consequences of limiting nutrients, the possibility of reproduction in comets and, of course, planets, and adaptations for the survival of microbes in space. To suggest a plausible source of Earthlife, Hoyle calculated entry trajectories compatible with a "soft landing." Moreover, from a retrospective on the 1976–77 Viking mission to Mars, Hoyle concluded, contrary to general opinion, that "The balance of the evidence therefore is that life is indeed present on Mars."[29]

The most intriguing (if not compelling) evidence for extraterrestrial origins of Earthlife is the quality of carbon compounds and the presence of water in chondrite inclusions of carbonaceous meteorites. Hoyle restores credibility to the work of George Claus and Bart Nagy, in the early 1960s, on meteorites recovered from a spectacular meteorite shower near Orgeuil France in 1864 and the Ivuna meteorite which fell in Tanzania in 1938, but Hoyle relies most heavily on the analysis performed by Hans Pflug in the early 1980s on the Murchison meteorite. All these meteorites contained carbon coated, microscopic spheres, interpreted as microfossils of biogenic origin. Murchison contained even more life-like forms at the electron microscopic level of resolution, carbon-bearing casts in the form of tiny filaments. Hoyle sees these casts as spitting images (so to speak) of Earth-bound living things (the flower-like, iron-oxidizing bacterium, *Pedomicrobium*, the outer cell walls of methanogens, terrestrial viruses) and "clear evidence of extraterrestrial life."[30] Moreover, Hoyle speculates that a type of iron-oxidizing extraterrestrial colonizer (possibly of the *Pedomicrobium* variety), capable of shuttling oxygen from one substance to another, arrived at Earth's virtually oxygen free atmosphere approximately 3.8 Ga ago. "*Pedomicrobium* simply did not have time to evolve in a supposed primordial soup — it must have 'appeared' intact."[31]

Little can be added to speculation on directed panspermia except, perhaps, some consideration of the size of the inoculum. If one thinks about panspermia as something analogous to *in vitro* fertilization and embryo transfer,[32] then, a multiple inoculum is more likely to be successful than a single inoculum. The poor statistical outlook for the *in vitro* fertilization justified, as far back as the 1970s, multiple embryo transfers, as a "backup." Having induced polyovulation and fertilized several eggs *in vitro*, physicians transferred several blastocysts to the hopeful (if not expectant) "mother" (to be), limited only by the number of blastocysts available, common sense and laws governing such things. Would it not make sense, therefore, to expect that multiple inocula would have introduced life to a planet?

Mars and "All That Jazz"

Recently, the idea of panspermia hit the newspapers and politicians with a splash.[33] The news was the alleged presence of biogenic material and possibly microfossils in a meteorite probably from Mars. The possibility that life existed on other planets and could have traveled to Earth on meteorites suddenly became acceptable, and the idea that Earth had been inoculated with life from space took on new life.

The 1.9 kg, softball-sized meteorite, ALH84001, found in the Allan Hills region of Antarctica in 1984, is thought to be of Martian origin because of its similarity with 11 other meteorites thought to be of Martian origin. These meteorites are members of the Shergotty–Nakhla–Chassigny (SNC) class of meteorites. They are igneous rocks (i.e., crystallized from molten larva) which are thought to be of Martian origin for one reason. In the official words of the Earth Science and Solar System Exploration Division of the Johnson Space Center:

> The conclusive evidence that SNC meteorites are from Mars is the analysis of gases trapped in glass inclusions in EET·A79001, which chemically and isotopically match gases measured in the unique martian atmosphere by the Viking lander spacecraft.[34]

Meteoriticists suggest on the basis of radiometric dating that ALH84001 congealed 4.5 Ga ago from magma in the Martian crust. The rock shattered from a meteorite impact 4.0 Ga ago and was launched into space by another impact no less than 16 million years ago where it remained until crashing to Earth 13,000 years ago.

Claims for life on Mars are based primarily on two types of "biomarkers": polycyclic aromatic hydrocarbons (PAHs) commonly formed by the breakdown of living things; carbonate globules aged 3.6 Ga, found along fractures and in pore spaces in the meteorite containing shock-faults of nonterrestrial origin. The carbonate globules resemble bacterial-induced precipitates on Earth. Moreover, the carbonate spheroids are rimmed by layers of an iron oxide magnetite resembling "magnetofossils" and containing a core of similar magnetite-rich and iron-rich sulfide material. These materials and their distribution resemble deposits left in terrestrial sediments by bacteria and an iron monosulfide called pyrrhotite. While pyrrhotite could share a biogenic origin with magnetite, it is unlikely to precipitate in the same deposit abiogenically.[35] Finally, the spherules share some features of possible microfossils. Measuring 1 to about 250 μm across, the iron-rich rim of globules typically contain ovoids about 100 nm in longest dimension and 20 to 80 nm across,[36] while the centers of some globules contain ovoids about 20 to 100 nm in their longest dimension. These textures and the nanosized magnetite and iron-sulfides are possibly products of microbiological activity or even the fossilized casts of nannobacteria.

How quickly "News" changes![37] At the same time additional support for Martian life appears, "cold water" is thrown on the entire concept. Support comes from the proportion of carbon isotopes ($\delta^{13}C$ up to 42‰) in a carbonate globule from ALH84001. A carbonate with the observed ratio of carbon-13 to carbon-12 of acknowledged Earth-bound sources would be considered near the typical range of enzymatic methane fixation ($\delta^{13}C$ values of ~50 or ~60‰). On the other hand, "cold water" (literally) may well have seeped into the Martian meteorite during its long hibernation on Arctic ice fields. Jeffrey Bada, who analyzed amino acids and PAHs in bits of another Martian meteorite, EETA79001, also collected from Antarctica, found the biomarkers matched those present in Antarctic ice. The latest word will not be the last word on the subject.[38] The question remains, are the biomarkers truly indigenous to the meteorites or are they tokens of terrestrial contaminants?*

Certainly, some sort of panspermic scenario is suggested by the alleged presence of extraterrestrial microbes on a meteorite.[39] The suggestion would be that much stronger if the extraterrestrial microbe had DNA twisted in a right-handed helix, like DNA on Earth. But such a finding would still leave panspermia in limbo. Logically, the possibility of panspermia (i.e., that Earthlife came from afar) is only of equal status with the possibility that extraterrestrial life originated on Earth, and a microbe on a meteorite was the descendant of one that originated on Earth (as the ejectum from Earth [following an impact of an asteroid on Earth]). According to Paul Davies, if Earthlife is "an inevitable consequence of the outworking of the laws of physics and chemistry, given the right conditions,"[40] then the possibility of terrestrial origins is virtually "[i]ndistinguishable from the panspermia hypothesis."[41] However, if extraterrestrial DNA had a left-handed helix, unlike DNA on Earth, or if an extraterrestrial microbe had some entirely different sort of informational molecule, it would suggest an origin independent of Earth. Such a microbe would not say much either way about the origin of Earthlife, but it would, "constitute powerful evidence against the theory that the origin of life was a freak event — a highly improbably random accident."[42]

If all these speculations, arguments and assumptions boil down to anything, it is that Earthlife is more likely to have had many origins than a single origin. This is not to say that the source of Earthlife was terrestrial or extraterrestrial, organic or inorganic, stochastic or determined. It is merely to say that prospects for multiple origins simply outnumber prospects for a unique origin. One can imagine, however, that most of these early origins occurred in isolation and became extinct. The future of life, it would seem, was cast when various forms of early Earthlife came together and "pooled resources."

Between Chemistry and Biology

Paleobiologists, physicists and chemists, while invoking as little magic as possible, postulate organic Earthlife coming about through a quasievolutionary process involving geochemically unproblematic starting materials, selection, cooperation (symbiosis), a chemical form of inheritance (auto-

---

*Skepticism ran high at a 2–4 November 1998 meeting at the Lunar and Planetary Institute, Houston, Texas, as reported in *Science*. See Kerr (1998).

catalysis), and, hence, growth and reproduction. Briefly, transformations from chemistry to biology would have occurred when simple organic compounds, of whatever origin, changed to the monomers of biologically relevant compounds (biomonomers such as amino acids and nucleotides) with rudimentary catalytic properties for polymerizing themselves into larger compounds.[43] Presumably, "hot spots" of polymerization of one type of biopolymer or another occasionally met and merged. In all likelihood, most of these cooled, but, presumably, some became hotter as a result of synergism set up by merging and mixing. The assimilation of reactions in these centers might then have led to new reaction centers by additional spreading or by fragmentation, sometimes taking a whole set of reactions along, sometimes taking a new mix of one or another reactions into new domains. Viruses and cells may be seen in the paleobiologists's snapshots of these mixes and fragments.

Ingredients

Origins are not necessarily linear, and every feature of life need not have arisen simultaneously. (for example, metabolism and reproduction, compartmentalization and cellularity). Alternative life-forms, spinning out in exotic directions (forms we might not recognize as living) might also have contributed directly (via cooperative exchange or symbiosis) or indirectly (via competition) to the origin of life (as we know it). For example, clay-mineral systems and crystal "genes"[44] may have preceded organic compound systems and nucleic acid genes. Mineral sheets, especially of double-layer hydroxides, which were probably abundant in the anoxic, Archean benthos, may have played a "vital" role in concentrating phosphate, providing scaffolding, and lowering the activation energy of reactions.[45]

The conditions on the early Earth and the requirements for life even suggest that spatial and temporal separation were most likely if not necessary for life's origin(s). In one corner of this world, a prebiotic oil slick of hydrocarbons might have formed at the air–water interface, possibly concentrated by wave action in intertidal pools, and, as a result of exposure to intense ultraviolet (UV) radiation, undergone a variety of chemical reactions. Adsorption (chemisorption) on mineral surfaces, might have provided the hydrocarbons and their derivatives with a surface-bonded layer and opportunity for polymerization into compounds with increased thermal stability. Minerals might also have imposed a biochiralic preference (i.e., the optical rotation of biologic compounds).[46] Elsewhere in this prebiotic world, other organic compounds might have suffered a similar fate, becoming concentrated from a prebiotic soup or accumulating by adsorption on the surface of minerals (i.e., catalytic substrates [templates?]) such as the pyrite crystal.

Release of a fluid membrane (surface layer) into the aqueous phase might have led to the formation of lipid-monolayer membranes by self-assembly, and mechanical agitation might have led to the formation of closed monolayers or micelles and, by condensation, to lipid-bilayer membranes and vesicles, or liposomes. Given enough unsaturated carbons or branches in the tail of the amphiphiles, the lipid membranes would be fluid, like contemporary plasma membranes, and given long enough tails, the membranes of liposomes would be both sufficiently permeable to acquire nutrients and ionic solutes and sufficiently impermeable to macromolecules to retain them. Repeated cycles of drying, fusing and wetting would have further concentrated the contents of liposomes relative to the external environment. Liposomes would also have provided a locus for the sequestration of pigment systems[47] able to absorb light energy and produce a proton gradient across the membrane, thus, a chemiosmotic energy source.

Mixes and Assimilation

Prebiotic chemistry probably produced many way-stations, "intermediates" and plateaus on the way toward life. Before the origin of life, phosphate would have to have been incorporated into organic moieties; energy-linked oxidation–reduction reaction of inorganic and/or organic compounds would have to have come under the aegis of chemiosmosis; light-induced formation of energy-rich phosphate would have to have coupled biological energy-conversion systems with ion channels and proton-selective pores; inorganic pyrophosphate (PPi) would have to have linked up with thioesters via the bioenergetically energy-rich metabolite acetylphosphate; primitive converters and couplers of chemical energy, such as simple proton-pumping pyrophosphatase enzymes with H^+-pyrophosphatase (PPase) and inorganic pyrophosphate (PPi) synthetase activity would have to have metamorphosed into complex, oligomeric enzymes capable of proton-motive force capable of producing H^+ gradients across membrane;[48] etc., etc.

In time, relatively stable biopolymers such as polypeptides and nucleic acids[49] with enhanced catalytic properties could have "gotten it together," and, in the first instance of symbiosis, nucleic acids and polypeptides, in the company of other polymers (lipids and carbohydrates), become a mutually enhancing, self-evolving system. Lipids might have offered the potential to organize membrane-enclosed compartments capable of separating (internal and external, intermembraneous and extramembraneous) spaces with the possibility of exchange between compartments. Membranes might also have provided sites for autotrophic energy transduction (e.g., starting with FeS, H_2S and CO_2), or (less likely) the external "soup" of organic molecules, no matter how dilute, might have provided the fuel for the evolution of fermentation-based, heterotrophic metabolism. Finally, while only derivatives of molecules, the system would have spread throughout the prebiosphere, processes called growth and reproduction by biologists.[50] Ultimately, the straightforward chemistry which allowed low-propensity reactions in a constant chemical environment, presumably, produced high-propensity reactions in fluctuating chemical environments, and, derivatives that catalyzed steps in their own production process resulted in the effect called "memory" by chemists or inheritance by biologists.

RNA-BASED GENOMES

Nucleic acid-based genomes, like other forms of early Earthlife, are likely to have had multiple origins. In his vastly under-cited book, Periannan Senapathy[51] suggests that nucleic acids were generated from the random polymerization of nucleotides in large numbers of pools around the globe. These nucleic acids could have encoded meaningful "words" (polypeptides), even if sentences (cistrons) contained meaningless syllables (introns). In Senapathy's scenario, the meaningless syllables would have been deleted by selection prior to the process of translation.

A great deal of speculation surrounds the issue of what the primeval nucleic acid actually was, RNA, DNA or both. The preponderance of opinion currently seems to favor RNA and an "RNA world"[52] in which RNA was both the code and the catalyst. This RNA world is thought to have preceded the "DNA/protein world" we live in today.[53]

Similarly, a great deal of speculation surrounds the issue of the presence of introns in transcribed regions of split genes. Introns,[54] or intervening sequences, are portions of the transcribed regions of DNA which are cut out of RNA transcripts (and not translated in the case of protein encoding genes[55]), while adjacent exons, or expressed sequences, are spliced together and retained in the transcript (and translated). On one hand, introns in DNA might be tenacious hitchhikers, junk or even deleterious DNA. On the other hand, introns might have been or may still be functional. Ideas on intron usefulness center on their history: If they were there at the beginning, they might have aided in exon shuffling (the introns-early hypothesis; see below); whereas, if they were absent at the beginning, they might have been added to genes in some regulatory capacity (the introns-late hypothesis).

The RNA World

The hypothetical RNA world is thought to have had a brief existence, lasting only about 400 My at the beginning of the Archean, somewhere between 4.2 (4.0) and (3.8) 3.6 Ga ago.[56] Inhabitants of the RNA world were not RNA-based life forms otherwise resembling today's DNA/protein-based life forms, although this plausible idea is not excluded among the melange of possibilities. Rather, inhabitants of the RNA world featured RNA "ribozymes,"[57] catalysts operating in the absence of protein and sequestered, at some point, in RNA-based cells, either floating in a "nutrient broth" (containing nucleotide substrata in an activated form) or fixed on a crystalline substratum. In the RNA world, RNA cells replicated an RNA "genome," albeit not especially accurately or efficiently, through the activity of RNA-dependent RNA polymerases of RNA and ancillary activities.

The RNA-world hypothesis is an extrapolation from the behavior of various forms of RNA in extant cells (which would suggest that the "RNA world" is still with us, albeit, only a satellite to today's DNA/protein-dominated world). Inspired by the many catalytic roles of RNA in modern cells, RNA-world enthusiasts

> come to an astonishing and unexpected conclusion. RNA apparently has a huge distribution of sizes and shapes, and a collection of RNA molecules, even a collection of rather short RNA molecules, could have been used for an enormous variety of chemistry prior to the invention of proteins.[58]

In particular, the inspirations for the RNA world hypothesis are extant ribozymes, or RNA with autocatalytic[59] activity, namely, (1) group I and group II introns[60] with "self-splicing activity" by transesterification,[61] described by Thomas Cech (b. 1947), and (2) ribonuclease P[62] whose "cut-and-trim" RNA catalyzes the hydrolysis of phosphodiester bonds (at least *in vitro*) described by Sidney Altman (b. 1939). Group I introns, by far the most widespread variety of introns, are present in T4 bacteriophage, prokaryotes, organellar genomes and the nuclear DNA of many eukaryotes. "Conserved" regions of group I, II (III present in *Euglena*) and eukaryotic nuclear introns, while not similar to each other, seem to be responsible for secondary structures in transcripts required for excision of the RNA (at least *in vitro*), although excision may also depend on products of external genes *in vivo*. Other small ribozymes (hammerhead, hairpin, and hepatitis δ agent[63]) of viruses and virusoids also undergo self-cleavage or play a role in cleaving repeated coding regions in (bicistronic) RNA. Possibly the greatest inspiration for the RNA world hypothesis, however, comes from the activity of RNA in the small nuclear RNA particles (snRNP), or snurps, in spliceosome complexes[64] which remove nuclear introns from intranuclear mRNA of eukaryotes and, presumably chloroplasts.[65] Although, in their extant form, these particles include protein, their RNA may be responsible for their cut-and-splice activity, dimly forecast by activities of precursors in an RNA world.

What is more, ribonucleotides (of which there are over 100 substituted varieties) and similar molecules are major transport elements between metabolic pathways and the electron transport system, as well as cofactors for many proteinic enzymes, about half of which cannot function without their cofactors. Conspicuously, RNA is the "primer" for DNA nonprocessive replication *in vivo* and the polymerization chain reaction (PCR) *in vitro*. In addition, an internal tRNA-like template provides the "primer" for telomerase during telomere regeneration.[66] RNA is even an extracellular receptor in bacteria and linked to lipid in the membrane component of the purple non-sulfur bacterium *Rhodopseudomonas acidophilia*, to say nothing of RNA's role as template for reverse transcription in retroviruses (or QB replicase, a viral RNA-dependent proteinic RNA polymerase). RNA's most important cellular role, however may be the catalytic activity residing in ribosomal RNA (rRNA) itself (not ribosomal protein) for forming the covalent bond between amino acids in elongating polypeptide chain.[67] Indeed, "[t]ampering with rRNA has dramatic effects on translation."[68]

One of the rationales for the RNA world hypothesis is that simple properties of oligoribonucleotides might have provided crucial attributes under early-Earth conditions (attributes not available to polypeptides). The number of strong interactions available to these nucleotides might have allowed strong associations into ordered assemblies, and their large temperature dependence for binding complementary sequences might have set up a catalytic cycle favoring the binding of short oligomers and elongation at low temperature and the release of elongated products at high temperature. Furthermore, tight substrate binding at low temperature would have enhanced the fidelity of the interaction.[69]

On the other hand, the chemical repertoire of extant ribozymes is quite limited, confined to phosphotransferases or phosphohydrolases, although *Tetrahymena* ribozyme has weak hydrolytic activity for the ester between a 3′-hydroxyl oligonucleotide and the carboxyl of an amino acid.

"The chemistries available to the five canonical bases may allow them to carry out a number of simple reactions, but their functionality is still quite limited when compared to the 20 amino acids of proteins."[70]

Critics of the RNA world do not paint a rosy picture of it. Criticism is directed first and foremost at the principle model for the primordial RNA enzyme, namely ribozymes. Each of these is capable of generating and supporting the cleavage and sometimes the ligation (and hence elongation) of preexisting RNA, but none of them support replication *sensu stricto*.[71] In terms of the bonds that are cleaved and the groups that are generated, the activities of these RNAs more nearly resemble the digestive activities of DNases, exonucleases and ribonucleases.[72] Members of this group catalyze nucleotidyl (phosphoester transfer) reactions whose specificity for complementary substrates is directed by the sequence of an internal template. Ribozymes do not, however, move processively along the template, and no extant ribozyme is known to be self-replicating. In fact, RNA-dependent RNA replication in extant cells is rare, best characterized by replication of plant viroids and so-called satellite RNAs of virusoids dependent on viral RNA coreplication. The DNA-dependent RNA polymerase of T7 phage, which can also utilize an RNA template and thus act as an RNA replicase, is not RNA at all but a protein![73] Likewise, proteinic enzymes catalyze RNA replication in negative-strand RNA animal viruses and even in positive-strand RNA coronavirus.[74] Retroviruses, with a single-stranded RNA genome, replicate through the action of proteinic reverse transcriptase and a DNA intermediate (hence inhibited by actinomycin D, an antibiotic which binds only to DNA). Retroviruses, therefore, provide a better model for transitions to DNA-based life forms than for the working part of an RNA world. RNA viruses are also inappropriate as a model for ribozymes, and extant ribozymes would hardly seem an appropriate model for self-reproductive particles in the RNA world, whether floating around in a dilute, prebiotic soup or even fastened to an absorptive substratum.[75]

Among several problems cited by molecular biologist Gerald Joyce, the RNA world's staunchest critic, the most cogent is the assumption that a ribozyme produced abiotically could replicate with fidelity:[76] very short RNAs would not have appreciable replicase (i.e., catalytic) activity, while longer RNAs are not likely to form spontaneously. According to Joyce,

> neither chemical self-replication nor RNA-catalyzed replication offers a plausible explanation for the spontaneous appearance of the RNA world in the prebiotic environment. The conclusion is that life did not start with RNA; RNA followed in the evolutionary footsteps of some other replicating molecule, just as DNA followed in the footsteps of RNA.[77]

Another point of contention is the purity of the RNA world.[78] Were simple peptides present and accounted for? One suggestion that they were arises from the ability of present-day aminoglycoside and peptide antibiotics, not only to influence translation by reacting with specific rRNA sites on ribosomes, but also to influence ribozyme activity by reacting with transcripts of presumably ancient group I introns. Thus, transcripts of group I introns and rRNA would seem to share specific binding sites for these antibiotics, and translation and splicing would seem to share an ancient relationship to aminoglycoside and peptide.[79]

Early Intronic RNA[80]

The introns-early hypothesis, also known as the "exon theory of genes," proposes that introns are relics from the RNA world. Walter Gilbert[81] proposed the hypothesis, suggesting that introns were present at the origin of exons and were instrumental in piecing them together. Mitiko Gō also suggested[82] that in the beginning were exons, minigenes, or the remnants of exons encoding modules (discrete, structural and functional units of globular proteins). Senapathy's view may be considered a "very early introns hypothesis" (accompanying the very origin of nucleic acids), while the Gilbert/Gō theory emphasizes a role for introns in transferring (shuffling) and fusing (splicing) stable exons.[83] In any case, early intron hypotheses place introns

> in the most primitive genes, where they accelerated the evolution of new functions by enhancing the recombination of protoexons encoding small stably folding polypeptides to yield mosaic (sub-)genes encoding functional domains. These domains could then be "reshuffled" to form genes encoding the basic enzymes of primary metabolism found in all cells.[84]

Are introns ancient? The extreme "conservation" of intron positions among chloroplast genes (tRNA genes) suggests that these introns are ancient and would have accompanied the endosymbiont into the eukaryotic cell.[85] "It is remarkable that introns in five chloroplast tRNA genes ... are all in the anticodon loop ... [I]ntrons in both eukaryotic nuclear and archaebacterial tRNA genes are also in the anticodon loop."[86]

A bit of evidence that makes the introns-early hypothesis slightly more plausible is the presence of coding sequences within some extant introns. Some group I and II introns encode a maturase that operates in excision from RNA or a similar endonuclease that initiates a double-stranded break in

DNA and allows the intron, much like a transposon, to be copied at the cleaved site. However, these coding sequences are not independent, since their transcription seems to depend on "read-through" mechanisms from the upstream promoters rather than anything intrinsic to the intron.

The introns-early hypothesis has more important implications than merely providing hosts for internal exons. Elaborating on the theme of "intron ribozymes" (the self-excising sequences transcribed from group I and group II introns), the introns-early hypothesis proposes that RNA introns in the RNA world were capable of specifically cutting and joining RNA sequences. In general, the spliced pieces are exons belonging to the same transcript, but not necessarily. In kinetoplasts (*Trypanosoma, Leishmania, Crithidia*), where no introns interrupt the coding regions for surface glycoproteins, messenger RNA is still prepared by obligatory splicing between (*trans*) primary transcripts and a "mini-exon-derived" (med) RNA polynucleotide (140 bases long) encoded by tandem arrays of larger units on one or a few chromosomes.[87] "[S]pliceosomal introns may eventually also be found in genes encoding transacting [small nuclear] snRNAs that are implicated in [nuclear] splicing process and interact with pre-mRNAs and introns within the spliceosome."[88]

Unbound introns may once have performed functions performed today by extant mobile genetic elements, "jumping genes," or transposons. Like transposons, group I and II introns occasionally have internal encoding sequences (exons), the products of whose translation are utilized, even required for splicing (yeast mitochondrial cytochrome *b* gene), and some group I introns with internal reading frames are known to be mobile (in *Physarum*, the plasmodial slime mold and T4 phage). Ancient RNA introns, thus, may have evolved from particular functional domains in ribozymes and devolved into genetic nucleic acid.

Genetic and/or Operational Codes

Martin Eigen and colleagues[89] estimated that the genetic code is approximately 3.8 ± 0.6 Ga old based on the amount of randomization in tRNA sequences and the assumption that the domains of life[90] branched off 2.5 ± 0.5 Ga ago. Questions of resolution notwithstanding,[91] the genetic code would seem to be at least as old as life itself if not older.

The code is written in codons of nucleotide triplets on messenger RNA (mRNA). The counterpart to codons are anticodons of nucleotide triplets on transfer RNA (tRNA).[92] Ribosomes are the venue where anticodons find their complementary codons, and tRNA transfers an amino acid to a growing polypeptide chain. The magic of the genetic code is that the tRNA carrying a particular amino acid to a ribosome contributes the "right" amino acid to a polypeptide sequence. Each tRNA has the key to each codon, and the sum of tRNAs, is the "code book" for deciphering the genetic code. The enigma is how tRNA became the cipher of life's secrets.

tRNA

How does life achieve the near-perfect coordination of amino acids with tRNAs having the right anticodon? Answers posited to the question differ regarding the degree to which the "identity," or the specificity of each tRNA for its particular amino acid, is determined by an "operational RNA code" residing outside the anticodon. Three proposals are on offer for how tRNA combined operational RNA codes with anticodons capable of recognizing specific codons: Francis Crick's "frozen accident,"[93] in which chance mated specific ribonucleotide structures with particular amino acids; Carl Woese's[94] specific physicochemical interactions linking anticodons and amino acids; Michael Yarus's hybrid proposal that both "a highly specific physicochemical basis and an element of chance [lie] behind existing coding assignments."[95]

Nominees for operational RNA codes consist of as little as a sequence in the acceptor stem[96] or as much as the single-stranded loop regions at the ends of helices, single-stranded bulges within helices, and modified bases.[97] The mystery is heightened by the sheer physical separation of the amino acid acceptor stem at one end of the L-shaped tRNA molecule and the anticodon loop at the other end.[98] The D arm, a variable extra arm, and the TψC[99] arm all separate the acceptor stem and anticodon loop, but their histories are obscure.[100] Identity may also be influenced by the anticodon, since a tRNA can be "recruited" from one isoacceptor group (a group tRNAs capable of binding the same amino acid) to another by changing a base in its anticodon.[101] This host of possibilities suggests that the identity of tRNAs evolved from several structures and mechanisms.[102] Whatever its source, each extant tRNA's identity is recognized by part of a cognate aminoacyl synthetase, while another part of the same enzyme catalyzes the binding of the tRNA to the appropriate amino acid.[103] The synthetases come in two classes of enzymes, but these employ the same mechanism to charge their cognate tRNA with its amino acid.[104] Since sequence similarity is remarkably high for some of these enzymes (isoleucyl-tRNA synthetases) from archaea, bacteria, and eukaryotes,[105] the enzymes, or at least the aminoacylating part, would seem to be very old proteins (old enough to challenge the RNA-purity of the RNA world).[106] tRNAs and their synthetases would seem to have had a long history of coevolution, each mutually feeding the fitness of the other.[107]

mRNA

As long as the "universality" of the genetic code was an established principle, scientists maintained that the code was engraved in the progenote(s), the common stem(s) of noncellular and cellular life, before the separation of life's domains.[108] The idea of a single, ancient and indelible code is not quite as universally accepted today, however, since the genetic code no longer seems quite as "universal" as it was once. Of course, one does not know whether the "nearly universal" code is the product of merging among several original codes or of variation springing up in the course of time, but it is gratifying to have alternative hypotheses to think about if not to test experimentally.

The first indications of idiosyncratic genetic codes came from the analysis of anticodon mutations in *Tetrahymena* tRNA[109] and changes in a *Paramecium* gene required to allow its efficient expression after transfer to *Escherichia coli*.[110] Initially, only ciliates seemed to have any unique codons, but, soon, a large body of sequence data on mitochondrial tRNAs and tRNA genes suggested other significant lapses in the universality of the "universal" code:

> (i) The genetic code in mitochondria and in some organisms can vary in some ways from the "universal" genetic code. (ii) Mitochondria and organisms with highly compacted genomes such as *Mycoplasma* spp. utilize a codon reading pattern different from that of most organisms and requiring a smaller number of tRNA to read all the codons of the genetic code. (iii) tRNA gene transcripts may in some cases undergo editing.[111]

Some variation occurs within a species, in the form of "codon bias," for example, where genes begin with different "start" codons. The amino acids assigned to certain codons may also have changed as indicated by variation within groups, for example, the AG(plus any purine) codon in animals.

Other variations cross species lines, with some codons having strange bedfellows. For example, yeast mitochondria share an otherwise unique codon with *Mycoplasma*; protozoan and cnidarian mitochondria use the universal code except for the UGA[112] codon; AAA and AUA codons are read differently in the mitochondria of various Metazoa; in plant, bryophyte (*Marchantia polymorpha*) and *Acrasiomycota* spp. mitochondria, UGA is a termination codon rather than a tryptophan codon (CCA is used instead); echinoderm and platyhelminth mitochondria use AUA as the codon for the amino acid isoleucine instead of methionine as in most Metazoa and yeast; AGA and AGG are termination codons in all vertebrate mitochondria, but AGA is the codon for the amino acid serine in the mitochondria of several insects and echinoderms; etc., etc.[113]

Scenarios for building a genetic code while incorporating some variety are invented quite easily. In the RNA world, an amino acid bound to an RNA sequence of a self-replicating ribozyme might have increased the fidelity of its replication and thereby allowed it to "out replicate" other ribozymes (increasing its "fitness" in the Darwinian sense of the word). Under the rubric, "if a little is good, a lot is better," the addition of more amino acids might then have been favored, and preferential relationships among particular amino acids and RNA sequences in ribozymes might have set the stage for establishing incipient operational RNA codes. Were stability to increase fitness further, polypeptide linkage among the amino acids, whose formation was catalyzed by RNA, might also be selected for and an evolutionary count down begun between the translation of polypeptides and the replication of ribonucleotides. What began as the "reproduction" of amino acids in the service of RNA replication might have ended as a translation system encoding polypeptides. Additional RNA codes might have been established, and codons added either by linkage to yet other ribozymes (the more probable route) or modification of portions of the same ribozyme. Anticodons would then have preceded and prescribed codons, and tRNA would have been invented first,[114] but, in the end, mRNA emerges as the logical extension of competition.

> In this way, information needed for the development of a translation system, such as a set of coding assignments, could accumulate slowly and be integrated piece by piece. The genetic code and the apparatus of translation could thus have appeared gradual, hand in hand.[115]

Or, to put it another way,

> the basic selection pressure for the origin and stabilization of the primitive translational apparatus was the enhancement of the catalytic activities of these RNA-based cells in order to increase their dynamic stability and reproductive fitness.[116]

Ribosomes

The RNA world was supposed to solve "the problem of proteins having to invent their own synthesis, but it does not answer to what extent they reinvented the process after it first got started."[117] It is one thing to suggest that primitive ribozymes made themselves (autocatalysis) and quite another to suggest that they "read" the triplet genetic code while providing a "read-out" in the form of a colinear polypeptide chain. Autocatalytic activity of replication is hardly comparable to the heterocatalytic activity of translation. After all, these processes proceed in opposite directions: Whereas RNA polymerases (and DNA polymerase, for that matter) proceed from the 3′ to the 5′ ends of their template (elongating the incipient strand in the 5′ to 3′ direction), all modern ribosomes move from the 5′ to the 3′ ends of the mRNA template strand (building polypeptides in the amine to carboxyl direction). Suggestions that "peptide-bond formation was catalyzed by progenitors of ribosomal RNA"[118] would seem unlikely, although it would seem that RNA catalyzes the peptidyl transferase activity of ribosomes.[119]

Possibly, a primitive ribozyme "was a predator living high on the food chain of the RNA world, with its prey being oligoribonucleotides that it consumed during its self-replication."[120] In such an event, the ribozyme might have been a progenitor of the small ribosomal subunit, providing a device to unwind duplexed RNA and separate strands during translation. Other primitive ribozymes might have been progenitors of the large ribosomal subunit, providing a device for moving (translocating) the mRNA and tRNA complex in lockstep, thereby preventing strand slippage especially on homopolymer or short-repeat sequences.

Another problem for ribosomes in the "RNA world" is that "RNA genomes are notorious for their low fidelity of replication."[121] Hence, primitive ribozymes would tend to be small and show frequent mutations, while rRNA structures are large and stable. This problem would be solved if compact, protodomain progenitors of ribosomes gathered into larger assemblies and became ligated into permanent associations with the help of ribozymes.[122] These processes might resemble parts of self-cleavage in viroid RNAs, RNase P activity in bacteria, transcripts of group I introns, and the capacity for recombination (copy-choice) and *trans*-splicing.[123] Likewise, something like transcripts of group II introns might have been the precursors of spliceosomal small nuclear RNAs of the eukaryotic nucleus.[124]

DNA-BASED GENOMES

At some point in Earthlife's history, DNA entered the scene that it would ultimately dominate. Presumably, early DNA was double stranded and circular, since DNA in life's three domains is double stranded, and it is predominantly circular (lacks ends) in two of the three domains (Bacteria and Archaea).[125] Moreover, the DNA of viruses of both bacteria and eukaryotes is predominantly circular, although linear DNAs are well represented among viruses. The telomeres of eukaryotic DNA would seem to have evolved later.

The first question confronting genomic evolutionists is, "Why would RNA have been so completely replaced in its genetic role by DNA (ignoring the role of RNA as the gene in RNA viruses)?" Part of the answer might rest in the potential nucleophilic activity of the 2′ hydroxyl of the pentose in RNA. By forming additional interactions, the 2′ hydroxyl of RNA confers versatility on RNA at the expense of stability. "Just as structure in mRNA affects translational frameshifting, hopping, and read-through, structures in genomic RNA are likely to affect replicase-associated recombination and probably mutations."[126] In addition, the deamination of cytosine to uracil, which is corrected by uracil-DNA glycosylase in DNA, is not corrected in RNA. Furthermore, UV irradiation produces far more photochemical changes in RNA than in double-stranded DNA. Finally, RNA is more easily hydrolyzed, especially in the presence of transition metal ions than is DNA.[127] In other words, double-stranded DNA is more suitable as a medium for long, sturdy genes.

The second question might be, "Where did DNA come from?" One possibility is the RNA world. If an RNA world ever flourished, however briefly, at some point, it might have given rise to a DNA/protein world. The turning point may have been the appearance of enzymes that transcribed DNA from RNA rather than replicated RNA from RNA. Some sort of reverse transcriptases[128] would have used RNA templates to transcribe single-stranded DNA (ssDNA), which, in turn, would have been replicated by some sort of DNA polymerase into double-stranded DNA (dsDNA). In the pure-RNA world hypothesis, these enzymes would themselves have been RNA, and the synthesis of DNA was either an error or a subversive activity by a fifth column of RNA with DNA-leanings. In an impure-RNA world, contaminated by protein, the enzymes could have been protein, having no loyalty to RNA whatsoever and prone to use any substrate available. DNA nucleotides might just have been the only thing available.

A second alternative posits the arrival from abroad of proteinic nucleases (enzymes that digest nucleic acids) with an appetite for RNA. RNA might have fought back by converting itself into more resistant, dsDNA. The devolution of reverse transcriptases and DNA polymerases from earlier ribonucleases is also convenient for explaining the origin of split genes in the eukaryotic genome. In the introns-early hypothesis, part of the RNA available for reverse transcription to DNA would have contained intronic

RNA as well as intronic exons. These intronic RNAs would represent an expendable portion of RNA sequences which, in the service of more central exonic RNA, could be sacrificed (much like telomeric DNA is expendable in extant cells). The prevailing genes of bacteria and organelles lacking spliceosomal introns would have lost all their introns while preserving only their exons. Other bacterial genes seem to have been lost completely (*Mycoplasm*[129]).

For a third alternative, imagine that a DNA/protein world (or, perhaps, a region or even an asteroid) was flourishing somewhere on Earth at the same time as the RNA world. DNA polymerases in this world would then, somehow, somewhere (or, more likely, many times and in many places) have slid over to the RNA world, mingled and assimilated as reverse transcriptases. The idea of a single world evolving in a totally linear direction might seem simpler than simultaneous worlds (or what-have-you) meeting, but the idea of proteinic enzymes entering the scene and foreclosing on the RNA world is a little less of a *deus ex machina* than the creation of reverse transcriptases and DNA polymerases *de novo*. Although negative evidence can not be conclusive, no evidence whatsoever eliminates this (these) alternative possibility(ies).

Genes: Stable Conveyers of Heredity

Genes are hardly homogeneous. One can only speculate that early genes, likewise, came in a variety of shapes and sizes. Most hypotheses involving early genes, however, imagine them to be simple and small, and limit consideration to structural genes encoding transcripts of RNA. One must "read between the lines," so to speak, to find the complexity in models of early genes.

Exon Shuffling

The exons-shuffling (alias introns-early) hypothesis[130] posits that early exonic RNA comprised "minigenes"[131] or "domains" (also motifs) of small, open reading frames (ORFs) encoding 15 to 20 amino acids, corresponding to extant exons encoding domains of 20 to 80 amino acids.[132] At the ends of early ORFs were noncoding regions corresponding to portions of present-day introns. Larger genes formed with full-length introns when sequences with encoding ORFs were linked together through their adjacent noncoding regions. Novel genes emerged by the shuffling of exons via their associated introns. Efficient translation, or RNA-dependent protein synthesis, required mRNA processing, however, to removed introns, either by excision from transcripts or by imparting signals that allowed the translation machinery to skip introns (as leader sequences).

The exons-shuffling hypothesis would not have engendered so much interest if modules, such as the helix-turn-helix (HTH) were not considered so important (as DNA-binding domains[133]), if exons and introns were not so widespread among eukaryotic coding sequences, so abundant, and frequently, so impressive.[134] The chief evidence supporting the exons-shuffling hypothesis is drawn from sequence databanks, namely, that, frequently, nuclear introns do not occur at random positions in the transcribed parts of genes. A random distribution of nuclear introns would be expected if introns were added to coding sequences in the course of eukaryotic evolution (given a particular set of assumptions about how introns might be added). Another tantalizing, if more speculative, argument on behalf of the introns-early hypothesis takes account of the position in a codon at which an exon begins and ends. Numbering the position between adjacent codons as phase 0 and consecutive positions within a codon as phases 1 and 2, the frequency of intron intrusions in phase 0 is 48%, 30% in phase 1 and 22% in phase 2. These frequencies are significantly different, indicating that the distribution of introns is not random. Likewise, a preponderance of exons begin and end in the same phase (symmetric exons). Gilbert et al suggest that the relatively high frequency of exons in phase zero and the preponderance of exons beginning and ending in the same phase are adaptations for making it "easier"[135] to move exons, the theory known as exon shuffling. What is more, 30% of the exons observed in databases are symmetrical around phase 1, which is hard to explain as a consequence of the random insertion of introns but consistent with the repeated shuffling of exon.

Other arguments on behalf of the introns-early hypothesis refer more directly to the cognate hypothesis, the exon theory of the gene. The idea is that if introns accompanied exons early in the evolution of proteins, then exons should bear some fundamental relationship to the structure of proteins and the position of introns should respect this relationship. The evidence, in any case, also comes from the sequence databanks. Basically, the position of nuclear introns is related to protein structural elements in two ways. (1) Exons correspond to functional domains. The prime example is the immunoglobulin proteins in which nuclear introns punctuate variable, joining, hinge and constant domains of both heavy and light chains, when exons are brought together by alternate splicing. Furthermore, exons encoding similar modules, such as the epidermal growth factor (EGF) module, occur in genes for many different proteins and broadly across species lines.[136]

The widespread absence of introns in prokaryotic genomes and in "intronless" eukaryotic genes presents a problem for the exons-shuffling hypothesis, but not an insurmountable obstacle. These types of genes could have formed by intron deletion (unaccountably more prevalent toward the 3′ end of genes than toward the 5′ end of eukaryotic genes), by a systematic error in replication, or possibly by reverse replication from messenger RNAs coupled to recombination and gene conversion.[137]

The chief problem for the exon shuffling theory is that even in genomes where introns are common, they are not uniformly distributed, suggesting that introns were not uniformly distributed among genes or that exon (domain) shuffling (or modular behavior) was not an ancient phenomenon.

> There is essentially no evidence in metabolic enzymes for modular behavior (e.g., domain shuffling), either for developing primitive catalysts or for altering the function of advanced enzymes. Rather, modular behavior is observed primarily in proteins involved in advanced regulatory systems in advanced organisms; in particular, protein involved in the immune system, blood clotting, and other regulatory pathways that emerged only within the last 400 million years. The particular attention paid to these types of proteins by contemporary molecular biologists has, we believe, created the illusion that domain shuffling is more widespread (and more ancient) than it actually is.[138]

Introns-Late Hypothesis

The contrary, introns-late hypothesis posits that, at the end of an RNA world, if there was such a place, the relevant RNA contained only exons. The invention of introns would have followed the origin of DNA rather than preceded it and then only after purely exonic DNA had already formed chromosomes of some sort. Enthusiasts for the introns-late hypothesis claim that, after the separation of eukaryotes from the progenote(s) (the last common ancestor(s) of all life), contagious introns selectively invaded contiguous nuclear genes, possibly by DNA transposition.

Several lines of evidence argue on behalf of the introns-late hypothesis: (1) The position of nuclear introns in similar genes is not necessarily "conserved," for example, the Gα nuclear intron's positions among genes within the GTPase superfamily expressed in eukaryotes (*Arabidopsis GPA1*, human *Gnas* and the G_i class genes from *Drosophila* and mammals).[139] (2) The inferred phylogenies of introns are sometimes incongruous with the species harboring them, suggesting that introns are gained and possibly lost (presumably by horizontal gene transfer via a vector such as a virus [see below]). Indeed, even introns occupying the same position in a sequence are not necessarily similar in sequence, suggesting that mobile introns "home" onto particular locations in coding sequences.[140] (3) Introns sometimes appear with a random distribution, may even intrude on protein structure (notably, in globin[141]), and genes which are broadly the same in their coding regions may not have similarly distributed introns (paralogous fibrinogens in the rat; orthologous actins across several taxa). (4) Exons encoding some proteins do not have a size distribution corresponding to elements of the protein's secondary structure.[142]

At the moment (1998), both "early" and "late" hypotheses have to be conceded merit because the evidence supporting one hypothesis does not excluded the other. Furthermore,

> Introns that are widely conserved within a multigene family may have been acquired early in eukaryotic evolution while the less well conserved introns may have appeared more recently. In addition, some sequences may favor intron acquisition during evolution of homologous genes in separate eukaryotic lineages.

Indeed,

> The simplest explanation for the dissimilarity in intron position in many Gα genes is that introns were both acquired and lost during eukaryotic evolution. A similar conclusion was reached following comparison of serine protease and α and β actin and tubulin genes.[143]

Conceivably, in the "great unknown" some forms of "progenotes" had introns and others lacked introns, although they existed simultaneous and met occasionally. Possibly the best argument for a joint hypothesis is evidence that argues both ways, namely high degrees of similarity in sequences as evidence for very ancient genes, such as pyruvate kinase (PK) and triosphosphate isomerase (TIM), which, nevertheless lack introns in prokaryotes and have similarly placed (nonrandom) nuclear introns in animals, fungi and plants.[144]

Chromosomes: Repositories of Genomes

Like genes, chromosomes are not homogeneous (with respect to GC content, dinucleotide frequencies, codon usage, etc.). Chromosomes, thus, offer several challenges, especially regarding the distribution of genetic sequences and nongenetic sequences.

Coding Sequences

In genetic maps of bacteria, neighboring genes are frequently responsible for related, but not identical, functions, and genes performing reactions within particular biochemical pathways are frequently arranged in the order of their biochemical reactions.[145] These genes are also, but not necessarily, organized in operons, units of contiguous coding regions under the control of a single promoter (bi- and polycistronic genes).

> Virtually all bacterial operons are composed of genes that show no obvious homology; many are closely related to unlinked genes encoding proteins that catalyze mechanistically similar reactions (e.g., dehydrogenases, kinases, etc.). Few cases of gene duplication and divergence within an operon have been demonstrated in bacteria.[146]

Gene clusters in bacteria, therefore, would seem to offer some advantage albeit not necessarily with regard to the position of genes in the chromosome. Selection of coadapted gene complexes might provide benefit to individuals if recombination would "generate new combinations of logical groups but would not disrupt the individual clusters of genes whose products must work together most intimately."[147]

Alternatively, benefits for the individual may be derived from coordinated coexpression and facilitated coregulation. Clustering may also (or alternatively) be advantageous to the genes themselves rather than the individual organisms. "The Selfish Operon model predicts that clusters of functionally related genes can colonize naive genomes or genomes having lost two genes of a single pathway. Clusters of coadapted, functionally related genes can invade genomes from which only a single gene has been lost."[148]

Horizontal Gene Transfer

Horizontal gene transfer,[149] or the movement of genes through the environment, may be a major mechanism for introducing novel genes into a genome. The phenomenon of horizontal gene transfer in bacteria was first demonstrated experimentally in the pioneering work of Frederick Griffith on transformation, and, today, genetic engineers routinely transfer genes horizontally among bacteria, yeast and cultured mammalian cells with recombinant DNA technology. On the "down side," at least for human health, resistance to antibiotics in pathogenic bacteria seems to have been transferred horizontally following an "unintended release of the rDNA [resistant DNA] production strain from the fermentation plant."[150] On the basis of incongruities between the inferred phylogenetic trees of genes and their "host" species, at least seven genes appear to have been transferred between bacteria and the nuclei of eukaryotes.[151] For example, sequence data suggest that the II form of the CO_2-fixing enzyme Rubisco moved from anaerobic protobacteria *Rhodospirillum* to plastids of dinoflagellates of the genera *Gonyaulax*, *Symbiodinium*, and *Amphidinium*.[152] Likewise, lateral gene transfer is held responsible for the massive movement of genes from mitochondria and chloroplasts to eukaryotic nuclei.[153] Eukaryotes's genes may also move horizontally. In wild *Drosophila*, the movement of the *P*-element (hybrid dysgenesis) seems to be responsible for horizontal gene transfer via a mite vector.[154]

Many statistical methods are now available for assessing the significance of incongruities between phylogenetic trees and sequence sets,[155] and a great deal of circumstantial evidence for a history of horizontal gene transfer among bacteria has recently accumulated in the form of atypical base compositions in DNA sequences, different codon usage patterns for particular genes, or genes with anomalous features compared to most genes in the resident species but resembling features of genes in a remote species. By these criteria, the chromosome of *Escherichia coli* seems to contain more than 600 kb of horizontally transferred, protein-encoding DNA, and at least 10% of the genes in "[*Salmonella*] *enterica* sv. Typhimurium were acquired by horizontal transfer."[156] Since the characteristics that distinguish transferred genes would probably be "ameliorated" over time, as genes accommodated to prevailing conditions in the host, horizontal gene transfer presumably represents an even larger phenomenon than that which is presently detectable. *Salmonella* and *E. coli* may have "gained and lost a total amount of DNA corresponding to 60% of the size of its current genome."[157]

Duplication Followed by Divergence

In eukaryotes, most genes with related functions are not even on the same chromosome, no less clustered, but where gene clustering occurs, it generally involves multiple forms of similar genes (homeotic genes; hemoglobin genes) or multiple alleles. Duplication followed by divergence may very well be a major source of clustered genes in eukaryotes. Certainly, duplication has been posited as a major instrument for genomic growth even at the origin of chromosomes[158] (although bacterial gene clusters probably did not originate primarily *in situ* by gene duplication).

The main problem with "duplicated" genes in a single genome is that they often do not resolve back to a single gene without throwing into the calculus the additional assumption that divergence among the "duplicates" occurred at different rates. The problem is only exacerbated when different types of genes are considered. Molecular taxonomists are at particular pains to explain "contradictions between the rRNA phylogenetic frameworks and those inferred from [other] genes"[159] The "way out" is to invoke some sort of duplication "before the separation of the eubacterial, archaebacterial, and eukaryotic lines of descent" or an "aboriginally duplicated" gene.[160] The "chimeric origin of the eukaryotic nucleus"[161] is even more difficult to rationalize along classical, evolutionary lines.

Another problem with "duplicated" genes in eukaryotes is their frequent dissemination among chromosomes. One would expect that duplicates would appear near to each other, as they seem to do in *Hox* clusters and hemoglobin sites, but most "related" genes in eukaryotes are highly dispersed. Horizontal gene transfer, rather than duplication, would seem a more likely explanation for the dispersed pattern of "related" genes, since it places no inherent restriction on the point of gene insertion anywhere in the genome.[162]

Noncoding Sequences

The vast amounts of nongenetic DNA in some eukaryotic chromosomes is even more difficult to explain. To suggest that it is the disposable remnants of transposable elements (TEs) or mobile elements would seem the sheerest of "cop-outs," but that is basically what is available by way of explanation. In any event, only TEs with some apparent (if not real) function have attracted attention, leaving the rest of the noncoding DNA to the imagination.

TEs fall into at least two classes with subclasses: those associated with euchromatin (the portion of the eukaryotic genome implicated in unique gene expression) and those with heterochromatin (the portion of the eukaryotic genome surrounding the centromere and telomeres of chromosomes or sequences thought to be permanently repressed in a cell).[163] The first class includes class I mobile elements comprising retroelements or retroids found in all types of cellular organisms and reputed to be very ancient.[164] Retroids transpose by reverse transcription of RNA intermediates (transcription/reverse transcription/integration mechanisms) and are broadly distributed among life's domains, retroviruses, some DNA viruses, mitochondrial group II introns, yeast's Ty elements, and *Drosophila*'s gypsy and copia-elements. Retrovirus-like elements (retrotransposons) with long terminal repeats (LTRs) represent one subclass of retroelements.[165] These elements appear to be integrated near euchromatic genes where they may modify regulation. A second subclass consists of retroposons or nonLTR retrotransposons (lacking LTRs). Members of this subclass, also known as LINE-like elements (long interspersed nuclear elements [LINEs]) and poly(A)-type retrotransposable elements (because of a poly[A] site in R2 retrotransposons) are broadly distributed (insects, mammals) and integrated independently (into specific sequences of 28S rRNA genes).

Class II TEs transpose by DNA to DNA transposition (not utilizing an RNA intermediate) and are associated with heterochromatin. This class includes an apparently ancient, broadly distributed subgroup of elements with short, inverted terminal repeats (ITRs), including *P*-elements, or the jumping genes of *Drosophila*. A second subgroup has terminal repeats of heterogeneous sizes.

Perhaps the singular lesson to be taken away from transposons is that the genome is not nearly as static as classical genetics would suggest (and the molecular biologist might like to think). The acquisition of transposon is not necessarily an ancient phenomenon. "[A] variety of evidence shows that *Drosophila melanogaster* has acquired at least three new families of transposable elements (P, hobo, and I) within the last 50 yr."[166]

Superfamilies of Genes

Gene families have been defined in various ways and discussed in Chapters 2 and 3. Here, attention focuses on superfamilies or gene systems, groups of allegedly ancient genes with representatives found among the three domains of life (or at least eukaryotes and prokaryotes) and sometimes in viruses as well. The importance of superfamilies to the narrative of life is that they might represent genes present in the progenote(s), the last common ancestor(s) of all life prior to the separation of noncellular and cellular Archaea, Bacteria and Eucarya. On the one hand, if more or less homogeneous classes of genes can be placed in superfamilies, their existence would not only testify to the existence of the progenote(s) but would clarify its character. On the other hand, if superfamilies of genes breakdown upon closer inspection, then either (1) the progenote(s) did not exist or (2) the progenote(s) performed its vital functions with materials that have long since disappeared or dissipated.

Historically, a commonality of function played a part in identifying superfamilies, but since the Human Genome Project, the "acid test" for superfamilies is similarity of nitrogenous base sequences in DNA or amino acid sequences in the proteins in life's three domains. Most work has centered on superfamilies concerned with informational processing, with bioenergetics and intermediary metabolism, and receptor or signaling proteins.

Informational Processing

RNA superfamilies exhibiting sequence similarity begin with ribosomes (70S and 80S), including their two (or three) large subunit (50S and 60S) with LS rRNAs (23S and 23S-like [28S], and 5S [except in mitochondria] or 4.5S in chloroplasts and 5.8S [similar to the 5′ terminal of the 23S RNA] or 7.8S in Eucarya) and their small subunit (30S and 40S) with SS rRNA (16S and 16S-like [18S–19S]). In prokaryotes, the genes encoding 16S, 23S and 5S rRNAs are clustered in the RNA cistron and transcribed as a single unit which is then reduced to the mature ribosomal RNA species. In eukaryotes, the genes encoding 18S, 5.8S and 28S RNAs are also clustered and transcribed as a unit, but the 5S rRNA coding region is usually not linked and is even transcribed by a different RNA polymerase.

Sequence similarity extends to group I introns of tRNAs[167] and to the 7S sequence recognition particle (SRP).[168] Comparisons should not be stretched, however, and the 16S rRNAs, while comparable in size in Archaea and Bacteria, are not especially comparable in sequence or pattern of base modification in Eucarya. Furthermore, the "group-invariant rRNA characteristics"[169] of eukaryotic rRNA differs significantly from the common characteristic structure of bacterial and archaeal rRNA. Introns also constitute RNA superfamilies, but spliceosomal RNAs are absent in prokaryotes.[170]

Many of the enzymes associated with the replication of DNA are similar among representatives of the three domains. Likewise, the superfamilies of elongation factors associated with translation, have been considered the bedrock of ancient genes, sufficiently well established to serve as the basis for rooting the universal tree of life.[171] Some ribosomal proteins associated with translation also constitute superfamilies, notably ribosomal A protein,[172] although other ribosomal proteins are found in Archaea and Eucarya but not Bacteria, while still others are found in Archaea and Bacteria but not Eucarya.

Bioenergetics and Intermediary Metabolism

The list of bioenergetic and intermediary metabolism enzymes with at least one representative among each of the domains of life is quite varied. In 1993 it included 19 classes of 1 to 8 or more (various) enzymes.[173] For example, the superfamily of aminotransferases and synthases included aspartate aminotransferases, tyrosine aminotransferases, aminocyclopropanecarboxylate synthase and histidinol phosphate aminotransferases. By 1998 the list had expanded so much as a result of sequencing that 271 proteins could be pulled out of databases on the basis of only two ATPase motifs in *Methanococcus jannaschii*, an archaean.[174] These proteins included DnaA, proteins with chaperone-like activity important to protein folding, including FtsH (a zinc-dependent protease), enzymes with ATP-dependent proteolytic activity, proteasome components and yeast HSP78 (heat-shock protein 78). The archaean sequence also seems to include two other ATPases containing metal-binding domains (ferredoxin-like domains).

Some bioenergetics superfamilies are extremely variable (K^+ channel) constituting an "extended family" rather than a superfamily.[175] The adenosine triphosphatases (ATPases) will serve here as a representative for bioenergetics superfamilies. ATPases constitute a broad superfamily, with two major families of transporters: ATP-dependent transbilayer lipid transporters and primitive or P-type ATPase simple ion transporters, the latter including ATPase metal ion transporters (the bulk of the family), bovine ATPase II (chromaffin granules), yeast *DRS2*, and other ATPases of unknown function.[176]

Vacuolar ATPases (membrane-bound), H^+-ATPase (= F_0F_1-ATPases), are central to the energy metabolism of aerobic bacteria and eukaryotes, to the maintenance of internal pH (by consuming ATP) in anaerobic bacteria, to the production of ATP in archaeans (in association with methane generation [proton-translocating ATPase] in methanogens and electrochemical motive force [bacteriorhodopsin] in halophiles).[177] In addition, as a result of the complete sequencing of *Methanococcus jannaschii*, a novel family of 16 proteins has been found which appears to belong to a broad class of proteins with ATPase activity.

Sequence data are not entirely straightforward, however. The F_1-ATPase catalytic subunit of bacteria and mitochondria is only moderately similar to the solubilized archaean ATPase, suggesting a distant relationship and justifying placing it in a different family. The subunits of eukaryotic vacuolar membranes and archaean ATPase have greater sequence similarity. Specifically, the amino acid sequences of the 70- and 60-kd subunits of vacuolar H^+-ATPase belonging to the eukaryotic endomembrane system are about 25% identical to the β and α subunits of bacterial F-type F_0F_1-ATPases while about 50% identical to the β and α subunits of the archaean, sulfur-metabolizing *Sulfolobus acidocaldarius*. Furthermore, an 88-amino acid sequence near the NH_2-terminal end of the 70-kd eukaryotic subunit is absent from the bacterial-type F_0F_1-ATPases but present in the *Sulfolobus*-type.

The problem is that if the gene family arose monophyletically through divergence among orthologues,[178] and the gene in all three domains was acquired from a common ancestor(s), dissimilarity should be about as great between bacteria and archaeans as between bacteria and eukaryotes. The eukaryan ATPase does not seem to have been acquired from bacteria, like other energy/metabolic genes, via lateral gene transfer, since the eukaryan enzyme resembles the archaean version more than the bacterial version. Nothing whatsoever in the data excludes the possibility of polyphyletic origins for the genes or their subunits and spread among the three domains by horizontal transfer.

Superfamilies of enzymes involved in intermediary metabolism may also have been constructed by polyphyletic means. The tricarboxylic acid cycle, pentose phosphate cycle, and glycolysis contain enzymes with multiple "homologies" and degrees of specificity. The glyceraldehyde-3-phosphate dehydrogenases (GAPDHs) gene system, for example, belongs to a metabolic enzyme superfamily. GADPHs are common to both glycolytic and photosynthetic pathways.[179] GAPDH Class I comprises the widespread, phosphorylating GAPDHs present in archeans, bacteria and eucaryotes. Bacterial GAPDHs (*gap1*, *gap2* and *gap3*) are explained by multiple gene duplications. Eukaryotic nuclear *Gap* genes resemble *gap* genes in both cyanobacteria and purple bacteria (*Escherichia coli*). Specifically, the *GapC* genes of plants, animals and fungi (the gapC or purple bacteria–mitochondrial line) resemble *gapC* of *E. coli* more closely than they resemble *gapC* of cyanobacteria, while *GapAB*, encoding photosynthetic GAPDA, of Metaphyta, *Chlamydomonas*, red algae (*Chondrus*) and *Euglena gracilis* (the gapA [*gap1*] or cyanobacteria–chloroplast line) resembles *gapA* of the cyanobacteria *Anabaena* and *Synechocystis*. Possibly, three members of the family (the gapC or purple bacteria–mitochondrial line) persisted in the genomes of cyanobacteria and protobacteria, while another member of the family (the gapA [*gap1*] or cyanobacteria–chloroplast line) was donated to plants by the ancestor(s) of plastids.[180]

Receptor Proteins

Receptors are a heterogeneous groups of proteins that operate within the cell and may transport signal-molecules to the nucleus or chromosome of cells (steroid receptor family) or operate on the cell surface and participate in transducing environmental information into an intracellular informational cascade. The latter would seem, potentially, a more ancient class, and G-protein–coupled receptors will serve here as a representative for receptor protein superfamilies. G-protein–coupled receptors constitute a large superfamily (over 200 sequences in the superfamily with over 1000 receptor types expressed in human beings alone). The superfamily is so large that, "it is possible that recombination events may have played a major role in generating and scrambling the receptors over the millennia, such that it may prove difficult to resolve their relationships and evolutionary history."[181]

The receptor is heterotrimeric, consisting of three subunits, each from a multigene family. In Eukaryotes, α, β and γ subunits, each with several isoforms, have coevolved as receivers of intercellular communication (developmental signals) and components of sensory transduction systems regulating a variety of specialized functions in terminally differentiated cells. The α subunit, itself constituting a superfamily of GTPase, provides specificity in the signaling pathway while binding and hydrolyzing GTP (guanine triphosphate); the β and γ subunits are required for effector regulation.

Sequence comparisons leads to the identification of three major groups of Gα genes: GPA (known mainly from fungi, plants, and slime mold; 14 genes), Gα-I (consisting of G_o, G_t, G_x, and G_i genes), and GαII (consisting of G_q, G_{12} and G_s genes) groups. Four classes of the GA-I and Gα-II genes (G_i, G_q, G_{12} and G_S) are expressed in vertebrates but not other eukaryotes. Five amino acid motifs, G1–G5, are essential components of the GTP-binding pocket, critical for GTPase activity. These motifs must be represented in any nominee for a common ancestor(s) of the GTPase superfamily. Although members of the classes are often similar, members are "probably exclusively found within metazoan organisms. Lower eukaryotes also express Gα subunits that regulate adenyl cyclase or phospholipase C..., but these sequences appear to be distinct from any of the mammalian Gα class genes."[182] In addition, "it appears that the branch lengths from the common ancestor(s) to the Gα-I and Gα-II Groups are different."[183]

EARTHLIVES

In its devolution, life would seem to have collapsed somewhere around noncellular (viruses and retroviral elements) and cellular (Archaea, Bacteria, Eucarya) forms. The noncellular forms, it would seem, collapsed further into parasites, requiring a cellular host cell for their reproduction. Some sequence similarities suggest a common origin or a commonality to noncellular and cellular life. For example, sequences in domains of the RNA directed RNA polymerase and chymotrypsin-like proteases, are similar in eukaryotic and prokaryotic viruses, suggesting an origin for the enzyme earlier than the separation of cellular life forms.[184] Other sequence similarities suggest abundant transfers of genes between viruses and their hosts. Thus, the historical relationship between noncellular and cellular life remains ambiguous.

Devolution of Noncellular Life

The origin of noncellular life is steeped in mystery. Inasmuch as viruses are thought of as simple in structure, they would seem to have originated prior to more complex cells, but, inasmuch as all extant viruses are obligate parasites of cells, they would have to have evolved after the origin of their cellular hosts. The alternative is that noncellular life had a life of its own at one time, and that this (these) form(s) of life devolved into today's cell-dependent viruses and retroviral elements.

Three mechanisms for viral origins suggested by Macfarlane Burnet in 1944 have each been recast in modern terms by molecular biologists (notably, Joshua Lederberg, André Lwoff, Salvador Luria, and Howard Temin),[185] and, although each has enjoyed popularity at one time or another, the mystery of noncellular origins remains unsolved. None of these mechanisms is mutually exclusive and the different mechanisms might even have operated independently, accounting for different forms of noncellular life.

The mechanisms are as follows: (1) Viruses arose as degenerated life forms that lost functions while retaining the genetic information required for a parasitic life (the regressive evolution model); (2) viruses evolved from self-replicating molecules in the prebiotic RNA world (the independent entities model or retroviral hypothesis); (3) viruses originated from subcellular particles or functional assemblies of ("runaway") macromolecules that escaped from cellular controls (the cellular origin model).

Regressive Evolution Model

The regressive evolution model is consistent with present-day examples of parsing out life in small, viral-sized quantities. For example, a mutant *Escherichia coli* K-12 forms minicells containing RNA and protein in normal ratios but devoid of DNA.[186] Fragmentation of this sort might have given rise to negative (–)stranded RNA viruses with lipid envelopes. The eukaryote-like features of RNA viral genomes, such as 5′ methylated caps and 3′ poly(A) tails, are also consistent with the regressive evolution model. The genomes of DNA viruses are especially eukaryote-like, and the plasticity of the phage genome and other viral genomes that integrate and separate from their host genome fit the regressive evolution model.

If viruses originated from cells as such, however, one would expect a consistent relationships with host-cell genomes. Instead, one finds two (Rhabdoviridae and Bunyaviridae) of the five families of negative (–) stranded RNA viruses containing parasites of both animals and plants. Indeed, both types of Rhabdoviridae, lyssa-like and vesiculo-like viruses, occur in both animal and plant varieties. Furthermore, in positive (+) stranded RNA plant and animal viruses (picorna-like and alpha-like),

> scoring criteria based on the degree of similarity for the pol [polymerase] homologies showed that the plant cowpea mosaic virus (CPMV) was more closely related to the animal picornaviruses such as polio or foot-and-mouth disease, than to the other plant viruses such as alfalfa or brome mosaic. The CPMV shares many more structural features and infective mechanisms with the picornaviruses. Thus, viruses are first likely to have developed their strategies along given paths and, once effective, spread across various hosts and species. Apparently, structural and functional development superseded development in and for particular host types, such as plants or animals, and required much more time and evolutionary effort.[187]

What is more, plant comoviruses (CPMV, dandelion yellow mosaic, anthriscus yellows and parsnip yellow flect viruses) and animal picornaviruses (poliovirus) also share capsid architecture, folds and β-barrel motifs as well as the alignment and order of capsid-protein encoding genes.

Another version of the regressive origins model is reserved for DNA viruses, especially those with the "large" variety of dsDNA genomes. Based on the similarities of domains in viral-encoded proteins and sequences in cellular-encoded proteins, many nonstructural (but not structural) viral genes are thought to have cellular origins. In the case of herpesviruses, a "core set" of viral/mammalian genes is present in all sequenced mammalian herpesviruses where it is responsible for viral replication and virion structure. This set would seem to have ancient nonviral sources. Other, "sporadic" viral/mammalian genes are not present throughout the mammalian herpesvirus family and, therefore, are presumably refinements of recent origins.

The problem with ancient and recent DNA genes is that genomic DNA sequences suggest that the viral genes are probably about the same age:

> [V]iral genes with non viral homologues resemble other genes of the same virus in attributes such as nucleotide and dinucleotide frequencies, and differ from their cellular counterparts; all have thus probably been in the viral genome for a long time.[188]

In mammalian herpesviruses, "core set" genes have diverged by nucleotide substitution, by large scale rearrangement of gene blocks (scrambling the order of genes), and adding large repeat elements. Outside the "core set," genes have been added and multigene families developed, greatly expanding the genomic size. Consequently,

> This would imply that great differences previously noted among mammalian herpesviral genomes trace only the latest part of the evolution of this one sub-group of herpesviruses, and further that we cannot expect to use sequence comparisons, even of whole genomes, to follow herpesvirus evolution back to a pre-herpesviral or pre-viral ancestor(s).[189]

Independent Entities Model

This "cross-dressing" of RNA viral families with both plant and animal hosts suggests that viral origins took place from entities existing before the "universal" phylogenetic tree had become a permanent part of life, supplying viruses with different, specific hosts. Supporting the independent entities model are similarities in sequence segments encoding RNA-directing DNA polymerases (reverse transcriptases) and sequences in transposons suggesting "that RNA viruses are as old as or even older than retroelements and that nonLTR retrotransposons are the oldest group of retroelements."[190]

Life in the RNA world may not have resembled life in the extant world. What sort of ancient proteinic enzymes would have been produced by the RNA world? The presence of tandem repeats between 88 and 440 amino acid residues in polyproteins of RNA virus has suggested a latent 11-meric principal periodicity. Possibly, "the ancestor of RNA viral genomes was a short gene sequence that amplified multiple times throughout evolution."[191] This possibility becomes all the more attractive given the plasticity of enzymatic domains found in viruses and similar sequences in cells. Indeed, the very idea of enzymatic specificity is challenged by the multiple functions of helicase domains in small DNA viruses and the protease activity of similar sequences in *Escherichia coli*.

Life in the RNA world could have been much more fluid, with genes or their equivalents moving more or less freely among life forms. For example, although traditional criteria (nonsegmented [monopartite] versus segmented genomes, host specificity and antigenic cross reactivity) link picorna- and alpha-like viruses, a phylogenetic tree based on RNA viral helicase links alpha-like viruses to corona-like viruses, traditionally coupled to flavi-like viruses, while the same tree divides picorna-like viruses into two groups, one, a sibling (sister) group to all the other (+)ssRNA viruses, the other, linked to the flavi-like viruses. Two RNA helicases, thus, may have originated from a source in the RNA world and moved into separate supergroups of viruses prior to the advent of DNA and the collapse of noncellular and cellular life forms.[192]

The presence of other retroelements across a vast spectrum of cellular life also suggests their ancient origins. Such independent entities, thriving in the RNA world, would have antedated species as such, rendering moot questions of whether viruses are monophyletic or polyphyletic.[193] DNA, RNA and retroviral elements ("switch" viruses along with retrotransposons) may thus be nonphyletic, entirely lacking a common ancestor in the form of a species. Monophylety among viruses and retrotransposons might have arisen after their enzymes.

The period before the collapse of life into noncellular and cellular forms also provides opportunities for speculation. For example, since both nonLTR- and LTR-retrotransposons (Copia-Ty1 and Gypsy-Ty3 varieties) are present from fungi and protozoans to plants and animals, the origins of retrotransposons can be placed before the origins of cells. Indeed, phylogenetic analysis of sequences of reverse transcriptases, the pivotal enzyme for retroviral genome replication, indicates that reverse transcriptase "emerged only once in the course of evolution"[194] and that "all examples of elements from the same class are clustered on the tree ... indicating a monophyletic origin for each of these classes or retroelements."[195] Were elements of these sorts freely picked up by the emerging life forms or might they have moved freely among these forms?

Assuming an ancient origin for retroviral-like elements, the question becomes, "Which came first?" Did retrotransposons give rise to retroviruses or were retrotransposons derived from retroviruses by losing the ability to leave their host cell? On the one hand, retrotransposons are present in a larger variety of organisms than retroviruses, suggestive of ancient origins. On the other hand, the sheer abundance of endogenous retrovirus-like elements in mammals (VL30 and IAP), suggests that some sort of trap mechanism has been at work accumulating these elements within cellular genomes for a very long time.

In the run-off between LTR and nonLTR retrotransposons for the position of oldest retrotransposons, the nonLTR appear to be the more primitive (but not necessarily closer to retroviruses). The primitiveness of nonLTR retrotransposons is also suggested by the phylogenetic map drawn for sequences of the reverse transcriptase.[196] Moreover, the insertion of LTR-retrotransposons into the host-cell's DNA is most often random (Ty3 elements in yeast being exceptional), suggesting a process that evolution has not yet refined. In contrast, the insertion of nonLTR retrotransposons is frequently site-specific (in 28S rRNA gene; spacer region of ribosomal DNA; spliced leader exons; other mobile elements; 5′ tRNA genes) even if most nonLTR retrotransposons exhibit random insertion, depending on random breaks in host chromosomes and host-supported DNA repair for second-strand DNA-synthesis.

Cellular Origin Model

The idea of viruses as "escaped" genes might seem to resurrect the long-defunct debate over spontaneous generation, but the complexity of viruses and even retrotransposons forecloses such a debate. Similarities in the genetic maps and gene sets across viral family lines, especially in ssRNA viruses, would be hard to explain as a collection of "escapees" randomly thrown together by transduction. The cellular origin model only makes sense at the level of parasexuality, the transfer of genetic complexes, on the order of plasmids.

The lesson of biotechnology, and the discovery of a spate of antibiotic resistant bacteria, indicates how easily plasmids can move among cellular species, but viruses are, in general, far more selective, and the origin of this selectivity would have to be earned if it were not inherited. Ironically, the plausible place to look for evidence for the evolution of viral specificity is among viruses with ambiguous identities.

Coronaviruses are sometimes nominated for the position of intermediates between positive and negative ssRNA viruses: The coronavirus polymerases occupy an intermediate position in the phylogenetic tree generated from subsequences; coronaviruses share similar genome polarity, and proteinase and helicase motifs with other positive (+)stranded RNA viruses; coronaviruses exhibit particle morphology, gene order, genomic termini and transcriptional expression with negative (–)stranded RNA viruses. However, picornavirus "appear to possess genes that originated from at least four different genetic sources"[197] which, along with numerous other "intermediates" are explained by interviral recombination and copy-choice mechanisms during replication.

Consensus

In any event, all three models for viral origins gravitate toward a primitive (+)ssRNA viruses, whether they originated as degenerate cells, retroviral elements, or plasmids. Double-stranded (ds) RNA viruses, conceivably, evolved as an adaptation to stabilizing the RNA and lengthening it. Negative (–)ssRNA virus might then have devolved as a spin-off of dsRNA viruses, coupled to the advent of segmented genomes. Viruses with smaller, more compact genomes (as well as segmented genomes) might have evolved as an adaptation to replication at faster rates. The (mainly) antisense, ssDNA viruses, which resemble (+)ssRNA viruses by way of undergoing replication before transcription, may have devolved with the collapse of the RNA world. Similarly, lysogenic bacteriophage, which, like retroviruses are inserted into the host genome and packaged as dsDNA genomes in virions, may have devolved by skipping the RNA intermediate, while the pararetroviruses devolved by skipping the DNA-insertional step.

Devolution of Cellular Life

Cells come with a complete repertoire of nucleic acids but otherwise with formidable differences in their contents. Cells come in "single" to "multicellular" varieties, and in forms that change from single to multicellular forms or vice versa; as prokaryotes (anucleate) and eukaryotes (nucleate). Prokaryotes exchange DNA through various processes but shunt replicated

DNA into new cells via its attachment to cell membranes. Eukaryotes may exchange DNA through sex (eggs and spermatozoa) or some other form of cellular copulation and exhibit a variety of types of cell division. Some eukaryotic cells have single nuclei, some are oligonuclear, and some are polynuclear syncytia (with nondividing nuclei) or plasmodia (with dividing nuclei).

Cells probably originated from the progenote(s) by processes of assimilation, something like minicells taking up plasmids (drug resistant factors). The "segregated plasmid transforms the inactive anucleated cell into one that can also synthesize RNA and protein."[198] Such a merogenote would then have to acquire ribosomes before it were a full fledged cell, but such a process is not out of the realm of possibilities.

Prokaryotes: "Which Came First?"

Contrary to prevailing prejudice that Archaea must be more ancient than Bacteria, the rooted "universal" phylogenetic tree based on sequences of 16S-like rRNAs supports Cavalier-Smith's idea of an ancient Gram negative-type bacteria exceeding Archaea and Eucarya in antiquity. Phylogenetic branching patterns, such as those inferred from 5S rRNA,[199] also show Bacteria emerging first, followed by the separation of Archaea (Metabacteria) and Eucarya.

The issue will not be decided, however, until the question of Archaea's monophylety is settled.[200] On the one hand, James Lake and other microbial taxonomists[201] argue that the Archaea are paraphyletic, consisting of two separate branches: a *Sulfolobus*-like branch of extreme thermophiles (or sulfur-dependent Archaea) with the name Eocyte which shares affinities with eukaryotes (an Archaea/Eucarya synapomorphy) and a branch of extreme halophiles (*Halobacterium* and *Halococcus*) and methanogens.[202] The best evidence for this split is the presence in the *Sulfolobus*-like branch and eukaryotes of a shared, unique, eleven-amino acid insert in their elongation factors (EF-1s) which is not present in EF-2 or any bacterial EF-1.

On the one hand, Woese claims that ribosomal rRNA profiles merely show a greater affinity of Archaea to Eucarya than to Bacteria, and that phylogenetic trees inferred from small subunit rRNAs, 5S RNA, and RNA polymerases, are consistent with the monophyletic view of Archaea.[203] Moreover, bootstrap resampling, which assigns an estimate of reliability on branch points, does not support the relationship of the principle Archaea groups to the Eucarya,[204] and eocytes are not resolved in the universal phylogenetic tree obtained with large subunit rRNAs.[205]

By no means is the situation settled. Even Lake's strongest critics cannot erase all doubt:

> According to the phylogenetic tree inferred from the small rRNA sequences, the branching pattern of archaebacterial species is monophyletic, all archaebacterial species form a single cluster. Phylogenetic trees based on the large rRNA , 5S RNA ... and the largest ... subunit of the RNA polymerase ... also show the monophyletic branching pattern.... The phylogenetic trees based on the catalytic and noncatalytic subunits of the ATPase ... and the phylogenetic tree based on the noncatalytic subunit..., however, show polyphyletic branching patterns among the three major groups of archaebacteria and eukaryotes. Also, the tree based on the second largest ... subunit of the RNA polymerase does not show the monophyletic branching pattern.[206]

Eucaryan Origins: Assimilated and Assimilating

Where do eukaryotes come from? Surprisingly, conjecture on the origins of Eucarya rarely mention an ureukaryote or protoeukaryotic cell. Those concerned with the issue have presented the ureukaryote as either archaeal (the Archaea hypothesis)[207] or an archaeal product of the fusion of Bacteria and Archaea.[208]

Instead of speculating on ureukaryotes, biologists interested in the origins and evolution of Eukarya emphasize eukaryotic organelles (see Chapter 3 for a discussion of organelles).[209] Two reasons seem to dictate this preference. In the first place, organelles make the eukaryote. Eukaryotes are defined by their organelles, and eukaryotes lacking one or another organelle, such as mitochondria, are considered primitive. Advanced eukaryotes, on the other hand, sometimes referred to by Cavalier-Smith's term, "Metakaryota," have the full complement of endomembranes and cellular organelles (mitochondria, peroxisomes and the golgi apparatus or dictyosome).[210] The second reason (more compelling if not less tautological), is that organelle acquisition is an idea "whose time had come" if only because it proved inescapable in the light of sequence data.[211] As a result of this concord, "[T]he conception of the eukaryotic cell as a consortium, a chimera, a mosaic entity, seems to be as firmly established within biological science as the chromosome theory of heredity,"[212] and mainstream biologists routinely speak of eukaryotes as holobionts, integrated assemblages of genomes originating from symbiotic partners. What is less certain is the mechanism through which such holobionts formed and, it would seem, kept forming.

•

SET

Thirty years ago, Lynn Margulis's serial endosymbiosis theory, known as SET, took root and grew until it dominated thinking on mechanism of organelle assimilation.[213] SET began by positing the origin of mitochondria from endosymbiotic bacteria, and, presently,

> In the case of mitochondria, there is now a satisfying concordance of structural, biochemical, and genetic evidence not only supporting an origin of this organelle from eubacteria, but, indeed, tracing this origin to a specific subdivision, the α-purple bacteria.[214]

From there, SET expanded to the origin of chloroplasts from endosymbiotic cyanobacteria, cilia or undulipodia[215] and other organelles from yet other prokaryotes.

In tracing the history of SET,[216] Margulis discovered that K. S. Mereschkowdky (1855 [1865]–1921), coined "symbiogenesis," and seems to have been the first to propose an endosymbiotic origin for chloroplasts.[217] Others shared similar views along the way, leading to the present consensus. Symbiosis is "the prolonged association of organisms that are members of different species,"[218] and endosymbionts are the special case of symbionts from one taxonomic lineage living permanently and compatibly within cells of another lineage. Symbiogenesis is the "emergence of new species with identifiably new physiologies and structures as a consequence of stable integration of symbionts [or endosymbionts]."[219] That "stable integration" depends on lateral gene transfer of genes from the endosymbiont to the partner's nucleus.

Endosymbiosis is an evolutionary stable strategy (ESS in the lexicon of sociobiologists) for many life forms: the archaea in gills of clams and the gut of tubeworms in geochemically driven communities around hydrothermal vents;[220] the chlorophytes (or zoochlorella) and xanthophytes (or zooxanthella) within ciliates and radiolarians, sponges, cnidarians, flatworms and mollusks.[221] Some of these endosymbionts can be separated experimentally from their partners, and most are acquired in each generation afresh, since they are not transferred to offspring through the egg.[222] An extraordinary case of endosymbiosis occurs in cicadas, in which some endosymbionts (intracellular bacteria) are separable, while others are not only inseparable from the host, but the host is inviable without them.[223] The case in point is the leafhopper's egg where a membrane-bound "symbiont ball" containing reduced endosymbionts (endocytobionts, protoplastoids or intracellular procytes), resembling mycoplasm or rickettsiae, is inherited and passed onto specific cells in the embryo. This ball would seem to exercise morphogenetic control, since its experimental displacement anteriorly from its normal posterior position (followed by anterior ligation) may reverse the polarity of the developing embryo or induce the development of a double-abdomen monstrosity lacking a head at either end.[224] Endosymbiosis, thus, covers a large range of cell-to-cell interactions, and nothing about endosymbiosis should be taken for granted. The presence or absence of walls surrounding algal endosymbionts, for example, which grades so beautifully between separable and inseparable species (zoochlorella among the green *Hydra*), and chlorella sending cytoplasmic processes between cells of its *Convoluta* host may have as much to do with the mode of transmission as with the intimacy of the cells's interaction.[225]

One can easily imagine endosymbiotic partners in the Precambrian "age of symbiosis,"[226] transferring nutrients and ATP, coupling up into permanent partnerships and sealing their bond with a more or less one way transfer of genes to the dominant partner's nucleus (although, "massive" trafficking among genes seems to have occurred between chloroplasts and mitochondria,[227] and an occasional movement from nucleus to mitochondrion as well[228]). This congenial picture is not entirely complete, however, since it does not suggest when the "couple" were joined or why some couples divorced. Evidence for mitochondrial-like genes in deeply branching amitochondrial protozoa suggest that the "wedding" took place earlier than generally expected, and the cytotoxic effects of cytochromes outside the confines of mitochondria suggest that the "marriage" was not entirely happy. Type II amitochondrial eukaryotes divorce their mitochondria, and type I amitochondrial eukaryotes divorced both mitochondria and hydrogenosomes.

Alternatives to SET

Inevitably, SET has faced challenges. First, other routes of organelle assimilation have been suggested as competitors with endosymbiosis, "including horizontal gene transfer, ephemeral or transient symbioses with organisms related to those that finally generated mitochondria, or by the original fusion between eubacteria and archaebacteria that has been proposed to generate the first eukaryotes."[229] SET has withstood this challenge, or so it seems, because endosymbiosis suggests the only way that genomes could be combined en masse, namely through lateral gene transfer.

A second sort of challenge centers on different organelles or the nucleus[230] residing in the cytoplasm of one or another alleged protoeukaryote. The entire census of organelles was considered in a "the more the merrier" mode. Thus, "the archezoan eukaryote that first successfully converted bacterial symbionts into mitochondria and probably also peroxisomes (and perhaps chloroplasts), and therefore became the first metakaryote, [probably] was very similar to the retortamonad *Chilomastix*."[231] These and other alternatives do not seem to dull the point, however, that mitochondria evolved from endosymbionts.

Other recent findings, however, have "blasted apart our ideas about the earliest eukaryotic cell," according to Geoffrey McFadden, although Michael Gray believes "It's still a debatable story."[232] The bombshell would seem to be the discovery of DNA sequences resembling those of mitochondrial genes but found in amitochondriate eukaryotes (*Entamoebar*[233] and *Giardia*[234]) lacking organelles (type I amitochondriate eukaryotes), although most speculation concerns the other amitochondriate eukaryotes (type II) containing an organelle known as a hydrogenosome (because it evolves molecular hydrogen).[235] Previously thought of as a degenerate mitochondrion, the hydrogenosome has now been raised to the status of a mitochondrial precursor.[236]

Present in ciliates and chytrid fungi lacking mitochondria, the hydrogenosome is a double membrane-bound organelle that converts pyruvate or malate to acetate, carbon dioxide and hydrogen gas with a net yield of two additional mol ATP. According to Martin and Müller's "hydrogen hypothesis" (HH) "a cell that is organized in a manner strikingly similar to the amitochondriate eukaryote *Trichomonas vaginalis* [was the first eukaryote, and n]ot a single evolutionary invention was necessary to deduce this organelle-bearing cell."[237] The HH emphasizes the bacterial-like qualities of metabolism in amitochondriate eukaryotes, down to the bacterial-like quality of DNA sequences encoding their glycolytic enzymes.[238] An α-protobacteria which was to have contributed to the ancient symbiosis relieved itself of H_2 and CO_2 waste products from its anaerobic fermentation of organic compounds acquired from the external environment. The archaean proto-eukaryotic cell, like extant methanogens and synthrophic hydrogen-producing organisms of marine sediments, used these waste products as the source of its energy and carbon metabolism.[239] An uncertain world and the potential of mutual reliance is thought to have provided a selective nudge toward permanence in the relationship between the cells. What is more, were the proto-eukaryotic archaean to have been an autotroph to begin with, reliance on its symbiont for nurture may have turned it into a heterotroph, able to use complex organic compounds for metabolism and growth. The hydrogenosome might then have evolved into a mitochondrion.

In any event, HH deviates considerably from SET, although both rely on the transfer of genes as the major device for making acquisitions permanent (and the major source of evidence potentially consistent with their position). Where they differ is primarily on the originary organelle and the ecological interactions that might have dominated at the time of an endosymbiont's incorporation. SET features the origin of mitochondria from a protobacterium resembling extant α-purple bacteria (modeled on *Paracoccus*) and a heterotrophic host already equipped with cytoskeletal elements and membranes capable of endocytosis and sequestering an endosymbiont. HH features the origin of a hydrogenosome from an anaerobic autotrophic α-protobacterial H_2/CO_2 producer (modeled on *Rhodobacter sphaeroides*) and a lithoautotrophic, anaerobic archaean methanogen dependent on H_2 for ATP production, and capable of utilizing acetate as a carbon source but still lacking a cytoskeleton. Because they differ, molecular biologists are hopeful of differentiating the hypotheses experimentally.[240]

Lateral gene transfer in the evolution of Eucarya

The lateral transfer of genes from the proto-organelle into the host's nucleus is central to most theories on the transformation of endosymbiont to organelle, but lateral gene transfer is, by no means, limited to organellar genes, although the term, "horizontal gene transfer," seems to be used more often when referring to genes from other sources than proto-organelles. Southern blots of DNA samples and systematic screens of sequence databases routinely turn up candidates for horizontal gene transfer. For example, the transfer of sequence motifs seems to have occurred between glucose-6-phosphate isomerase of the plant, *Clarkia ungulata* and glyceraldehyde-3-phosphate dehydrogenase of the bacterium, *Escherichia coli*. Likewise, the fibronectin type III domain, widely distributed among animals but absent in plants and fungi, is found in bacterial carbohydrate-splitting enzymes. Other examples of apparent gene transfer seem quite reasonable, in retrospect. For example, the transfer of the Ti plasmid from *Agrobacterium tumifaciens* (*A. rizogenes*) to plant cells, "appears to be just one aspect of an evolutionary long-standing and complex symbiosis between the two partners."[241]

The evidence that lateral gene transfer has taken place is largely sequence data, and, thus, subject to all the problems of alignment, sequence size and heterogeneity, and metricizing discussed in Chapters 2 and 3. In general, these difficulties are no more attended to regarding lateral gene transfer than regarding the usual monophyletic assumptions. Lateral gene transfer is, however, adopted as the "default" solution after phylogenetic trees based on these data are found to be too incongruous with other phylogenetic trees based on other data. In addition, three types of evidence are marshaled to "test" the hypothesis of lateral gene transfer: (1) Sequences encoded in the nucleus of the organelle-bearing cell are present in the genome of an allegedly "primitive" organelle; (2) sequences encoded in the nucleus of the organelle-bearing cell are present in an extant representative of an alleged ancestral type; (3) bizarre divisions of related genomes are traced through alleged ancestral types.

For example, the *TufA* gene encoding elongation factor EF-Tu, present in the cpDNA of green algae and many Charophytes (thought to be closely related to land plants) is unearthed in the nucleus of land plants (*Arabidopsis thaliana*).[242] Furthermore, nuclear genes encoding the chloroplast CS-RBD-type RNA-binding proteins (28–33-kd) in tobacco share the RNA-binding consensus sequences with genes in cyanobacteria (*Synechococcus* sp.).[243]

> [I]t appears that the cyanobacteria-like endosymbionts contained CS-RBD-type [RNA-binding] protein gene(s) in their own genomes, and that this gene(s) was subsequently transferred to the nucleus. Consequently, multiple genes encoding either one or two RNA-binding domains could possibly have been produced by duplication of this gene or by fusion to genes already present in the nuclear DNA. The two domain-type protein genes may have evolved to chloroplast RNP genes after acquiring a DNA region encoding a transit peptide.[244]

Differences in the whereabouts of organellar genes in plants and algae suggest that lateral gene transfer has occurred independently at a slow but continuous rate. Variation in gene content in the cpDNAs of rice, tobacco and liverwort is also suggestive of continuing lateral gene transfer. The most bizarre distribution of genomes, however, occurs for the small and large monomers of the ribulose bisphosphate carboxylase/oxidase (RUBISCO) holoenzyme:

> In the cyanobacterium *Anabaena*, the small subunit coding sequence is located immediately upstream from the large subunit sequence and both are transcribed as a dicistronic message.... [Likewise,] the small subunit is encoded by genes adjacent to the large subunit gene on the cpDNA in some algae and in the cyanelle genome of *Cyanophora paradoxa*.[245]

In all land plants and some algae, however, the RUBISCO holoenzyme's large subunits are encoded in cpDNA (the *rbcL* gene), while the small subunits are encoded by a small gene family in the nucleus (the *rbcS* genes). A transit peptide targeting the small subunit proteins to chloroplasts is cleaved following importation into chloroplasts.

Evidence, thus, warrants taking lateral gene transfer seriously and fleshing it out with mechanisms. Several possibilities are easily imagined for the transfer of genes: xenologous recombination in which a resident gene is replaced by a similar foreign gene; a copy-choice mechanism requiring replication; DNA synthesis through reverse transcriptase followed by a device for gene copy; hybridization followed by translocation and independent assortment. A composite of these possibilities would have mobile genetic elements, resembling retrotransposons, performing the transfers with the help of reverse transcriptase within introns.[246] Introns are not uniformly distributed in the genomes of organelles. For example, land plants and *Chlamydomonas* have no introns in their large subunit RUBISCO genes (*rhc* L) in cpDNA, while *Euglena*'s gene has nine introns, and photosystem I's P700 chlorophyll apoprotein genes (*psa* A) encoded in cpDNA has no introns in land plants, but it has two introns in *Chlamydomonas*, while *psa* B has no introns in either land plants or *Chlamydomonas*, and both *psa* A and *psa* B of *Euglena* cpDNA contain multiple introns. If introns are involved in lateral gene movement, therefore, one would expect significant differences in the rate of gene movement. The rarity of such elements in the genomes of extant organelles would suggest, however, that organellar genomes are more or less stable, and sequencing suggests that more or less the same genes are transferred across the board, indicating directed rather than random movement.[247]

Another problem posed by lateral gene transfer is an apparent contradiction with common sense: the transfer of genes would seem to place the endosymbiont at a disadvantage with respect to maintaining itself outside the partner's cells. Why then would any endosymbiont transfer genes?[248] The simple response to this problem, namely, that the organelle's requirement for products of the partner's genome presents no obstacle as long as the organelle remains within the cell, does nothing more than beg the question. A more satisfactory response requires a better understanding of endosymbiosis.

Presumably, those endosymbionts which became chloroplasts and mitochondria were integrated in unique ways into their partner's physiology. Today, that integration is reflected in the major contributions of the products of the nuclear genome to the organelle (ribonucleoprotein, other proteins, and phospholipids), and the complex mechanisms for importing and sorting components. Apparently, however, there are limits that chloroplast and mitochondrial genomes do not cross, as suggested by the extraordinary convergence in chloroplast and mitochondrial genomes: both genomes encode the rRNAs and usually most of the tRNAs of their translation systems, specific components of their electron transport chains and several (mitochondria) to many (chloroplast) ribosomal proteins. Those limits may be the requirements of differentiated double membranes linking chemiosmosis to the electron transport system and photo- and oxidative phosphorylation. The inner membrane of these organelles (the inner mitochondrial membrane and thylakoid membrane of chloroplasts) would be especially sensitive to cooperation inasmuch as it is a mosaic of organellar and nuclear gene products.[249] Alternatively, the requirements for division in DNA-bearing organelles may set limits on the transfer of genes.

Other organelles, hydrogenosomes, peroxisomes, undulipodia (generally), lacking nucleoids, may reside within eukaryotic cells under rubrics that allowed the transfer of their entire genome (assuming that they once had a genome). What crucial difference(s) may account for these alternative "ways of life" among organelles is unknown, since even organelles reduced to membranes within a eukaryotic cell retain the property of self-perpetuation, although they may not divide like some mitochondria and chloroplasts with the help of constricting rings.[250] Like prions, it would seem, "gene-less" organelles control the products of nuclear genes. This is not to say that these organelles are not differentiated, but their differentiation may have a larger component of self assembly than organelles retaining a genome.[251]

Finally, the movement of genes from any proto-organelle to a nucleus has implications for the study of "eukaryotic" evolution, and the movement of genes from several proto-organellar sources to a nucleus has even greater implications. If at least part of the eukaryote's nuclear genes are derived from a cellular organelle (mitochondria, chloroplasts or whatever other organelle was assimilated), then eukaryotic genomes are not monophyletic (having a source in only one taxon), and all cladistic analyses incorporating monophyletic assumptions (such as maximum parsimony and distance matrix methods) are inevitably in error. This problem for eukaryotic systematics becomes all the more profound with each compounding of the genome (if genes were transferred from several proto-organelles to nuclei). Lateral gene transfer boils down to polyphylety in the eukaryotic genome, and polyphylety will not go away simply because one would like to assume the "simplest case scenario."

Monophylety Among the Organelles

An entirely different question is whether each organelle is itself monophyletic. Many have affirmed the monophylety of cellular organelles, but the affirmations smack of special pleading rather than conviction. For example:

> Certainly, a monophyletic origin for each organelle would most easily explain the striking amount of conservation of chloroplast and mitochondrial genes in different organisms even given that this apparent gene conservation partially reflects the inordinate past focus on chloroplast and mitochondrial genomes of a limited set of eukaryotes (animals, mostly mammals, a few fungi, and flowering plants).[252]

Despite testimonials of this sort, monophylety among the organelles remain uncertain.

Cavalier-Smith invokes "[T]he transfer of chloroplasts from one metakaryote to another"[253] to explain the difference between chloroplasts with two membranes (land plants, Chlorophyta, Gamophyta and Rhodophyta), three membranes (Euglenophyta and Dinoflagellata), and four membranes (chromophytes).[254] Despite multiple "infections" of eukaryotic cells, however, all these chloroplasts may still have had a monophyletic origin. The results of sequencing are ambiguous, consisting of similarities between *gapA* (alias *gap1*) genes in free-living cyanobacteria (*Anabaena* and *Synechocystis*, hence, a "sister" group to chloroplasts), and the *GapAB*, present in all chloroplasts of eukaryotes (with the exception of the lineage leading to trypanosomes, after the divergence of *Euglena*, which lost *GapA*).[255]

> Taken together, the results support a monophyletic origin of oxygenic photosynthesis derived from cyanobacterial ancestors but suggest that *the acquisition of prokaryotic phototrophic endosymbionts in the evolution of eukaryotes may have been a multiple event*" (emphasis added).[256]

Morphological distinctions among the "cristae" of mitochondria argue for a polyphyletic origin of tubulocristates (organisms whose mitochondria contain tubular cristae) and platycristates (organisms whose mitochondria contain flattened cristae).[257] Furthermore, "it is sometimes observed that mtDNA lineages in two different species are more closely related to each other than to other lineages in their own species."[258] Recently, however, the idea of monophyletic mitochondrial origins from bacteria-like endosymbionts was reinforced by the discovery of proto-mitochondrial DNA in the protozoan *Reclinomonas americana* (ATCC 50394)[259] resembling a bacterial genome in miniature.[260] The proto-mitochondrion's closest "relatives" would seem to be the rickettsial group of α-protobacteria, but the genome is also (if not better) reflected in mitochondrial genomes. The genome of *R. americana*'s endosymbiont contains 44 protein-coding genes typically found in mitochondrial mtDNA, two unidentified open reading frames resembling sequences in plant and some protistan mtDNA, and three rRNA genes, including a 5S rRNA gene previously found only in mtDNA of land plants and unicellular chlorophytes. The same genome also contains, however, the RNA component of RNase P, rarely encoded in mtDNA, and no fewer than 19 proteins otherwise unknown in mtDNA, including four genes encoding a eubacterial-type of RNA polymerase with multiple subunits. Other features of gene content and characteristics of genomic organization and expression are also indicative of a close resemblance to bacteria. It would seem, thus, that *R. americana* represents an early mitochondria-containing protistan, and its primitive mtDNA is as much that of the ancestral endosymbiont as that of a mitochondrion.

What about other organelles? Differences among ribosomes and their RNA are so often ignored that one tends to assume that ribosomes are monophyletic, but were ribosomes also derived from endosymbionts? Do the universally present genes for small and large subunit rRNAs (16S and 18S squeezed into the category of 16S-like, and 23S and 28S squeezed into the category of 23S-like), found throughout the Eukarya, owe their presence there to lateral gene transfer? Did the eukaryan nucleus originate from plastids (the "nucleomorph" of cryptomonad flagellates from red algal cytobionts)? Did an archezoan ancestor successfully convert bacterial symbionts into peroxisomes?[261] Certainly, the most audacious extension of SET is Lynn Margulis's own symbiogenetic derivation of eukaryotic locomotory organs, her "undulipodia" (centrioles, kinetosomes and blepharoplasts [internal members], cilia and flagella [external members]) from the highly motile spirochetes of the genera *Mobilifilum* or *Spirochaeta*.[262] The question of the undulipodia's source, however, will have to remain unanswered until a satisfactory demonstration to its own DNA and sequencing provide the molecular biologist with an answer.[263]

The concept of "secondary endosymbiosis" derives an endosymbiont by ingestion of a eukaryote that itself harbors an endosymbiont without the original host remaining as an endosymbiont itself except, perhaps as a vestigial membrane. "Max" Taylor's extension of secondary endosymbiosis takes "cells-within-cells"[264] to new depths by placing "secondary symbioges" or eukaryotes within eukaryotes and marking other distinguishing characters of the "cytobiont" (the partner within) from the host. Going even further, one finds symbiogenesis recommended as the explanation for, what is arguably, the most complex cellular organelles known: the apicoplast (a nonphotosynthetic plastid) in parasitic apicomplexans[265] and the cnidocysts in Cnidaria.[266] The cnidarian cnidocysts are especially interesting, since tracing their morphological character against a presumptive phylogeny suggests that cnidocysts are polyphyletic in the Medusozoa, traceable to two sources, but monophyletic in the Anthozoa.[267] The molecular biologist have their work cut out for them, not only sequencing the genomes of all these esoteric species, but sequencing as well their organellar genomes (if they exist).

CRITICAL ISSUES FOR SAMENESS AND DIFFERENCE

Sameness and difference have not "gone the way of all flesh." They remain determinants in the biologist's formulary and guide to life. This is not to say that they must continue to have the same impact they have had in the past on the way biologists think. Raising the shades of sameness and difference should allow some intellectual light into the biologist's future. The glimmering is devolution.

Varieties of living things can be created, it would seem, either by starting from varieties or by acquiring various components in the course of time. The former idea is implied by the concept of life's multiple and separate origins, whereas the latter is implied by the concept of parsing out life into various forms. The two ideas are not mutually exclusive, since multiple origins could conceivably have been followed by mixing and sorting. In any case, the signature properties of present-day life may well have been represented by long-ago-lost elements and "strange bedfellows" in early Earthlife.

Under the rule of samenesses and differences, the properties of life are generally reduced to single entities. For example, prevailing prejudice dictates that memory is predominantly the property of nucleic acid/protein-based chromosomes. One is surprised, therefore, to find properties of life in places where they are not expected, such as memory attributed to nonchromosomal cellular constituents. For example, in ciliates, the inheritance of experimentally induced patterns of kinetosomes (ciliary patterns) such as "Siamese-twin" or double-bodied cells, and "inherited" mutants such as "swimmer" and "snaky.", are cortically (cytoplasmically) determined.[268]

Indeed, "We must not forget that membranes have genetic continuity through their growth and division and that cell heredity involves both DNA heredity and membrane heredity."[269] Moreover, prions, the vectors of several mammalian spongiform encephalopathies, would seem to be proteins with a memory. The mixing of life's properties in life's entities comes to light when the rubric of sameness and difference is set aside.

What has been learned from the search for ancient gene pedigrees? Ever since the Human Genome Project went into "high gear," superfamilies of genes have expanded exponentially. At the same time, the basis of comparison has shrunk, sometimes to just a few amino acids in a sequence, or a short "consensus" of "conserved" sequences with a broad distribution and criteria only vaguely relevant to function. On the plus side, as the sense of superfamilies has grown, faith in the specificity of enzymes has been eroded. Biologists have become increasingly aware that proteins, or RNA, for that matter, with similar sequences do more than one thing, and even more things at the same time than would have been thinkable under the old rubric of enzymatic specificity. New concepts of building metabolic pathways by recruiting enzymes only weakly active in a prior step have come into vogue. On the minus side, as the sequences relied upon to demonstrate "homology" have become more homogenized, more homologies are demonstrated with less basis in similarity. Similarities among members of superfamilies are sometimes so attenuated that horizontal gene transfer and exon shuffling cannot be distinguished from orthologous genetic evolution. Instead of presenting molecular biologists with a vast sameness pointing to an explanation in orthologous genetic evolution (sameness through common ancestry), ancient genes confront molecular biologists with vast difference challenging any simple explanation.

One should probably not assume that all the properties of life were assembled at the same time and in the same place during life's origin if only because these properties are not present in all present-day living things. Nothing whatsoever precludes the possibility of life's components being assembled in different times and places, and, what is more, nothing excludes their assembly in different forms than those predominating in extant living forms. Life organized around nucleic acids may very well be all we have to go on for understanding extant life, but it is not all that is imaginable for life's possibilities.

At present, life is broken down primarily into cells and viruses (to which should be added viroids, virusoid, and prions and probably other forms even more consistently ignored by biologists). Did these forms devolve from the parsing and separation of life's materials and properties as well as from their mixing and assimilation? Did some of the combinations that "worked" survive to the present while others ceased?

Possibly, life was once a more or less homogeneous mixture to which complexity was added here and there through disequilibrium and asynchrony. Possibly life broke up into small, cooperative communities in some places and into localized, competing individuals in other places. Possibly, under some conditions cooperative communities prevailed and under other conditions competitive individuals prevailed. Given the revolutionary success of the free market economy in the closing decades of the 20th century, one can have no doubt which form of organization currently prevails among human beings. One may wonder, however, what became of collectivity in life.

CHAPTER FIVE

ALTERNATIVES: EVOLUTION VERSUS DEVOLUTION

Darwin's great novelty, perhaps, was that of inaugurating the thought of individual difference. The leitmotiv of The Origin of Species *is: we do not know what individual difference is capable of? We do not know how far it can go, assuming that we add to it natural selection.*

Gilles Deleuze (Difference & Repetition, *p. 248*)

Karl Popper, the philosopher of science, prescribes a hypothetical-deductive method to cure science of its chronic ills, its descent to dogma and intolerance.[1] The method requires scientists to posit alternatives to their hypotheses against which the "truth" of an argument is evaluated. The Popperian method assumes that truth is not an absolute virtue, but provisional truth is fostered by contrasting or testing the relative merits of hypotheses.

For "ordinary" Kuhnian science,[2] inventing alternative hypotheses, performing the experimental tests and evaluating results are all of a piece. For example, practicing biologists frequently design hypotheses and experiments to whittle away at other hypotheses on the edge of Occam's razor, the proposition that simplest is best. Statistics and computer-assisted methods, likewise, offer numerous opportunities for designing experiments and testing hypotheses against a standard, a null hypothesis or a matrix of probabilities. In the long run, or, in retrospect, the Popperian method works in these instances if only because uncertainty and the circulation of alternatives is constitutive of rhetoric, fashionable methods, weighting evidence, and mathematical legerdemain. Presumably, biologists will continue to solve "mop up" problems by working in their traditional way. But what about "revolutionary science"? Where do alternatives come from to the extraordinary scientific idea? What, for example, is the alternative to evolution?

"Alternative to evolution"? Indeed! Evolution is so deeply etched in modern consciousness that even asking for alternatives produces shrugs of bewilderment. One might as well ask, "What is the alternative to free-market economics for providing the greatest common good?" Furthermore, asking for an alternative in substance, not merely for the substitution of details, is likely to be construed as a subversive activity (like asking for an alternative to motherhood).[3]

The biological sciences, nevertheless, desperately need an alternative to evolution. The need is acute if only because the cornucopia of molecular data, acquired through the Human Genome Project, overflows with rash evolutionary hypotheses such as those on consensus sequences and gene superfamilies. In general, the data garnered through cloning and PCR, from sequences and computer-assisted alignment protocols are squeezed into conformity with the notion that every taxon is monophyletic (having but one source) and neither paraphyletic (including descendants from more than one source) nor polyphyletic (including descendants having combined sources). The dynamics of life, the coming and going of individuals, speaks volumes about the unreliability of evolutionary theory based on the current vogue in simplifying assumptions.

Evolution's predicament is traceable to none other than sameness. Ironically, sameness, and hence stability, is so ingrained in evolutionary theory that it is well nigh impossible to eliminate. Life without sameness (the individual) would hardly be life as we know it, and evolution without sameness (the species) would hardly be evolution as biologists teach it. The same reliance on sameness spreads from top to bottom of the systematist's hierarchy, from domains, kingdoms and phyla to genera, species and individuals. Likewise, evolutionary theory is indifferent to difference. How would biologists conceive of evolution if they did not consider difference a matter of common sense and the intuitively obvious? One might wonder what biology would be like without its fundamental attitude toward sameness and difference for framing its ideas. What would evolution look like without the biologist's reliance on the sameness/difference dialectic?

In Chapter 4, devolution was reconstructed around life without sameness and difference. Chapter 5 goes further, by positing devolution as an alternative to evolution. The object here is not to re/present devolution and certainly not to represent it as the solution to evolution's problems but to show where alternatives are needed to current evolutionary theory. The chapter begins, thus, with a description of evolutionary studies and goes on to mainstream evolutionary concepts in the categories of genes and mutation, heredity, and natural selection. In each instance, the mainstream concept is followed by alternative concepts based on devolution. Finally, a brief reprise is devoted to reasserting the need for alternatives to Darwinian evolution as biology moves into the 21st century.

EVOLUTIONARY THEORY: BIOLOGY'S HORSELESS CARRIAGE

One might think that in this golden age of molecular systematics, evolution would be one of the hottest subjects in biology. Certainly, the idea of evolution percolates through the kinds of studies reviewed in Chapters 2–4. But evolution as subject (ends) rather than object (means) is not "hot"; it is not even warm, and may be cooler now than it has been for a century. Evolution as such is not taken seriously.

For example, instead of questioning evolutionary theory, the premier problem of biologists contemplating evolution seems to be finding the right metaphor. The Nobelist, Christian de Duve, for example, attempts to explain the difficulties of tracing biological descent by describing the struggles of linguists:

> By careful comparative studies based on the assumption that words change only gradually with time, he [the linguist] might even succeed in reconstructing the ancestral Latin, as well as the manner in which the four languages [French, Italian, Spanish, and Romanian] evolved from it.[4]

Biologists playing linguist should bear in mind that what "evolves" from one language to another is more than words: it is grammar, syntax, vocabulary, indeed, culture. Language also folds back on itself, covers itself, erases and copies, expands and unfolds. It takes one route in oral traditions and another in literature, movies, music, theater and television. Contemporary evolutionary theory reflects none of this complexity.

Evolutionary theory is biology's horseless carriage. For nearly one hundred and fifty years, the idea of evolution has not changed, or not very much. Thus, evolutionary theory gets biologists to where they want to go, to an explanation of sameness and difference, and no further. The biologist's horseless carriage has all the essentials of a modern vehicle but only in rudimentary form: Natural selection is the engine; heredity is the transmission; genes and mutations are the fuel. The irony is that in its contemporary incarnation, the horseless carriage is purely a virtual machine, existing only in interactive software and on the circuits of the Human Genome Project.

What is Evolution?

The dearth of studies on evolution (as opposed to the use of evolution as rationale for studies) at our centers of learning was exposed by Michael Ruse,[5] the doyen of paradigm hunters among contemporary historians of science. He has searched in vain for an evolutionary paradigm in the history of biological sciences. After massive fiddling with folios and footnotes (at least in English-language libraries) he is well qualified to tell his readers that he has not discovered a professional standard for studying evolution in evolutionary studies. Ruse assumed that an evolutionary paradigm would have surfaced in the wake of Darwinism but found that evolution remained submerged while mainstream biology flowed toward the molecule.

In the beginning, Thomas Henry Huxley ("Darwin's Bulldog," *The Devil's Disciple*, and *Evolution's High Priest*[6]) failed to incorporate evolution into his educational reforms, preferring to buttress the medical sciences with physiology and morphology.[7] Even where evolution was taken seriously by professional biologists, such as (W. F). Raphael Weldon (1860–1906), no school of evolutionary studies emerged, and, according to Ruse:

> Three decades after the *Origin*, one of the leading evolutionists in North America [Alpheus Hyatt, 1838–1902], a man of proven moral courage, would not have students exposed to evolutionary theorizing. If this does not point to the peculiar status of evolutionism in the world of professional science, I do not know what would.[8]

Even Edmund B. Wilson, the dean of American cytologists,

> regarded the study of evolution as beyond professional bounds.... [He] simply presupposed evolution as the metaphysical background, and then he got on with what interested him.... Wilson assumed evolution as fact; he rather looked down upon efforts to discern evolution as path; and he was professionally uninterested in the question of evolution as cause.[9]

Perhaps, Ruse's concept of evolutionary studies is too narrow for him to appreciate that Wilson and the other "egg rollers" at Woods Hole, squinting through lenses at developing marine embryos, were studying (or, at least criticizing) evolution in the mold of Haeckelian recapitulation. Perhaps their concept of evolution was too unformed to be included in Ruse's pantheon of biological paradigms, but evolutionary studies in the Haeckelian mold were probably never more popular than between the *fin-de-siècle* and the

Great War. Moreover, the flowering of evolutionary theory during the nineteen thirties and forties, known as the "new synthesis," neoDarwinism, or microevolution, was certainly a period of ferment in evolutionary studies.

Ruse's point, nevertheless, must be granted: evolution as subject has hardly gotten off the ground in the second half of the 20th century. Of course one finds here and there a department or subdepartment of biological sciences that incorporates the word "evolution" into its title ("evolution/ecology"), but at most universities, evolution is rarely represented by *bona fide* faculty, and at most granting agencies, proposals on evolution are hardly ever funded. What is more, with few exceptions (Sewall Wright, was probably the last one), individuals rarely spawn "schools" of evolutionary thought. What then is it that goes by the standard of evolution at universities?

Pervasive Ambiguity

Ambiguity in evolutionary theory did not start in the electronic age. The habit of illustrating evolution with pictures drawn from metaphor goes back to Huxley, and even the composite portrait is hardly realistic, evolution being so many things to so many people. Indeed, one hardly knows what anyone (present or past) is talking about as evolution in the first place. Take, for example, the simple concept of evolution as change in life's forms. This concept was anathema to many Englishmen in the 19th century but is widely accepted in the English-speaking world today if only because it is sufficiently ambiguous to pass without notice.

Evolution as change has many lacunae, frequently filled with unstated assumptions. Change itself is assumed to be directional, although the mutations producing the change are assumed to be random. Furthermore, evolution is assumed to affect individuals, but individuals themselves are not permitted to evolve! Indeed, any effort to reduce this ambiguity and search for a sense in which individuals can acquire evolutionarily significant samenesses and differences runs afoul of Western biological law and is labeled ludicrous, a laughable throwback to Lamarckian concepts of use and disuse and the inheritance of acquired traits.[10]

The abundance of ambiguity attached to evolution is nowhere better illustrated than in the failure of contemporary biologists even to agree on what changes during evolution. The molecular biologists, who long ago abandoned the study of life in individuals, now extend their belief in the "life" of molecules to the evolution of molecules.[11] The molecular heartbeat would seem to be the ticking of "molecular clocks,"[12] while the molecular id, ego and libido take the form of "selfish genes," pursuing their immortality in future generations through the organism they occupy. Mutations, blithely accumulating in DNA, provide the variation, and natural selection picks and chooses which molecule will wax and which will wane.

For biologists in general, however, the idea is anathema that life exists in molecules or anywhere else outside living individuals. Retaining their belief in the individual's monopoly on life is, possibly, an extension of the idea of a "self," or essence of oneself, and abandoning this belief, therefore, is too threatening to most Westernized individuals for contemplation. Life simply dissipates both above and below the level of living individuals. Life is not attributed to the communities of living things making up the "living" part of the environment nor to the chemistry of molecules and the physics of organization (with the exception of DNA). What then is left to evolve?

Some biologists will erode the individual's boundaries in order to speak of the individual as evolving. For example, the geneticist who argues that Darwinian fitness is the likelihood of an individual passing on its genes can only make the argument by ignoring the usual requirement of breeding for two individuals and the demands of rearing for a whole deme of individuals. A biochemist who claims that regulatory mechanisms evolve in individuals ignores the individual's dependence on communities (interacting individuals of different species) and neglects the community's reliance on a host of biogeological worlds, the hydrosphere, geosphere and beyond (Gaia).

Other biologists cast populations of living things in the role of the evolving entity. These biologists take their lead from chemists, seeing the behavior of populations as the aggregate of the individual's behavior, but the chemist's model does not accommodate biological change: biological individuals cannot be identical members of a class (the chemist's model of atoms and molecules) and at the same time acquire hereditable differences (the requirement of evolution). Indeed, so-called Hardy–Weinberg equilibrium, the model of living individuals in populations that most closely approximates the chemists ideal of atoms or molecules in groups, is precisely one in which evolution does not take place. Evolution occurs only when the model breaks down: when some force enters that violates the equilibrium, usually a biological force such as emigration, immigration, or selection. Disequilibrium can also be caused by nonbiological means that reduce population size, such as density independent "acts of God," but these only exacerbate the inadequacies of uniformitarianism as a model for evolution.

Many biologists offer repetition as a rationale and compensation for the inherent contradiction between the unchanging individual and the requirement of evolution for change. Typically, the biologist enlarges on the individual both temporally and spatially. For these biologists, the individual is recycled recurrently from one generation to the next, especially at its relatively stable plateau (adult) stage of life. The individual is also thought of as repeated in individuals with similar appearance in the same generation. Thus, biologists attempt to dilute the problem of changing individuals by generalizing from one to many, both across and within generations.

In a further generalization, statements about populations are enlarged both diachronically to species and higher taxa across their evolutionary lifetime and synchronically to a species across its range.[13] Taxonomists seem to believe implicitly that species and higher levels of taxonomic hierarchies[14] correspond to descendants of ancient individuals and constitute repetitions of individuals. Likewise, the physiologist enlarges on the biological individual to a view of species.[15] Thus these biologists see in a human being living fourscore and ten years a model for the aggregate of human beings, known as *Homo sapiens* (presently somewhat less than a million years old), surviving for a few million years before becoming extinct.

These biologists would argue, furthermore, that evolution studies are no different from studies in any other branch of biological research, whether the biologist is examining the behavior of molecules or cells *in vitro*, cells, organs or systems *in vivo*, island populations in the midst of an ocean, or everything in-between. Biological research is uniformly predicated on the belief that repetition is a fact of life. The concept of development presupposes repetition in embryos (larvae or fetuses); physiology presupposes vast chains of regulated repetition; endocrinology and immunology study repetitive cycles exhibiting memory; genetics is nothing if not the study of repetition among individuals (between parents and offspring); and neurobiology is devoted to periodic events, from activation potentials to bad habits. One should not single out evolutionary studies for criticism, therefore, if the assumptions made by biologists studying evolution are no different from the assumptions made by biologists laboring in all other subdisciplines.

One may agree that evolutionary studies share problems of repetition with all areas of biology, but working through these problems would seem preferable for scientific investigation than ignoring them. What is more, biology is not alone in having these problems, but others are confronting them. For a start, one may consult Gilles Deleuze's painstaking unpacking of repetition's problematic:

> Returning is being, but only the being of becoming. The eternal return does not bring back "the same," but returning constitutes the only Same of that which becomes. Returning is the becoming-identical of becoming itself. Returning is thus the only

> identity, but identity as a secondary power; the identity of difference, the identical which belongs to the different, or turns around the different. Such an identity, produced by difference, is determined as "repetition." Repetition in the eternal return, therefore, consists in conceiving the same on the basis of the different. However, this conception is no longer merely a theoretical representation: it carries out a practical selection among differences according to their capacity to produce — that is, to return or to pass the test of the eternal return.[16]

Or, more succinctly, "Difference inhabits repetition."[17]

The problematic of studying evolution, thus, is stagnation around notions of sameness and difference. Whatever alternative one pursues to evolution must come to grips with the sameness/difference dialectic.

Devolution is the alternative advanced here. From a devolutionary point of view, difference disappears when difference devolves from sameness and no longer gives rise to difference. Sameness disappears when sameness devolves from difference and no longer gives rise to sameness. Above all, in contrast to natural selection, which drives the evolution of sameness and difference through the competition of individuals, mechanisms of devolution, from assimilation to xenobiosis, operate through cooperation and interactions among members of life's boundlessly complex communities.

Virtual Evolution

Nowhere in the biological sciences has the computer had greater impact than in the acquisition of sequence data and sorting out its evolutionary implications (see Chapters 2 & 3). From beginning to end, the treatment of data demands simplifying assumptions known mainly to the cognoscente and frequently passed over without comment by the critic, unwilling to "take on the establishment."

Some problems are inherent in computation. Because DNA sequence data have only four possible character states (nucleotides), convergence onto the same state will occur frequently in rapidly evolving sequences, resulting in significant underestimation of genetic distance especially among distantly related strains. Corrections may be made for "multiple hits" at the same nucleotide (Jukes–Cantor; Kimura 2-parameter models), although using distance methods and averaging across sequences (BIOPAT) will not work where the position of a nucleotide influences the probability of substitution.[18] The alternative, adopted all too often, is to separate the analysis of rapidly and slowly diverging sequences and of synonymous (nucleotide substitutions not affecting an amino acid) and nonsynonymous (nucleotide substitutions affecting an amino acid). In this case, the problem is that as the size of sequences get smaller, the number of "false positives" gets larger.

Genes change at different rates. Mutation rates in lytic RNA viruses can be 300-fold higher than in DNA-based microbes, and retroviral genomes may evolve over a million times faster than nuclear genomes, if only during brief periods.[19] Faster rates of change do not, of course, mean faster evolution. Indeed "Muller's ratchet" drags down competitive fitness as slightly deleterious mutations accumulate. Furthermore, increased numbers of mutations per genome expected as changes accumulate would not result in an overall increase in rate of phenotypic change, since genomic redundancy would dampen opportunities for the introduction of novelty. Rates of recombination, reassortment and *de novo* gene acquisition would also, presumably, change during a species's evolutionary history.[20]

Beyond these computational problems, the molecular enterprise faces problems inherent in the assumption of molecular families and stands or falls upon their virtual relatedness. Computer-assisted searches have identified several dozen superfamilies of molecules accommodating as many as 1000 families[21] on the basis of clustered motifs and modules. These families raise all the issues raised by families in general, such as, "How does one establish 'paternity'?" Of course, convergence, or the evolution of similar traits from unrelated sources may confound efforts to trace gene families through similarity alone. For example, "Recently it has been shown that chlorophyll *b* has evolved independently at least three times in the prokaryotes ... requiring relatively little change from chlorophyll *a*."[22] On the other hand, "members of a family can diverge to a point where they have very few sequence identities and half their structures have different folds."[23]

Beyond "paternity," there is pedigree, with all its uncertainty:

> [U]sing sequence similarity alone to determine a transcript's identity can be misleading, as similarities may not necessarily impart the true relationships. Indeed, an inference of the evolutionary relationships of a gene family must be based on validated homology or common origin, just like the deduction of organismal phylogeny.[24]

The tracer of molecular pedigrees, thus, typically attempts to reconstructs evolutionary histories as phylogenetic trees (or maps) for putative paralogues and orthologues. Sequences in DNA (or RNA [edited] transcripts) or sequences in polypeptides provide the data, and a tree is constructed on the assumption that the sequences are linked by the fewest number of changes, or the most likely changes in a sequence (parsimony or maximum likelihood methods).

Despite these concerted effort to place genes into families, phylogenetic patterns are not ascertained for the vast majority of sequences. For example, regarding COGs (clusters of orthologous groups), only one-third can be placed into any phylogenetic context.

> This fact emphasizes the remarkable fluidity of genomes in evolution, revealed in spite of the fact that the analysis concentrated on ancient conserved families. Multiple solutions for the same important cellular function appear to be a rule rather than an exception, at least when phylogenetically distant species are considered.[25]

In other words, the prevailing prejudice that has led molecular systematists to a virtual reality of gene families and superfamilies defined by "computer-friendly" similarities has *not* led to the promised, "complete, consistent ... natural system of gene families from complete genomes."[26] The "systematic analysis of phylogenetic tree topology"[27] may be an exciting, scientific video game, but that is all.

GENES AND MUTATION: EVOLUTION'S FUEL

A good case can be made for mutations in genes powering evolution in the case of some retroviruses under epidemic conditions. A viral reproductive cycle on the order of minutes may simply respond to evolutionary pressures with entirely different dynamics than even the most actively reproducing prokaryotic or eukaryotic life form. For example,

> With RNA viruses' mutation frequencies averaging 10^{-3} to 10^{-4} substitutions per nucleotide..., we must consider whether the suppression of all but a limited set of variants is, in fact, not equivalent to a positive-selection event.[28]

In other instances, the faith biologists place in genes as the determinants of heredity and mutations as the fuel of evolution is misplaced, especially since genes and mutations are narrowly defined as classes of DNA sequences.

> Closeted as they are in their concept of bean-bag genetics and allelic bookkeeping, advocates of the Modern Synthesis still fail to realize that genes and genomes themselves evolve and, while so doing, change the ground rules for their own subsequent operations."[29]

The gene is, if nothing else, not simple. It is contradictory, exhibiting both static and dynamics properties, discriminating and mechanical behavior, contingent and determinist characteristics. The very idea of mutations fueling evolution flows from failing to distinguish their various facets.

Genes

What is the gene? Above all, genes are the instruments of hereditary continuity. Thus Watson and Crick sealed their own place in history by attaching their description of the secondary structure of DNA to a possible mechanism for its continuity.[30]

Arguments for continuity are most urgently pressed today by those insisting that the passage of nuclear genes through egg and sperm during sexual reproduction determines everything about the offspring for the rest of its life. This throwback to Weismann's germ plasm is as popular as it is ill-conceived. Nothing could be more absurd than attributing so much to so little (a stretch of nucleic acid or sequences of nitrogenous bases). Nevertheless, precisely this modern version of medieval determinism is touted by professionals in both biological and cultural studies as the great breakthrough in hereditary and evolution, development and physiology, social behavior and humanity's burden. The geneticist and genetic counselor who successfully predict things as trivial as the inheritance of blue eyes (as opposed to brown) in offspring have been elevated to the role of scientific soothsayer, capable of predicting everything precious and enduring in human nature. Nothing in life is that simple, and not even molecular biologists can justify adherence to these tenets of rampant geneticism.

Other arguments for continuity support the notion of gene duplication as the explanation for all similarities among DNA and polypeptide sequences. Duplicate genes within the same species are supposed to give rise to paralogues genes (homologues associated initially with functionally related molecules). Changes that occur in any gene, including paralogues, following descent of a separate species become orthologues.[31] Orthologous duplications, thus, are paralogous duplications in sibling species evolving from a common ancestor. Duplication may, of course, also follow the separation of incipient species in which case they are serial homologues.[32] One expects that "Phylogenetic trees of gene families from the same and different species can delineate the number of gene duplications and sometimes the relative timing of these events."[33] Both putative paralogous and orthologous duplications are used to trace gene families, paralogous families being within species and orthologous gene families being among species.

Beyond their structural identity as DNA and their arguable role in continuity, genes are difficult to define. They can be as small as a nucleotide, inasmuch as the loss or gain, or even the substitution of one nitrogenous base for another can sometimes upset the translation of an entire sequence. More often, however, the gene is somewhat larger, on the order of a codon/anticodon triplet of nitrogenous bases encoding an amino acid. A larger gene is the open reading frame (ORF) containing the sum of codons for an entire polypeptide chain, and a still larger gene is the transcribed region containing exons and introns. In the case of genes exhibiting alternative splicing, the series of exons and introns constitutes several genes or several "alternative splice sites" within one gene, and, in the case of genes subject to rearrangement, such as immunoglobulin genes, the alternative splice sites must be considered several genes or several "alternative rearrangement sites" containing alternative splice sites.

The sameness within any of these concepts of genes is not always sameness, however, and the differences are not always different. Indeed, a great deal of ambiguity is ordinarily offset by "gaps" in alignments, "synonymous substitutions" and sequence translations from DNA to polypeptide. The famous "wobble rules" of Francis Crick also redeemed ambiguity mainly in the third base of codons or the first base of anticodons. Originally, the genetic code of 64 triplets contained 61 codons specifying the 20 amino acids (the three remaining triplet sequences specified stop signals). Often, however, one tRNA recognized more than one codon. Crick contrived his solution to the problem of redundant recognition by lumping codons into "two-codon sets" (two codons specifying the same amino acid) and "four-codon family boxes" (four codons specifying the same amino acid). Placed into these receptacles of sameness, most codons were seen to require complementary nitrogenous bases in only the first two positions of anticodons,[34] and most of the remaining codons were only slightly more fastidious about pairing in their third base: uracil (U) pairing with either purine (adenine [A] or guanine [G]), and G pairing with either pyrimidine (cytosine [C] or U).

"Wobble pairing" allowed the recognition of more than one codon by a single species of tRNA. by accommodating pairing between G and U. Thus, U in the first position of an anticodon could pair with adenine (A) or G in the third position of codons, and G in the first position of an anticodon could pair with cytosine (C) or U in the third position of codons. C in the first position of the anticodon, however, paired only with G in the third position of a codon, and A in the first position of the anticodon paired only with U in the third position of an codon.

"Wobble" did not turn out to be the solution to the problem of redundant tRNAs, however. Another sort of ambiguity resided in the modification of the first base in anticodons. U in the first anticodon position of two-codon sets was found to be modified in most cases to 5-methyl-2-thiouridine (bearing a methyl group on the 5 position and having sulfur substituted for oxygen in the 2 position), or something similar. The modified U paired strongly with A and only weakly with G thus countering the effect of wobble. In addition, A in the first position of an anticodon was found to be modified to inosine (I), which paired with A, C or U. Furthermore, queuosine (Q), which was sometimes exchanged for G in tRNA, paired with C but more readily with U. Thus, modification increased redundancy, increased specificity, or altered specificity.[35]

Still more ambiguity keeps cropping up regarding the "gene": editing of mRNA, codon biases, and, conspicuously, deviations from the universal genetic code. Exceptions occur in the initiation or termination signals, especially in *Mycoplasma*, protozoa, ciliates, and mitochondria. For example, AUA is the codon for the amino acid methionine in yeast and most metazoan mitochondria except for echinoderms and platyhelminthes in which it is the codon for isoleucine. Mitochondria of protozoans and Cnidaria use the universal code except for the UGA stop codon. With the exception of plants and *Acrasiomycota* spp., UGA is the codon for the amino acid tryptophan and not a stop signal.[36] In higher plants and the Bryophyta (*Marchantia polymorpha*), UGA is a stop codon, and the codon for tryptophan is UGG. Presumably, in these cases, A in the first codon position is modified to I, adding to the mystery: Why do biologists advocate a determinist gene when genes in cells and their organelles are so prone to vacillation?

Mutations

"Gene change" is the standard definition of mutation, but many changes can affect the same gene. Genes may change "by substitution of bases; by deletion of bases; by recombination, domain shuffling, and gene conversion events that recopy sequences to produce hybrid genes; and by duplication and divergence events that generate families of related genes."[37] This list of changes can be extended easily: genes change by transformation, transposition (through the action of transposons) and transduction (through the action of viruses), etc. Moreover, some changes are transient, while others are stable; they may represent shifts within a sleeve of variability (microevolution) or drifts in one or another direction (under the influence of selection). And these possibilities are hardly exhaustive.

Mutations may occur at phenotypically unimportant sites (synonymous polymorphism) or at important sites (nonsynonymous polymorphism).[38] Changes in genes produce isoforms, which may be used to trace the gene's evolutionary history, while, phenotypically, mutations may be recognized as mutants. Mutations may be amorphs, when a mutant gene is thought to be utterly "turned off," hypermorphs, when a mutant gene is thought to be "working overtime," and hypomorphs, when a mutant gene is "lazy." Some mutations affect other genes, turning them "on" or "off," while other mutations "tune in," "speed up," and "slow down" other genes. Mutations of either sort are routinely attributed to changes in a DNA sequence, hence the justification for cloning genes and sequencing them. Single-base substitutions are, most commonly, transitions (changes from one purine to another or one pyrimidine to another) or, less frequently, transversions (changes from a purine to a pyrimidine or vice versa). Mutations demonstrated by sequencing to affect coding sequences are further identified by the consequence of the change: nonsense mutations introduce "Stop" codons prematurely in a coding sequence, thereby truncating a polypeptide and possibly destroying its physiological efficacy; "frame shift" mutations alter the sequence of codons and hence the likelihood of a "misbehaving" polypeptide; and missense mutations introduce the wrong amino acid at a site in a polypeptide.

The vast majority of sequences in the DNA of complex eukaryotes are noncoding, however. In mammals, such as human beings, of the 3 billion base pairs making up the genome, only 60 million would seem to be involved in coding sequences, given the presence of only about 60,000 genes and an average of 1000 nitrogenous bases per coding sequence. The remaining DNA consists of noncoding sequences such as introns, which interrupt the expressed sequences of coding regions, short tandem repeats and long tandem repeats. For example, in human beings, more than 5% of the total DNA consists of a so-called *Alu* repetitive sequence (present in more than half a million copies).

Presumably, changes in noncoding sequences are extremely more common than changes in coding sequences (if only because there are so many more) and one can assume a small likelihood for harmful effects following changes in individual base pairs of noncoding regions.[39] What is more, given "wobble" or ambiguity around the third base in codons, one can assume that as much as one third of the changes in coding regions would have no affect on heredity. All these changes would be "neutral" or "near neutral" (see Chapter 3) in the sense of neither adding to nor taking away from the host's inclusive fitness (likelihood of leaving offspring). Presumably, a molecular clock, stochastically ticking away individual changes in DNA's nitrogenous bases would produce mainly neutral or near-neutral mutations.

Other mutations, like other "genes," involve larger units than mere base pairs, and "It should be realized that evolution has been a process of the recombinational dynamics of modules, with the environment being permissive and the evolving genomes being proactive."[40] Transposition, or the insertion of new genetic elements into chromosomal sites, and excision, or the loss of elements from old sites, are other types of genetic change. Excision is not necessarily excision per se, since it can be mimicked by homologous recombination, for example, when a long terminal repeat in DNA (associated with a retrovirus-like element), leaves a copy behind as it "excises" as an extrachromosomal circle. Ectopic exchange, or crossing over between similar elements located at different chromosomal sites, resulting in chromosomal rearrangements (such as inversions), can also result in the deletion of DNA. DNA changes of this magnitude are likely to represent deleterious mutations, jeopardizing chromosomal physiology and lowering the fitness of individuals.[41]

DNA is not uniformly disposed to change, containing "hot spots" as well as "cold spots" and heterochromatin, or condensed chromatin, is not subject to the same changes as euchromatin, or uncondensed chromatin. For example, transposable elements are frequently inserted in telomeric and centromeric heterochromatin. Moreover,

> P elements [a type of mobile element] frequently altered the number of repeats in a tandem array of heterochromatic sequences, by inducing unequal gene conversion; suggest that a flux of transposable element insertions and excisions has the capacity to rapidly and nonrandomly modify heterochromatic sequences dispersed at multiple chromosomal sites; transposable elements maintain genomic heterochromatin in a state of dynamic equilibrium and drive its rapid evolution.[42]

The type of heritable transmission occurring in a species also plays a role in the type of genetic changes experienced by the species. For example, where hereditary transmission is uniparental, the genomes of mitochondria and chloroplasts do not contain transposable elements and, therefore, do not support transposition. Furthermore, different frequencies of hereditary exchange are likely to be reflected in different types of genetic exchange, for example, in bacteria with genetic exchange mediated by plasmids or transducing phage, and in self-fertilizing plants with extensive, nonrandom associations between alleles. Indeed, insertions may have fitness effects comparable to those of spontaneous mutations in sectors of the genome such as coding sequences (where genetically offsetting factors are absent or present in negligible frequencies).

Devolution

Alternatives on the Fringe of Genes and Mutations

Common sense predicts that if you wait long enough, every idea dropped at one time will reappear (the pendulum swings; what goes around comes around). Thus biology has witnessed a recycling of thought so that "Earlier ideas of genomes as constant, stable structures have been replaced with the realization that they are, in fact, dynamic and rather fluid entities."[43] In part, the explanation for this recycling phenomenon is that ideas are not so much lost in biology as ignored or left floating in the miasma of history. For example, the popularity of colinearity between DNA and polypeptides more or less deadened enthusiasm for noncolinearity due to RNA modification. A popular textbook may report that "These and other pairing relationships make the general point that *there are multiple ways to construct a set of tRNAs able to recognize all the 61 codons representing amino acids*" (emphasis in original), but the same book then goes on to warrant the point of view that "The predictions of wobble pairing accord very well with the observed abilities of almost all tRNAs."[44]

One is not surprised, therefore, to discover a no-man's-land between gene and mutation: they look the same to those in the trenches, but the generals know they are different. For example, different polypeptides are encoded by the same gene under a variety of circumstance: in RNA reassortment, genomic segments are exchanged among viruses with segmented genomes; in gene rearrangement, segments of DNA, ostensibly from different eukaryotic genes, are collected in coherent sequences of DNA; in alternative splicing, alternative introns are cut out of nascent transcripts as exons are spliced together. Mechanisms for these types of change are uncertain, but, together, they suggest the possibility of module or motif shuffling or intron-mediated (or intronic recombination-mediated) exon shuffling. The functional parts of proteins may recombine as a result of ligation of coding sequences or, in the event of replication, by a gene copying mechanism.

Repeated structural motifs or **modules** (see Chapter 2) are loosely defined as functional or structural units of proteins (modules generally limited to globular proteins or parts), while domains are their overarching coding regions in DNA (include several modules). Small (28 Å[45]), mechanically stable, regions of contiguous polypeptide segments of 10–40 amino acid residues in a compact conformation may be encoded by a single exon.[46] "Modular behavior" seems to be confined to proteins involved in advanced regulatory systems of higher eukaryotes, such as the immune system and blood clotting, and absent among metabolic enzymes.[47] Nevertheless, most eukaryotic genes contain multiple exons and may have experienced intronic recombination,[48] and proteins constituting the extracellular matrix (ECM) are virtual mosaic proteins of different motifs,[49] leaving open the possible role of motif shuffling in the evolution of new genes.[50] Are modules "the evolutionary building blocks of proteins"?[51]

Differential gene amplification, or the replication of specific genes during a period of cellular differentiation, occurs in the ribosomal organizing center of amphibian oocytes, in the *Drosophila* genes for egg membrane proteins, and vertebrate genes involved with methotrexate resistance in tumors, and plant mitochondrial genomes. The extra genes are frequently capable of replication as well as transcription, and, while most are shed into the cytoplasm and lost, those associated with methotrexate resistance, behaving like bacterial episomes, may become permanently incorporated in chromosomes (forming homogeneous staining regions known as HSRs).[52]

Two related hypotheses in particular have gained some attention among mainstream biologists: recombination and *de novo* origins of genes from preexisting sequences. Recombination is the end-to-end bonding of separate genetic molecules into one molecule. Recombination on the order of magnitude of whole genes is horizontal gene transfer if it occurs between cellular forms of life and lateral gene transfer if it occurs between endosymbionts and eukaryotic genomes. In the case of viruses, the simplest explanation for the incongruities among shared genes and their different combinations in viral genera "is that the progenitors of existing viral genera acquired such genes from more than one ancestral virus by recombination."[53] For example, the dsDNA baculoviruses and (–)ssRNA orthoacariviruses share sequences in virion glycoproteins. Recombination may also account for some of the large virogene families and for the presence of virogenes in hosts and host genes in viral genomes.

Viral/host recombination is also implicated in pathogenicity.[54] For example, in cytopathogenic forms of bovine viral diarrhea virus and hog cholera virus, flaviviruses are associated with a virally encoded 76 amino acid polypeptide resembling a truncated cellular ubiquitin (a widespread protein functioning as an intracellular protease cleavage signal and a chaperone in ribosome assembly) in all but two amino acids. Similarly, a gene in the insect virus *Autographa californica*, a baculovirus, shares 76% of its amino acids with insect ubiquitin.[55] The movement of *v*- and *c*-oncogenes to and from retroviruses and hosts is especially pernicious.

Mechanisms for the *de novo* origin of genes through the use of preexisting sequences rely on some sort of frame-shifting. "Overprinting," the use of unused or the reuse of used portions of an open reading frame (ORF), may explain some of the abundant instances of overlapping genes. In other cases, such as (+)ssRNA picornaviruses whose genomes "print" a single polyprotein, an original gene may have been copied (duplicated or obtained by recombination) before overprinting (frame-shifting), and the "palimpsest gene" and the original may then have evolved separately.

Biases in nucleotide usage have some interesting consequences for frame-shifts as an devolutionary mechanism. Biases in the third place, where they are most abundant, would switch to the first or second codon position of the new ORF where they would result in an overrepresentation of amino acids with the biased nucleotide in its new position.

> For example, the two codons for histidine and four of the six for arginine have cytosine in their first positions as do four for leucine; thus the overlapping protein of tymoviruses is a very basic and hydrophobic protein.... This suggests that genes produced by overprinting may first produce proteins whose usefulness results from their composition, and only later, after further evolution, will they acquire functions determined by their sequence and structure.[56]

Overlapping genes also occur in the genomes of cellular life forms, both prokaryotic and eukaryotic and in organellar and nuclear genomes of eukaryotes, for example, in yeast and their mitochondria. Novel genes may thus have devolved in cellular life as well as viral life by "overprinting" preexisting sequences of coding regions and noncoding regions. "Overprinting" may also account for similarities of the antisense (nonreading frame) strands of rRNA and tRNA genes with flanking regions and the "virtual" amino acid sequences of some eukaryotic genes.

The Special Case of Virogenes and the Viral Genomes

Viruses have been in the forefront of sequencing, since it started, but sequence data on virogenes and viral genomes are often squeezed into molds prepared from cellular genes and genomes. This is not to say that these data are wasted. On the contrary, predictions from sequences have been enormously useful for analyzing the structure of peptides and their function. Moreover, sequences have frequently supported the results of site-directed mutagenesis or side-group chemical modification. The point is that, when viewed on their own, viruses challenge monophyletic views of evolution as well as traditional views of viral taxonomy and systematics. For example, the viral RNA directed RNA polymerases (pols) do "not cluster according to plant, animal, or fungal cell categories [of hosts], although grand divisions according to prokaryotic and eukaryotic hosts were maintained."[57] Furthermore,

> [Phylogenetic] Trees were not constructed across the DNA and RNA pols because the two motifs suggested to be held in common by the two major pol types were considered too distant to reach any reasonable conclusion.
>
> ... Surprisingly, none of the exonuclease subsequences, taken individually or in combination and applied over all the clustering procedures listed, resulted in any interpretable tree and, often, stood in opposition to the required major classifications.[58]
>
> ... In any event, caution must be exercised in using motifs to reconstruct evolutionary development paths.[59]

With all the avenues open for the exchange of genetic material in and among viruses, viral phylogenies are likely to split and fuse as rhizomes rather than branch directionally as trees. However, proposals for ongoing viral evolution are often restatements of one or another principle of adaptation, and prospects for rhizomatic interactions are ignored most often. For example, the compacting of viral genomes, the presence of uniquely viral forms of differential transcription, overlapping genes, alternative splicing and differential expression are all advertised as adaptations to parasitism rather than consequences of gene recombination. The appearance of isolated "plant" genes within the genomes of negative single-stranded RNA ([–]ssRNA) plant viruses that otherwise resemble the genomes of animal viruses (bunyavirus, rhabdovirus), and extra genes (especially in nonstructural protein-encoding regions but also in structural genes) in positive single-stranded ([+]ssRNA [alphavirus]) animal viruses or in similar plant viruses (tobamovirus) are considered in terms of their facilitating some function of host–parasite interaction rather than their acquisition as "add-ons" to the viral genome.

Insects are cited as the culprit that extended viral parasites to plants or vice versa. Coevolution is invoked as the shaper and mover of genes in the shared host ranges of numerous plant and insect viruses and the plant-like segmentation of the genomes of Nodaviridae, a (+)ssRNA insect virus. The potential to mix genomes in infected plant cells is thought to be reflected in the segmented genomes of positive-stranded plant viruses.

In order to explain the failure of some viruses to respond to drugs in a consistent manner, molecular biologists frequently invoke "hypermutability" (frequent errors in proofreading–repair systems during replication by error-prone polymerases), "biased hypermutability" (errors in a "defective interfering" sequence changing adenine to guanine or uracil to cytosine), and "dislocation mutagenesis" (changes from guanine to adenine attributed to slippage of the primer relative to the template during reverse transcription) as a viral property. Similarly, gene exchange in potyviruses, "exon shuffling," recombination and gene copy are invoked to explain the capture of extra genes (comoviruses and picornaviruses) and other adaptations to plant and animal hosts.[60]

Alternatively, the possibilities for novel solutions to evolution's problems would seem to increase when viruses and retrotransposons are replicating thousands of times (or infinitely) faster than their cellular hosts. Under these conditions, recombination, gene capture, motif shuffling, the reassortment of sequence-modules, and horizontal gene transfer[61] might all be far more common mechanisms for genomic change in viruses than in cellular life forms.

Cross currents of genomic transfer are also readily imagined to have enriched DNA viruses and their cellular hosts, but subsequences of viral DNA directed and RNA directed polymerases (pols), proteinases, ribonuclease H II segments and helicases, also reveal a fluidity of virogenes that exceeds anything reasonably expected from mutation and accumulation in cellular genes. For example, DNA-directed DNA polymerases fall into four families which do not, however, cluster on a phylogenetic tree constructed by group-averaging: One family (A or I), including two domains with 3′–5′ exonuclease and polymerase activity, is found among bacteria, bacteriophage and yeast mitochondria; another family (B), with nucleotide-editing capability and 5′–3′ exonuclease activity, is found in bacteria, bacteriophages, viruses, yeast, mammals and plasmids; a third family (C) comprises the α-subunit of pol IIIs from bacteria; and a fourth family (X) consists of mammalian pol βs and terminal transferases.[62] Similar subsequences are also found in DNA-directed RNA polymerases, thymidine kinases (TK) of poxviruses, and in African swine fever virus and T4 phage.

New strains of viruses are said to come into existence virtually all the time (influenza B virus accounting for yearly flu epidemics) via high mutation rates, but in the life of pathogens, in the interpandemic years, antigenic drift, or serial changes in protective antigens coupled to the susceptibility of hosts accounts only for regional epidemics. Pandemics, such as those occurring every 10 to 30 years, in the case of influenza, follow an antigenic shift, that is, a gene reassortment in a reservoir species or the exchange of genes in the reservoir species with those in the terminal species resulting in a new strain able to reinfect a vulnerable, terminal host. Such a shift in influenza A hemagglutinin (from H2 to H3) of birds produced influenza A/Hong Kong/1968, responsible for the '68 pandemic in human beings.[63]

Viruses "jumping" between species are said to represent the origin of new viral species (HIV) when transmission among members of the new host species allows the virus to persist in the absence of the original host or reservoir species. With good reason, communicable disease fighters around the world are terrified by the possibility of pandemics of Marburg and Ebola viruses from monkey and arenaviruses from rodents.[64]

> Once a new virus emerges through (a) radical genetic change in another virus; (b) a change that allows the virus to cross a species barrier, such as the new morbillivurus infecting seals; (c) sympatric or allopatric speciation of old virus types; or (d) rogue oncogenes from a host, then the ability of a new virus to invade its host will depend [only] on the initial reproductive rate.[65]

Ultimately, however, and provided the host population is large enough to contain sufficient variability, a new equilibrium will come about through the seesaw of adaptation and susceptibility, of attack and counterattack characteristic of the pathogenic viruses and their host's way of life. Myxoma virus, for example, was introduced with devastating effect against wild European rabbits in Australia, but quickly lost virulence as it coevolved with genetic immunity in the rabbits.

HEREDITY: EVOLUTION'S TRANSMISSION

The transmission in biology's horse-less carriage connects the engine, natural selection, with the live axle of evolutionary change. This biological gear box furnishes the continuity that sustains biological change, the smooth transitions between the slow and powerful and between forward and reverse.

Genetic Continuity

Heredity is broadly understood as the consequence of chromosomal transfer, also called "vertical gene transfer." Heredity thus takes place between a parent cell and its offspring cells produced by cell division, by mitosis in eukaryotes, and during asexual reproduction. Genomic transfer also occurs from parent to offspring during sexual reproduction (and parthenogenesis) where it involves meiosis, the differentiation of gametes (spermatozoa and eggs), fertilization and embryonic development.

Heredity is highly homogenized by the concept of vertical gene transfer. For example, sex is blended with recombination in DNA viruses, copy-choice in RNA viruses, and segment reapportionment in segmented viruses. Similarly, asexual reproduction is combined with the reproduction of single-component viruses. The consequent debate over sex in viruses takes place notwithstanding the absence in viruses of sex cells (of cells at all), of eggs and spermatozoa, of meiosis and fertilization, of development into a diploid organism.[66]

Moreover, the extraordinary emphasis on chromosomal genes in heredity is not justified, because a lot more than genes is passed on to "offspring." For example, cytoplasm and mitochondria are passed on when cells divide, and fertilized eggs contain a variety of materials that take part in embryonic development. Moreover, many extraordinary powers rest in environments, and biological power is exercised by numerous features of reproductive structures (shells and uterus, extraembryonic membranes and placenta).[67]

Devolution

Heredity is the cusp over which one generation of cells or individuals collapses into the next. It is a line of escape between plateaus (attractor states?), and like every other biological transformation, heredity is not linear: its samenesses are not necessarily the same sameness and its differences are not necessarily the same difference throughout heredity's distribution space.

Discontinuous Transmission

The principle alternative to "vertical" gene transfer is "horizontal," "lateral" or "intergene transfer." These alternatives are invoked in the mainstream literature only when vertical gene transfer is confounded by incongruities in phylogenetic trees built on parsimony or maximum likelihood. Thus, various authors suggest that "intergene transfer may be responsible for the ambiguity in the tree,"[68] and "unique patterns ... might have evolved through horizontal transfer of typical eukaryotic genes into bacterial genomes."[69] In William Martin's words, "gene transfer throws a monkey wrench into the phylogenetics of early evolution,"[70] or, as Rudolf Raff explains,

> inference methods will produce a tree even if the version of a gene being sequenced for one of the included species was introduced to its genome by horizontal transfer from another species. That sequence would violate the model of a treelike evolutionary process for sequence evolution, but the tree-building procedure would incorporate it into a tree without blinking.[71]

Horizontally transferred DNA is probably not really rare, however. Beyond reliance on incongruities in phylogenetic maps, horizontally transferred DNA identified from sequence characteristics, such as base ratios, seems abundant. If horizontally transferred DNA is only gradually assimilated into the host's genome, through a process called "amelioration," the DNA only gradually acquires a base composition roughly that of the host. Based on this assumption, Jeffrey Lawrence and Howard Ochman estimate that the entire *Escherichia coli* chromosome contains more than 600 kb of horizontally transferred, protein-coding DNA which has accumulated at a rate of 31 kb per million years. This amount and rate suggests that horizontally transferred DNA introduces as much variation into *E. coli*'s DNA as point mutation.[72]

Sightings of horizontal gene transfer should, thus, be more common and should provoke greater interest in the mechanisms of transfer. For example, viruses as gene vectors are not confined to biotechnology. Arborviruses, in particular, appear to cross species barriers with ease, possibly carrying nuclear genes across kingdom barriers. In the case of the mite, *Proctolaelaps regalis*, carrying genes among *Drosophila* species,[73] a mechanical vectoring is most likely, but other mechanisms are not ruled out: the integration of fly DNA into the mite genome and secondary roles for viruses or bacteria.

Undoubtedly the most dramatic and threatening horizontal gene transfer has occurred among bacterial pathogens: the resistance factor.[74] In this case, the movement of genes would seem entirely at the behest of the donating and recipient species, but in another disastrous transfer, human beings may yet be the culprit:

> Ciba-Geigy has designed soybeans to be resistant to their Atrazine herbicide, while Dupont and Monsanto are developing crop plants with tolerance to their herbicides.... The problems associated with this strategy are so obvious[:] ... The herbicide- and pesticide-resistant genes introduced into crop plants are unlikely to stay in these species, and will be transferred by viruses, bacteria, and fungi to other species so that "weeds" will themselves become tolerant.[75]

Reticulate Evolution, Hybridization and Larval Transfer

Entire genomes are also transferred through biological processes other than reproduction, although these are not generally recognized as normal devices of heredity. Reticulate evolution is the mixing of genes introduced by recurrent hybridization among species and their sorting out, or introgression, by gene displacement. The problem of tracing sequences through reticulate evolution are the same as those of tracing sequences through endosymbiotic origins (see Chapter 4) and horizontal or lateral gene transfer, namely, the methods commonly employed for drawing phylogenetic inferences from sequence data do not accommodate readily to polyphyletism.

Reticulate evolution has, nevertheless, been followed through karyotypic analysis (quantitative and qualitative microscopic analysis of chromosomes in dividing cells) and found to be amazingly widespread in plants, probably accounting for the genotype and presumably the evolution of most monocotyledonous angiosperm. Reticulate evolution is rarely reported in animals, however, although the many instances of it documented by Michael Arnold suggest that hybridization in animals may not be as rare as commonly assumed. A lot of demystification will have to take place before the scope of hybridization and reticulate evolution is generally appreciated, and many biologists will be surprised to know that in some plants and animals

> hybrid classes were more fit than at least one of the parents.... This challenges the dogma that hybrids are uniformly less fit than their parents.... The surprise for some will come from the observation that hybrids can possess the highest fitness or be equivalent to the most fit class(es).[76]

Donald Williamson has also cast his nets in the troubled waters of hybridization and has come up with his theory of larval transfer. His thesis rests on the well-known observation that larvae of many animals live in different habitats than the adults that produced them and, while bearing scant similarity to these adults, resemble other larvae living in these habitats. Ordinarily, these observations are explained in a Haeckelian manner, assuming that the similarity among larvae is a legacy from common ancestors, the dissimilarity among adults is due to the addition of the adult stage at the end of development. Williamson, however, asserts that "these clues [from larval morphology] have been misinterpreted and that widely accepted groupings and trees are totally misleading."[77] As an alternative, he contemplates xenohybridization resulting in a reticulate formation followed by introgression.

Evidence for or against Williamson's theory is difficult to muster. However, several characteristics displayed by invertebrates exhibiting indirect development (having a larval stage) are compatible with Williamson's theory. For example, the notion of separate embryonic/larval genomes might be reflected in larvae that sequester cells destined for adult tissue and only produce adults upon metamorphosis and the complete destruction of embryonic/larval tissue. Well known since the 19th century and frequently documented,[78] this separation of embryonic/larval and "set-aside" adult cells only rarely receives theoretical attention.[79] In addition, some of the maternal genes influencing pattern formation in embryos may have no counterparts operating in adults. The widespread "homologies" said to exist among some of these genes in embryos[80] may ultimately prove the anvil for hammering out Williamson's ideas.

NATURAL SELECTION: EVOLUTION'S ENGINE

Natural selection is ordinarily supposed to power evolution by sifting variations in life's forms through an environmental filter: The filtrate evolves; the sediment expires. This maxim has been the received version of natural selection since the days of the "new (modern) synthesis." It requires no introduction or exegesis. It has been engraved in the consciousness of virtually everyone educated in the public (US) or national (EU and post-colonial) schools and most private (nonreligious) schools for the last fifty years. Natural selection is hailed as the epitome of reason's triumph over religious superstition.

Beyond an arguable consistency in data, natural selection is supported by the laws of genetics, the formulas of the modern synthesis, and the matrices of computational cladistics. In general, natural selection governs the emergence of difference from sameness in a common sensible way: by attributing difference to changes imposed on sameness. Two rules are fundamental: 1) life is continuous between generations; 2) life is capable of incremental, hereditary change over generations. Natural selection further demands that: 1) genes change; 2) changed genes accumulate in genomes (frequently as orthologous and paralogous gene families). Of course, gene changes and their accumulation are supposed to add up to qualitative changes in traits or phenotype, even if the calculus of qualitative change does not compute. Furthermore, phylogenies traced through genetic lineages are thought superior to phylogenies traced through any other sort of data. Indeed, molecular systematists define phylogenies as "the history of genetic connections through evolutionary time,"[81] even if phylogenies drawn for different sequences are incongruous.

In the less common sensible version of the modern cladist, natural selection is a purely power-driven mechanism of monadic descent. Genes retain their identity as they descend, although they can accumulate relatives. This convention is necessary, in order to make natural selection "computer friendly" (which is not to say "friendly" to data or to alternative hypotheses). Cladistics can treat natural selection as strictly arborizing evolution: it can branch but it cannot blend. The alternative rhizomatic pattern is ignored simply because it is too difficult for present computers and their programmers to contemplate.

The Roots of Natural Selection

Charles Darwin is credited with first conceiving of natural selection, notwithstanding Alfred Russel Wallace's simultaneous publication on the subject. By and large, Darwin's concept reflects 17th century nature philosophy stuck on a Leibnizian nature progressing in infinitely small, virtually continuous steps. Life was not supposed to proceed in great leaps and certainly not through discontinuities, and anything that smacked of leaps, even leaps between similar living forms, or gaps, even gaps between very different living forms, was heresy. Indeed, Lamarck, the great iconoclast and early 19th century revolutionary, was "driven from the temple," precisely because he took pincers to the Great Chain of Being and chopped away at continuity.[82]

Darwin realized that evolution as such was diametrically opposed to continuity, but, unless he succeeded in merging evolution with continuity, he would be faced with an insuperable obstacle. His version of natural selection undermined that obstacle even if it did not surmount it. Darwin imagined organisms producing minuscule variations in their progeny which were selected by the environment leading to quantitative changes. The accumulation of these changes then led to qualitative changes through some sort of *gestalt* switch.[83]

Modern biologists consider life's spatial discontinuities (individual members of hierarchically ordered classes) intuitively obvious. Even genes are supposed to be discontinuous (units of segregation) despite their presence in linkage groups. Life is not organized in an infinitesimally continuous array of forms, whether one speaks of the biological individual or jumps to the species, family, order, the class, the phylum or division, "*Bauplan*," kingdom or domain. Discontinuity, rather than continuity, describes how life is organized, but contemporary biologists, like their recent ancestors, continue to place natural selection at the antipole of continuity.

Michael Behe, for one, is preoccupied with the same conundrum that troubled Charles Bonnet (1720–1793) two centuries ago:

> Since natural selection can only choose systems that are already working, then if a biological system cannot be produced gradually it would have to arise as an integrated unit, in one fell swoop, for natural selection to have anything to act on.[84]

Behe goes on to argue that

> one couldn't take specialized parts of other complex systems (such as the spring from a grandfather clock) and use them

> directly as specialized parts of a second irreducible system (like a mousetrap) unless the parts were first extensively modified. Analogous parts playing other roles in other systems cannot relieve the irreducible complexity of a new system.... [T]he system can't be put together piecemeal from either new or secondhand parts.[85]

Ultimately, Behe elects an argument of design to explain the emergence of differences among living things,[86] but design's answer to continuity is probably not an improvement over natural selection's. Indeed, the very existence of imperfections in life's structures (such as the backward architecture of the vertebrate eye) is frequently taken as enough evidence to overturn any argument of design, especially with the added caveat, "To uncover the mechanisms by which novelty emerges from continued selection, we must turn to experimental studies."[87] Furthermore, Behe's reliance on argument by analogy leads him into conflict with Elliott Sober's[88] general criticism of design arguments.

What then is the modern dodge to the problem of continuity? Still neoDarwinian natural selection: The frequency of genes in a population will increase for a variety of reasons, and one of these provides an adaptive or selective advantage to particular genes or the organisms possessing them. Computer driven models of phylogenetic trees legitimize neoDarwinian natural selection, but several assumptions are required to prop it up. First, it works best with duplicated genes. In general, increasing the rate of transcription is supposed to provide gene duplication with an initial advantage. Second, added genes are assumed to be or to become somehow or somewhat different from the originals, for example, by any and all the mechanisms of mutation. The mere availability of duplicates for change is cited as a second (if not contradictory) adaptive advantage for gene duplication.[89] Changed genes are also expected to accumulate through some sort of displacement activity or competition. Finally, a duplicate gene will perform as an exaptation (or preadaptation), a "modification of a structure originally adapted for one function to serve another dissimilar function."[90] Were the new gene to influence the organism's phenotype in such a way as to promote fitness, natural selection will have done its job and served its role in evolution.

Devolution and the Problems of Natural Selection

Natural selection has many problems. In the first place, it is problematic because it is radically relative. It may occur imperceptibly slow, as in Darwin's natural selection,[91] or rapid, as in Eldredge's and Gould's punctuated equilibrium,[92] produce large changes, as in adaptation, or small changes, as in microevolution. Thus, one finds in the literature contradictory positions such as: "*It seems inescapable to us that the major conclusions from the fossil record are the very rapid appearance of functional morphologies inexorably followed by a period that is basically "morphological fiddling at the edges*" (emphasis original).[93] The problem is that lumping everything together under the aegis of natural selection may yield something too attenuated to support a scientific hypotheses. Possibly natural selection could account for "equilibria," or stabilizing selection, but it could not, at the same time, account for "punctuations."

Another great challenge to natural selection comes from Sewell Wright's concept of drift,[94] although it is generally "cleaned up" to represent the genetic setting for future natural selection. Drift is a natural change in gene frequency which, were it to occur experimentally would be attributed to "sampling error," the "luck of the draw." Drift occurs when small populations are somehow taken from large populations, and the small population's "gene pool," or census of genes, is not representative of the large population's "gene pool." The point is that this change in genetic composition of the population, the equivalent to microevolution, occurs without the filtering effect of the environment or the feedback of natural selection.

Drift may even be implicated in the explanation for one of the more enigmatic discoveries of the presequencing days of genetics: polymorphic alleles, alleles occurring in such great variation that individuals are more likely to have a different set of these alleles than the same set.[95] An example of polymorphism occurs at the major histocompatibility complex (MHC) loci (*H2* in the mouse and *HLA* in human beings) containing a multigene family some of which members are essential for the immune response to foreign antigens.[96] Several possibilities, in addition to drift, have been proposed to explain MHC polymorphism, some (1–3) would seem to operate without natural selection while others (4–7) imply an adaptive advantage and hence natural selection:

> (1) a high mutation rate ... (2) neutral mutations ... (3) gene conversion ... (4) maternal–fetal incompatibility [heterozygote advantage, but note: chickens have a high degree of MHC polymorphism even though they have no maternal–fetal interaction] ... (5) mating preference ... (6) overdominant selection [or heterozygote advantage: different MHC molecules bind different foreign antigens: therefore a heterozygote having two different MHC molecules would be more resistant to infectious

> diseases than a homozygote with only one type of MHC molecule] ... (7) frequency-dependent selection [host individuals carrying a recently arisen mutant allele have a selective advantage because pathogens will not have had time to evolve the ability to infect host cells carrying a new mutant antigen].[97]

How does drift enter the calculus of polymorphism at the MHC loci? Mutation would not seem to have the capability of adding alleles indefinitely if only because such an increase would inevitably impose a prohibitive burden on the organism. Instead, the number of polymorphic loci would seem somehow to be limited. Judging from the degree of divergence, some genes (classical) would seem older and more stable than others (nonclassical), the latter, possibly, arising independently in separate lineages. If classical loci duplicated and one of the duplicates became nonclassical as a result of mutation, drift might account for their presence in some individuals and their absence in others without positing any adaptive advantage to any combination of alleles.

Ever since data of molecular systematics began to accumulate, two additional problems have confronted natural selection: 1) Genes appear to change at different rates, and hence changes in different genes accumulate at different rates;[98] 2) some genes in different organisms are more nearly similar to each other than the organisms bearing these genes are similar to each other.[99] Clearly, a separate "genetic systematics" (a "gene genealogy") is required for genes as opposed to their organisms's phylogenetic systematics. For example, sequence data for modern, mosaic proteins of the extracellular matrix suggests that the separation of the human/mouse/frog and the human/mouse/chick lineages took place about 500 million years ago, near the beginning of the Cambrian, and long before any common ancestors of these groups existed![100]

Natural selection does not ordinarily accept as major instruments of evolution all sorts of phenomena implied by sequence data: transposition, transduction, transformation, horizontal gene transfer, hybridization,[101] reticulated evolution and larval transfer especially among otherwise unrelated organisms. Furthermore, because only monophyletic descent is permitted by natural selection, it requires devices for the recognition of "self" and the rejection of "nonself" as a prerequisite to fecundity. In its modern, synthetic version, natural selection also rejects the possibility of "miscible" hereditary material as opposed to or even in addition to the "immiscible" gene. Thus, nongenetic instruments of heredity such as maternal inheritance, cortical inheritance, and even otherwise genetic imprinting fall outside the province of natural selection.

What is more, natural selection is problematic because it does not explain the presence of vast amounts of noncoding and nongenetic DNA in eukaryotic genomes. This is the DNA of the "selfish" and "junk" varieties, but it is also the DNA of introns and the repeats so meticulously mapped by the Human Genome Project. All this "nongenetic" DNA seems to have accumulated without the permission or participation of natural selection.

"Hitchhiking" provides another route for introducing genes into a genome without natural selection. Hitchhiking occurs when a "hitchhiker" gene is tightly linked to an advantageous mutation (the driver). The hitchhiker may become fixed in the genome, while alternative alleles are eliminated, not by selection for it but by selection for the "driver." For example, the transposable elements, with non-long terminal-repeat retrotransposons, known as *mariner* elements, seem to be hitchhikers that have resided within the 28S rRNA genes of many insects and other animals for 500 million years.[102]

Finally, according to one scenario, genes themselves could not have originated had protogenes followed the prescriptions of natural selection. Genes arose when protogenes violated the rules.[103] Instead of the gradual accumulation of nucleotides into an effective sequence, "the present gene was created by a mosaic gene."[104] In this scenario, small polypeptides with some rudimentary enzymatic activity were somehow encoded by polynucleotide chains which were, however, larger than the encoded regions. These coding regions are today's exons or "minigenes," and correspond to domains that encode polypeptides of today's modules (not necessarily domains themselves), while the noncoding regions are today's introns. By working together, in contrast to competing, the original polypeptides (motifs) achieved some sort of advantage, either greater activity, greater stability or both, and this advantage was somehow reflected backward so that the polynucleic acids coding the polypeptide modules banded together permanently, thereby forming genes. Natural selection predicts that minigenes might later have sorted themselves out when one or another combination of polypeptides had a competitive (adaptive) advantage over other combinations. Natural selection does not predict the polymerization of minigenes, however, even if modules combined secondarily, since no competitive edge could be achieved through noncoding, intervening sequences. The widespread commitment among geneticists to evolution by single amino acid substitutions is also challenged by this "minigene" scenario, since much larger units of change, minigenes and modules, are projected as the lead players. The relative unimportance of substitutions in amino acids for many functions of proteins, especially those outside their active sites that do not alter secondary structure, suggest that modular changes are not only the big ones but the significant ones.

REPRISE

Evolution of Sameness and Difference was intended to be an analysis of sameness and difference in modern biology. The book moved through the problems, methods and solutions in the areas of recombinant DNA technology and sequencing, genomic mapping and molecular systematics, phylogenetic mapping and life's emergence, and, in each area, great differences in life and the materials of living things were seen to be stuffed into little boxes of sameness. In the end, evolution emerged not so much as a box of sameness, but as a black hole from which no differences were able to escape (hence the many claims for evolution being biology's organizing principle).

What is certain is that biologists have not come up with an alternative to evolution as required by the hypothetical-deductive method of science. As a minimum, an alternative would sharpen what is meant by evolution and may even suggest ways of testing evolutionary hypotheses. The Popperian challenge to all those advocating evolution as the explanation for life (sameness) in all its forms (differences) is to formulate an alternative to evolution.

Some alternatives to evolution are well rehearsed in biology's modern literature, although they are frequently relegated to the dust bin of Lamarckian hypotheses.[105] Indeed, two alternatives have recently risen on the horizon of legitimate scientific ideas. These ideas, determinist chaos[106] and complexity theory[107] are related and may ultimately illuminate each other,[108] but the existence of chaos in nature is still controversial[109] and complexity theory is largely descriptive, only hovering on the edge of explanatory mechanism.

A third alternative to evolution, devolution, is sketched out here. It is compounded from many previously published, if not well circulated ideas. Instead of ascendant evolution producing difference through intermediates, descendent devolution assimilates difference in chimeras and their fragmented products; in contrast to evolving samenesses stagnating in homology (biological inertia), devolving samenesses accumulate from redundant differences; whereas evolving differences emerge from mutations, devolving differences reassort into widespread samenesses.

Devolution, itself evolved from dissatisfaction with some planks in the evolutionary platform: Did life originate once, or might life have originated several times? Did samenesses and differences in living forms emerge entirely from the separation of life's branches, or might samenesses and differences have resulted from the joining of branches as well? Evolution and devolution might seem to approach these questions from opposite poles, but they are on different planes.

What is it that biologists, or anyone for that matter, would like to know about life? Is it an explanation for the boxes of sameness and their content of differences studied by biologists? These will, in all likelihood, be explained by evolution (guarded tentativeness is appropriate if only because some samenesses and differences may not have evolved in the usual sense of the word). Molecular systematics would seem to be as robust a discipline as science has produced and will continue to be as successful in the future as it has been in its short past at explaining biology's data. But is there more to life than these samenesses and differences, and does evolution hold out the possibility of explaining what remains beyond sameness and difference?

For me, lifting the veil of sameness and difference has exposed previously unimagined dimensions of life and content in biology. Free, life has reached new plateaus of cooperation instead of competition, of interaction instead of isolation, of convergence instead of divergence. Removing samenesses and differences has also unburdened biology of culturally laden metaphors, of its rhetorical legacy, and constraining mind-set. Evolutionary theory cannot come abreast of unbound life because evolutionary theory cannot escape sameness and difference. If devolution is to be an alternative, it must nurture life's possibilities and preserve biology's liberation.

ENDNOTES

CHAPTER ONE

1. Appel, 1987.
2. Deleuze, p. 50, 1994.
3. Deleuze, p. 126, 1994.
4. Deleuze explains (p. 120, 1994): "Resemblance is in any case an effect, a functional product, an external result — an illusion which appears once the agent arrogates to itself an identity that it lacked."
5. Aristotle, Book I: 815^{a}10, 1984.
6. See Rogers, 1996.
7. This sentiment has many sources. I take mine from Lampadeusa's, *Il Gatopardo*.
8. Various sources might be cited, including Shostak, 1998.
9. See Appel, 1987.
10. Appel, 1987.
11. Bateson would later become the first professor of genetics at Cambridge.
12. Boveri, 1964.
13. Desmond & Moore, 1991.
14. Lamarck, (1809) 1984.
15. An exception to the rule, Richard Dawkins frequently tries to define progress. For example, see Dawkins, 1997.
16. For example, Raff, 1996.
17. Baer, (1827) 1966; (1828) 1967.
18. See Darwin's autobiography in Darwin (1902) 1995.
19. Lamarck, (1809) 1984.
20. See Geison, 1995.
21. See Gould, 1977.
22. Olby, 1985.
23. Nei, 1987.

24. For a naughty, brief review see Rieppel, 1994.
25. See Mayr, 1982, 1988.
26. Mayr, 1994.
27. Maynard Smith, 1970.
28. Yomo et al., p. 261, 1995, further explains: "The properties of an enzyme are thought to have been improved and optimized based on the need of the living cell during evolution."
29. Moore, 1983.
30. Moore, 1986,
31. Which they shared in 1962 with Maurice Wilkins.
32. Codified, in part, by one of the principles: Watson, 1969. Also see Judson, 1979; Olby, 1994
33. Gunther Stent (p. 204, 1971) considered the statement "coy." Judson (p. 639, 1979), reported that "Crick says the statement was a claim to priority."
34. Watson & Crick, p. 737, 1953.
35. Watson, 1969.
36. Crick, 1958.
37. Medawar, p. 272, 1982.
38. "Genome" was coined in 1920 by the botanist H. Winkler, to designate the set of genes present in a haploid chromosome set, either that of a gametophyte or of gametes. Molecular biologist generally use the word to mean the sum total of genes and intergenic sequences in the chromosomes of a species. See Bernardi, 1996.
39. Shostak, 1998.
40. Senapathy, 1994.
41. According to the "protein-only" hypothesis, transmissible mammalian spongiform encephalopathies are caused by infectious protein particles known as prions (PrPSc) which are isoforms of the naturally occurring proteins (PrPC).
42. Bernard, 1957.
43. Lamarck, p. 112, (1809) 1984.
44. Lamarck, p. 113, (1809) 1984.
45. Lamarck, p. 108, (1809) 1984.
46. Lamarck, p. 108, (1809) 1984.
47. Lamarck, p. 58, (1809) 1984.
48. Lamarck, p. 70, (1809) 1984.
49. See Jablonka & Lamb, 1995.
50. Hershey, p. 100, 1966.
51. See Rogers, 1996. Hugh Montgomery, later commander-in-chief of the Royalist army in Ulster, experienced no sensation when his heart was palpated through a closed chest wound. Politically, the demotion of the heart because of insentience accompanied the promotion of blood to the rank of monarch,

and the elevation of the female element above the male element. Thus, with the English Revolution, not only had the king become unnecessary for government, but the male had become unnecessary for reproduction. Elsewhere, the institution of *Droit du Seigneur*, the customary right of the lord of the manor to inseminate any of his female servants on her wedding night, was severely under strain.

52. Schizophrenia genes are especially evanescent, fluttering between chromosome 5, 6, 9, 20 and 22. For schizophrenia see Marshall, 1995a; Restak, especially pp. 142–143, 1991, and for manic-depression genes see Morell, 1996a. Also see Steen, 1996 for a popular review of genes and behavior.
53. Cardon et al., 1994; Warren et al., 1995.
54. Barinaga, 1991; LeVay, 1991; Hamer et al., 1993; Pool, 1993; Marshall, 1995b.
55. For example, the successful synthesis of such biological building blocks as adenine from hydrogen cyanide and ammonia by experimentalists attempting to mimic early-Earth conditions is widely hailed as support for the abiotic origins of life, yet this synthesis bears no similarity whatsoever to cellular synthesis, and the ingredients employed would poison the multiple enzymes operating in cellular synthesis.
56. See Atkins, 1993.
57. See, e.g., Lovelock, 1990, 1996; Margulis, 1990b, 1996. Also see Shostak, 1998.
58. See Behe, 1996.
59. Aristotle, 1984.
60. Galen's term, conception, is incorrectly equated with fertilization, since it refers to fetal reception in the uterus. Fertilization takes place in a human uterine tube, and the blastocyst, produced by cleavage arrives in the uterus a week later. This preembryo (i.e., lacking an embryo, proper) then hatches from its egg membranes and begins implantation. A definitive "primitive streak," indicative of the formation of an embryo (although not the embryo itself), forms only after another week, or two weeks after fertilization.
61. Adelmann, 1942.
62. Harvey, p. 338, (1651) 1952.
63. "Epigenesis" was probably coined by William Harvey who attributes the concept to Aristotle.
64. Short, 1977.
65. Following the model of his teacher Fabricius, Harvey had set out to prove that rooster's semen could not enter the hen's uterus. Harvey could not so much as force air into the hen's uterus despite his utmost exertion at blowing into its external orifice.
66. Gray, 1970.
67. Baer (1827, 1828), 1966; 1967.

68. Translation quoted by Moore, p. 451, 1987.
69. The zoa of sperm (liquid seed) were named "spermatozoa" (zoa of sperm) in the 19th century by Karl Ernst von Baer.
70. Adelmann, 1966.
71. "*De minimis no curat les.*" Singer, p. 287, 1959.
72. Malpighi (1675), 1968.
73. According to Joseph Needham, p. 191, 1959, Bonnet declared that the overwhelming endorsement of his views by other scientists represented "one of the greatest triumphs of rational over sensual conviction."
74. Bonnet, (1762) 1985.
75. Driesch, 1892.
76. Spemann received the Nobel Prize in Physiology or Medicine in 1935 for the discovery of embryonic induction.
77. See Snow et al., 1981.
78. Wilmut et al., 1997.
79. DiBerardino, 1997.
80. See, e.g., Oppenheim, 1982; Oppenheimer, 1967.
81. Lamarck, p. 230, (1809) 1984.
82. Bichat is Michele Foucault's (p. 128–131, 1973) candidate for the parent of histology.
83. This divide is only recently breaking down with the discovery that "unicellular" myxozoans are closely related (by the criterion of the ribosomal RNA sequences) to multicellular cnidarians. See Smothers et al., 1994.
84. In what would seem an act of national chauvinism, Ernst Haeckel (p. 155, 1905) credits Hugo Mohl with designating "protoplasm" in 1864.
85. Schleiden was a botanist who had earlier suggested that the development of plant tissue depended on nucleated cells.
86. Schwann, 1839, quoted from Libby, p. 259, 1922
87. Golgi and Ramón y Cajal shared the Nobel Prize in Physiology or Medicine in 1906. For a description of the Golgi's unceremonious behavior at his Nobel lecture see Ramón y Cajal, 1989.
88. See Goldschmidt, 1956, 1960.
89. Margulis, 1991b; Margulis & Sagan, 1986.
90. See Rogers, 1996.
91. Bernard, 1957.
92. Even Bernard's close friend, Louis Pasteur had a posthumous tiff with Bernard. See Geison, 1995.
93. Ruth Hubbard (pp. 70–71; 1990) discusses contributions by all the "fathers" of genetics.
94. Sapp, 1994, p. 45, traces the ideas elaborated by Weismann back to the gemmules of Darwin, which de Vries called pangenes. Weismann (1882),

published an early version of his concept of germ plasm and developed it more completely ([1892] translated to English in 1912).

95. In a remarkable irony of biological history, the name, "germ plasm" is retained today only for "germ-line determinants" (also known as pole plasm in insects such as *Drosophila*), putative determinants of sex cells which are passed maternally through oocytes to embryos and concentrated during cleavage in primordial germ cells. This "germ plasm" is rich in mitochondria, RNA and a variety of specific proteins linked to signal pathways and cell adhesion, although the role of these components in sex-cell determination is presently unknown.
96. For an evaluation of Weismann's germ plasm, see Goodwin, 1984.
97. Oppenheimer, p. 161, 1967.
98. Muller received the Nobel Prize in Physiology or Medicine in 1946 for his work on the induction of mutation by X-rays.
99. See Garrod, (1909) 1923.
100. See Brøndsted, 1969; Mitmand & Fausto-Sterling, 1992.
101. Schiffmann, 1994.
102. For a discussion of the issues, the heroes, heroines and villains of the period, see Judson, 1979; Olby, 1994; Shostak, 1998.
103. RNA polymerase I transcribes most of the RNA of ribosomes [rRNA]; RNA polymerase II transcribes mRNA; RNA polymerase III transcribes tRNA, a small rRNA and several other small RNAs serving a variety of functions.
104. Given four nitrogenous bases (ACGU), 64 different codons are possible (i.e., 4 nitrogenous bases in groups of 3 [= 4^3]). Three of these codons are "stop" codons which signal the termination of a polypeptide chain, leaving 61 codons to match up with complementary anticodons on tRNA and encode amino acids in polypeptides.
105. Shiva, 1990.
106. Goldschmidt, 1940.
107. Baltscheffsky, 1997.
108. RNA is usually the culprit, as one might have imagined from RNA's role as a primer in replication.
109. Popper, p. 194, 1976.
110. Goldman, p. 14, 1977.
111. See Behe, 1996.
112. A representative example of the sentiment is found in Engel et al. (p. 40, 1994): "The evolution of genes ... may have been complicated by horizontal gene transfer."
113. Smith et al., 1998.
114. Cech (p. 251, 1993) calls this method of drawing conclusions "phylogenetic sequence comparisons" (or "phylogenetic covariation of bases method"), exemplified by success in drawing the cloverleaf secondary structure of tRNA.

115. Bergerat et al., 1997.
116. See Davies, especially p. 22, 1995, for the relevance of uniformly applied laws of nature to life elsewhere in the universe.
117. Monod, p. 81, 1971.
118. Feynman, quoted in Mehra, p. 490, 1994.
119. Geison, p. 37, 1995.
120. Lurie, 1988.
121. The phrase, "Secret of life," also appeared in Watson's first letter to Delbrück describing DNA's structure (Judson, p. 232, 1979).
122. Brian Goodwin (p. 225, 1984) makes a similar critique concerning DNA directly: "there's much evidence which makes it [the assumption that DNA's influence extends beyond protein structure] a proposition of dubious validity, to say the least."
123. Random mutation, recombination, and natural selection alter DNA, but these effects are easily accommodated to the DNA paradigm.
124. Aronowitz, 1988.
125. See, e.g., Bajaj, 1988.
126. Weinstein et al., 1997.

CHAPTER TWO

1. Codons are the triplets of nucleotides in messenger RNA (mRNA) that direct the incorporation of specific amino acids into a polypeptide chain during the process of translation. The "genetic code" refers to the relationship between specific codons and amino acids. Because codons were identified experimentally as sequences in mRNA, codons were traditionally listed as triplets of ribonucleotides, but, because DNA is said to be the genetic material, the genetic code is still considered to reside in DNA.
2. By serving as a template for the transcription of mRNA, DNA serves as the source of codons.
3. Kuhn, p. 24, 1970.
4. Kuhn, p. 36, 1970.
5. Zuckerkandl & Pauling, 1965. Similar ideas regarding cytochrome *c* evolution were expressed at the same time by Margoliash and Smith, 1965.
6. Macromolecules are or contain polymers consisting of monomers or the monomeric residues remaining after the condensing reactions involved in polymer formation.
7. Woese & Pace, 1993; Pace, 1997.

8. Hiding differences in boxes is frequently acknowledged in professional communications, if only in introductory remarks. For example, the introduction to the complete genome sequence of *Bacillus subtilis* explains that "Techniques for large-scale DNA sequencing have brought about a revolution in our perception of genomes. ... [I]t is now realistic to envisage a time when it should be possible to provide an extensive chemical definition of many living organisms." Kunst et al., p. 250, 1997.
9. See, for example, Boiker & Raff, 1996.
10. Mewes et al., p. 7, 1997.
11. For example, Mewes et al., goes on to say, "The significant number of gene duplications in yeast *must* reflect an evolutionarily successful strategy: gene duplications allow for evolutionary modifications in one of the copies without disturbing possibly vital functions of the other" (emphasis added; Mewes et al., p. 7, 1997).
12. See Eigen, 1992.
13. E.g., "Genecrunch" is available as an Internet service to "crunch" the complete yeast genome.
14. Svedberg received the Nobel Prize in Chemistry in 1929.
15. Tiselius received the Nobel Prize in Chemistry in 1948.
16. Polypeptides are the components of protein. They are made by the condensation of amino acids into a chain. Once incorporated, the amino acid is identified by its side group or residue. Any of the (typical) twenty amino acids can occupy positions in this chain, and several of these can be modified after incorporation. The polypeptide sequence refers to the identity and order of resides or amino acids in the polypeptide chain.
17. Chait et al., 1993.
18. Secondary structure refers to the presence of helical and ribbon-like portions of molecules.
19. DNA consists of two complementary strands of nucleic acids, each a polymer of nucleotides, which is to say, nitrogenous bases joined to a sugar-phosphate moiety. The nitrogenous bases in the two strands are paired (forming so-called base pairs) in such a way that for each adenine (A) in one strand, a thymine (T) appears in the other strand and for each cytosine (C) in one strand, a guanine (G) appears in the other strand. This is the famous A–T, C–G pairing discovered by James Watson and Francis Crick in 1953.
20. See Shostak, 1998.
21. Anonymous, 1998.
22. Eric Lander, of the Whitehead Institute for Biomedical Research at MIT, makes this point in his review of progress made in the first six years of the Human Genome Project. While others have compared the Project to the Manhattan Project and moon shots, Lander concludes that the Project "is best understood as the 20th century's versions of the discovery and consolidation

of the periodic table ... [by chemists in] the period from 1869 to 1889." Lander, p. 536, 1996.

23. Hamilton O. Smith (b. 1931), Werner Arber (b. 1929) and Daniel Nathans (b. 1928) shared the Nobel Prize in Physiology or Medicine in 1978 for their work on endonucleases capable of splitting DNA, especially, restriction enzymes. In general, enzymes derived from prokaryotes (bacteria and archaea) are identified by a three to four letter abbreviation, indicating the species, followed by a letter, indicating the type of enzyme, when more than one enzyme are derived from the species, and a Roman number, when that enzyme is represented by more than one variety. For example, *Eco*, is the three letter code for the bacterium, *Escherichia coli*.
24. The conditions are usually described as being highly stringent when numerous nitrogenous bases must match exactly and being of low stringency when mismatching is permitted.
25. Autoradiography uses the sensitivity of photographic emulsions to radiation as a device for detecting a source of radiation.
26. Named after its inventor, Edward M. Southern.
27. RNA or northern blotting (not named after anyone but thought of as the antipole of Southern blotting) is a similar technique employed for identifying specific RNAs. The comparable technique used for identifying specific polypeptides with monoclonal antibodies is called western blotting.
28. Pulse-field gel electrophoresis was invented by Charles Cantor (b. 1942) and collaborators at Columbia University.
29. The cosmid consists of a bacterial plasmid with phage λ *cos* sites, marking the ends of phage genomes. DNA intended for cloning is inserted within the plasmid, and the vector is packaged within an otherwise normal phage coat, as long as the distance between *cos* sites in the cloning vector is within 15% of the length of the normal λ phage DNA.
30. If the restriction enzyme's recognition site is still intact, retrieving the cloned DNA may be as easy as applying the same restriction enzyme used to create the hybrid phage or plasmid to extracted DNA.
31. Paul Berg shared the Nobel Prize in Chemistry in 1980 with Fred Sanger and Walter Gilbert.
32. Reverse transcriptase was discovered by David Baltimore (b. 1938) and isolated by Howard Martin Temin (b. 1934), for which accomplishment they shared the Nobel Price in Physiology or Medicine with Renato Dulbecco in 1975.
33. "Expression," usually refers to transcription, or DNA-dependent RNA synthesis, through which RNA is synthesized on a DNA template, but the term implies that the synthesized RNA is also employed by the cell in some aspect of protein synthesis and is not merely broken down.
34. ESTs are short sequences of bases in cDNA fragments of about 900 base pairs generally cloned from their 3′ end (3′ESTs). The 3′ESTs generally lack

polypeptide-encoding sequences, but, upon complete sequencing, can be brought into harmony with fragments cloned from their 5′ end (5′ESTs) containing polypeptide encoding sequences.

35. Schuler et al., 1996.
36. EST sequences are deposited into the National Center for Biotechnology Information (NCBI) EST database (dbEST) and hence into the GenBank database.
37. See Hartl, 1996.
38. See Gerhold & Caskey, 1996.
39. Schuler et al., p. 541, 1996.
40. Saiki et al., 1985.
41. The Nobel Prize in Chemistry in 1993 was awarded to Kary Mullis for the invention of PCR and to Michael Smith (b.1932) for discovering a mechanism for site-specific mutagenesis in 1978.
42. See Rabinow, 1996.
43. Nucleosides are covalently bonded nitrogenous bases (adenine [A], cytosine [C], guanine [G], thymidine [T] in the case of DNA; A, C, G, uracil [U] in the case of RNA) plus sugars (deoxyribose in the case of DNA [hence the D in DNA]; ribose in the case of RNA [hence the R in RNA]). Bound to three phosphate moieties, nucleosides are the active substrates for replication (in the case of DNA synthesis) and transcription (in the case of RNA synthesis). A nucleotide is a nucleoside plus one phosphate moiety (a monophosphate nucleoside).
44. Because nitrogenous bases are paired in DNA, the base pair (bp) and kilobase pair (kbp) have become the standard units for describing the length of DNA molecules.
45. See Gilbert, 1992.
46. Sanger shared the prize with Walter Gilbert and Paul Berg.
47. Actually the 3′–OH group is missing.
48. The theoretical physicist, Steven Koonin, points out that, "The debate on error rates should focus on the level of accuracy needed for each specific scientific objective or use of the genome data. [Furthermore,] The necessity of finishing sequences without gaps should be subject to the same considerations." Koonin, p. 37, 1998.
49. Gerhold & Caskey, p. 978, 1996.
50. Hartl, p. 1022, 1996.
51. Kevles, 1992.
52. Genomics is generally the study of complete genome sequences, but it is increasingly the study of structure and functional interactions between coding sequences and noncoding sequences or genome properties. See Bernardi, 1996.
53. See Botstein et al., 1997.

54. Behe, 1996.
55. The reader might like to check out some of these databases personally. The TIGR World Wide Web served (http://www.tigr.org), and other sites, offer copies of sequences already aligned for comparison. Updated constantly, new data is obtained from http://www.tigr.org/tdb/mdb/mjdb/updates/updates.html.
    European Bioinformatic Institute (mirrors NCBL data):
    http:// www.ebi.ac.uk
    GDB (Genome DataBase) collects and integrates genome mapping data from many sources including STSs (Sequence Tagged Sites); lists over 6000 STSs (in 1996) derived primarily from 3′ESTs (Expressed Sequence Tags) with thousands awaiting publication: http://gdbwww.gab.org
    Image (Integrated Molecular Analysis of Genomes and their Expression) Consortium provides detailed information on EST clones and vectors:
    http://www-bio.llnl.gov/bbrp/image/image.html
    Merck-WashU provides EST clones for a modest fee:
    http://genome.wustl.edu/est/esthmpg.html
    NCBI (National Center for Biotechnology Information), ENTREZ site on server lists updated comparisons of most EST to known genes:
    http://www.ncbi.nlm.nih.gov
56. See Collins & Galas, 1993. Also see Rowen et al., 1997.
57. See Marshall, 1998.
58. One Morgan is equal to 1 percent recombination. A centiMorgan, which has become the standard recombination or map unit, is 0.01 Morgans.
59. Cuticchia et al., 1993.
60. Collins & Galas, 1993.
61. NIH/CEPH Collaborative Mapping Group, 1992.
62. Cooperative Human Linkage Center et al., 1994.
63. Copeland et al., 1993.
64. The CA repeat contains cytosine and adenine as alternating nucleotides over a stretch of DNA. It is the most common short tandem repeat in the human genome.
65. EU Arabidopsis Genome Project, 1998.
66. Cooperative Human Linkage Center et al., 1994.
67. The assignment of a parental chromosomal phase and likelihood calculations were performed using the maximum-likelihood methods of the CRIMAP program package. Cooperative Human Linkage Center et al., 1994.
68. Schuler et al., p. 542, 1996.
69. See Hartl, 1996.
70. See Schuler et al., 1996.
71. See Hartl, 1996.
72. Copeland et al., 1993.

73. For example, the FACTURA program may be employed to "clean up" sequences. See Fleischmann et al., 1995; Fraser et al., 1997; Lang et al., 1997.
74. See Mewes et al., 1997.
75. Schuler et al., 1996.
76. *Saccharomyces* Genome Database, 1996; Goffeau et al., 1996; Clayton et al., 1997.
77. Deckert et al., 1998.
78. Kunst et al., 1997.
79. Fraser et al., 1997.
80. Blattner et al., 1997. The K-12 isolate chosen for sequencing was strain MG1655.
81. Fleischmann et al., 1995.
82. Tomb et al., 1997.
83. Cole et al., 1998.
84. Fraser et al., 1995.
85. Himmelreich et al., 1996.
86. Fraser et al., 1998.
87. Klenk et al., 1997.
88. Bult et al., 1996. See Edgell & Doolittle, 1997
89. See Mewes et al., 1997.
90. In yeast, centromeres are composed of three elements CDEI–III with the same motifs on each chromosome. CDEII, for example, consists of 79 to 86 bp, over 90% of which are A/T base pairs, but highly variable in actual sequence. See Goffeau et al., 1997.
91. In *Bacillus subtilis*. Kunst et al., p. 250, 1997.
92. In *Escherichia coli* K-12, Blattner et al., p. 1453, 1997.
93. The asymmetry in A–T and C–G switches polarity at the origin and terminus of replication in *Escherichia coli*, *Bacillus subtilis*, *Haemophilus influenzae*, and *Mycoplasma genitalium*. See Lobry, 1996.
94. E.g., Mewes et al., 1997.
95. scRNA is typically less than 300 nucleotides. As part of signal recognition particles, scRNA functions in the protein-translocating machinery of the endoplasmic reticulum.
96. snRNA is typically less than 300 nucleotides. It functions in the nucleus in small nuclear ribonucleoprotein particles (snRNPs), mediating and regulating posttranslational RNA processing
97. The subunits of ribosomes are identified by size measured in sedimentation or Svedberg units (S). The RNAs present in these subunits are also specified in S units.
98. Kunst et al., 1997.
99. Fraser et al., 1997.

100. Kunst et al., 1997.
101. Blattner et al., 1997; Fraser et al., 1997; Kunst et al., 1997.
102. French, 1992.
103. E.g., Blattner et al., 1997.
104. Operons are classically defined in prokaryotic chromosomes as clusters of two or more polypeptide sequences sharing a single promoter and operator sequence. Today they are often identified by a transcriptional termination sequence and may even be identified with only one coding sequence.
105. Promoters are portions of DNA that bind RNA polymerases and thereby "promote" the transcription of downstream nucleotides.
106. EU Arabidopsis Genome Project, 1998.
107. Collins & Galas, 1993.
108. See Sollner-Webb, 1996; Kable et al., 1996
109. See Bass, 1993.
110. Other programs may soon recognize nontranscribed regions involved in control or regulatory functions (enhancers, silencers, promoters, terminators) and other sequences corresponding to none of the above but containing long stretches of monotonously repeated short sequences that may affect the expression of the other sorts of genes (via imprinting).
111. "'Annotation' is the correlation of genome sequences with other "knowledge." For example, in the case of *Escherichia coli*, annotation is intended to (i) identify genes, operons, regulatory sites, mobile genetic elements, and repetitive sequences in the genome; (ii) assign or suggest functions where possible; and (iii) relate the *E. coli* sequence to other organisms, especially those for which complete genome sequences are available" (Blattner et al., p. 1454, 1997).
112. The Shine–Dalgarno-type groups is 5′–AAAGGA–3′ with not more than one substitution or insertion located 2–12 nucleotides upstream to an inferred initiation codon of polypeptide-encoding genes of bacterial and mitochondrial-DNA (mtDNA).
113. Gilbert et al., p. 243, 1995.
114. The codon adaptation index (CAI) is a measure of the extent to which codon usage agrees with a reference set from highly expressed genes. The CAI is considered a predictor of the extent of gene expression. See Blattner et al., 1997.
115. See Kunst et al., 1997.
116. Kunst et al., 1997.
117. Kunst et al., p. 252, 1997.
118. Fraser et al., p. 582, 1997.
119. Fraser et al., p. 581, 1997.
120. See Watanabe & Osawa, 1995.

121. FASTA is a program which measures similarities by comparing a similarity score of a candidate polypeptide sequence with that of the known polypeptide aligned with itself (a self-score). A score higher than one third of the self-score is considered a "strong similarity."
122. See Song & Fambrough, 1994.
123. Runnegar, 1994.
124. See Devereux et al., 1984.
125. See Woese & Pace, 1993. Raff, p. 105, 1996 is also unambiguous about "eyeballing": "In molecular phylogenetic analyses, homologous gene sequences from the species to be included are aligned to give the optimal match.... Despite the prevalence of computer methods for inferring trees, most alignments are done by eye.... Bases in a well-aligned set of sequences are presumably homologous in terms of position in the sequence."
126. See Bennett et al., 1994.
127. The Feng and Doolittle method packaged by J. Felsenstein, 1988. See Pollard & Quirk, 1994.
128. Benner et al., p. 31, 1993.
129. Applebury, 1994.
130. Doolittle, 1986.
131. Runnegar, p. 288, 1994.
132. Eigen, p. 56, 1992.
133. Eigen was awarded the Nobel Prize for Chemistry in 1967 jointly with Ronald George Wreyford Norrish (1897–1978) and George Porter (b. 1920).
134. Eigen, pp. 56–57, 1992.
135. In addition, a smaller class of transposed genes "Retrotransposons/Plasmid Proteins" is recognized by long terminal flanking repeats.
136. Mewes et al., p. 8, 1997.
137. Green et al., 1993.
138. Gojobori et al., 1995.
139. Gilbert, 1985; Doolittle, 1985; also see Fukami-Kobayashi et al., 1995; Gō 1981; Gō & Noguti 1995; Noguti & Gō, 1995; Oshima et al., 1995; Takahashi et al., 1995; Tateno et al., 1995; Yura & Gō, 1995.
140. See the discussion of clusters of orthologous groups (COGs) in Tatusov et al., 1997.
141. See Markert, 1996.
142. Clayton et al., p. 459, 1997.
143. Kunst et al., 1997.
144. Copeland et al., p. 60, 1993.
145. Mewes et al., 1997, for example, forget that "duplication" is not a mechanism but merely the recognition of similarity.

146. See Spradling, 1994. Gene conversion in eucaryotes (meiotic conversion), manifest as non-Mendelian segregation of alleles, is frequently explained by a nonreciprocal transfer and a DNA sequence's adoption of a complementary sequence from its partially base-paired partner. Micro-gene conversion may introduce multiple sequence changes to members of large gene families. Mitotic gene conversion may be responsible for some cases of gene rearrangement.
147. Schimenti, 1994.
148. See Kidwell, 1993; Syvanen, 1994; Smith et al., 1992
149. Hybridization and introgression of genes, or reticulate evolution, which assumes a sexual process, could also serve to double genes without duplicating them.
150. Blattner et al., p. 1454, 1997. Also see Lawrence & Roth, 1996; Lawrence & Ochman, 1997.
151. Clayton et al., p. 460, 1997.
152. See Goffeau et al., 1997.
153. Miklos & Campbell, p. 511, 1994. In contrast, all-atom modeling problems for RNA are frequently solved by an analysis of positional covariation with algorithms defining nucleotides as variables in the coordinates of three-dimensional space (developed to solve mathematical or topological constraint-satisfaction problems). Cedergran, 1995.
154. Hartl, 1996.
155. Riley, 1993.
156. Replication is DNA-dependent DNA synthesis.
157. Transcription is DNA-dependent RNA synthesis.
158. Translation is RNA-dependent polypeptide synthesis.
159. A so-called domain (or module) is a functionally distinct part of a protein or a sequence in a polypeptide with a specific three-dimensional configuration.
160. See, for example, Gerhold & Caskey, 1996.
161. See Hartl, p. 1022, 1996.
162. See Gilbert et al., 1995.
163. But also see Engel et al., 1994.
164. See discussion of COGs by Tatusov et al., 1997.
165. See Chothia, 1994.
166. See Green et al., 1993.
167. Tatusov et al., p. 633, 1997.
168. See Blackmore, 1994.
169. Hortsch and Goodman, 1991.
170. Erickson & Bourdon, 1989.
171. Gojobori et al., 1995.
172. Tatusov et al., p. 635, 1997.

173. Tatusov et al., p. 631, 1997.
174. E.g., Rudolf Raff, 1996.
175. Clayton et al., p. 459, 1997.
176. Eric Davidson quoted in Pennisi & Roush, p. 36, 1997.
177. "For nucleic acids, we call the sequence that arises from this procedure of alignment and superposition the *consensus sequence*." Eigen, p. 53, 1992.
178. Mewes et al., p. 7, 1997.
179. Hartl, p. 1022, 1996.
180. This method of geometrical representation and measure of separation was introduced in information theory by Richard W. Hamming and is known as the "Hamming metric." Ingo Rechenberg suggested applying the method to problems of evolution. See Eigen, 1992.
181. Felsenstein, 1984.
182. See Woese, pp. 7–8, 1991.
183. See Felsenstein, 1984; Lake, 1987a,b.
184. Felsenstein, 1988.
185. See Felsenstein, 1988.
186. See Nei, 1987.
187. Lake, 1987a,b.
188. Eigen et al., 1989.
189. See Bergström, 1994.
190. Taylor, 1994.
191. Polypeptides are components of proteins, although they may be joined in active proteins by prosthetic groups or coenzymes. Typically, polypeptides are said to be composed of amino (and imino) acids bonded together by peptide bonds. Historically, polypeptides were identified as long chains built of peptide bonds with projecting amino (and imino acid) residues.
192. Codons are units of the universal genetic code, a triplet of nitrogenous bases in messenger RNA (mRNA) that specifies which amino acid is incorporated at every position in a polypeptide chain.
193. Bult et al., p. 1068, 1996, would have you believe that "basic translation machinery is similar in all three domains of life."
194. For example, coenzyme A, biotin, thiamin, pyridoxalphosphate, adenosylmethionine, nicotinamide adenine dinucleotide (phosphate) (NAD[P]H), the flavins, the pterins.
195. Wächtershäuser, 1994.
196. For example, the authors of the *Escherichia coli* genomic sequence report, "We have not yet attempted to represent proteins with multiple roles that depend on physiological circumstances." Blattner et al., p. 1459, 1997

197. Blattner et al., for example, inform the reader, "The genes in all genomes are derived from a set of unique ancestral genes present in a progenitor of all extant organisms." Blattner et al., p. 1459, 1997.

CHAPTER THREE

1. Kuhn, 1970.
2. See Shostak, 1998.
3. Rudolf Raff, p. 54, 1996, for example, excuses contradictory data by explaining, "The problem is, of course, that despite the evidence for an overall parsimony in evolution, nonparsimonious things do happen in the real world."
4. See Ruse, 1996 for an exhaustive discussion of high and low in biological theory.
5. Taxonomy is only tangentially allied to systematics (based on proposed descent), since nothing whatsoever like common ancestors were identified by any taxonomic level within taxonomy's hierarchies of nomenclature. Phyla did not represent common ancestors of classes; classes were not common ancestors of orders, and so on down the hierarchy.
6. Copeland, pp. 7–9, 1956. The idea of naming organisms in natural groupings identified by relatedness through descent, the heart of modern systematics, did not get started until the end of the 18th century and did not become broadly accepted until neoDarwinism legitimized the new standard.
7. Lovejoy (p. 227, 1964) makes the point unambiguously: "No history of the biological sciences in the eighteenth century can be adequate which fails to keep in view the fact that, for most men of science throughout that period, the theorems implicit in the conception of the Chain of Being continued to constitute essential presuppositions in the framing of scientific hypotheses."
8. See Haeckel, 1901, especially Chapters 2 and 14.
9. See Gould, 1977.
10. See Haeckel, 1892.
11. Chatton, 1938.
12. See Copeland, 1956.
13. Whittaker, 1959.
14. Taylor, pp. 314–315, 1994.
15. See Margulis, 1990a and Margulis & Schwartz, 1982 for summary and discussion.
16. Cavalier-Smith, 1991.
17. Plastids refers to all the photosynthetic and nonphotosynthetic organelles resembling chloroplasts in terrestrial plants. Plastid is preferred by many

systematic botanists and algologists because it avoids the implication of "green" body, but, because "plastid" looks and sounds like "plasmid," the term is dropped here in favor of "chloroplast."

18. Sogin (1994) suggests that the Archezoa is a polyphyletic conglomerate.
19. Cavalier-Smith, 1991.
20. See Morell, 1997.
21. Woese & Fox, 1977.
22. See Woese & Fox, 1977; Woese et al., 1990, and elsewhere.
23. See Cavalier-Smith, 1992.
24. Yang et al., 1985.
25. Woese & Fox, 1977; Woese, 1987; Woese et al., 1990; Woese, 1991. Cedergren, p. 150, 1995, adds "the great legacy of the Woese school which turned the determination of the complex folding pattern of ribosomal RNAs into an art form."
26. Woese, 1987.
27. See Eigen, 1992.
28. The term, "rDNA," is often used for the DNA encoding rRNA, and it is, indeed, the DNA that is sequenced even when speaking of rRNA sequences. However, "rDNA" is also used for antibiotic "resistant DNA," a notorious plasmid spreading through prokaryotic pathogens. "rRNA" is, therefore, used here to avoid ambiguity.
29. Gu, 1997.
30. Lake, 1987b.
31. Swofford, 1993.
32. Hibbett, 1996.
33. Maddison & Maddison, 1992; Song & Fambrough, 1994.
34. Felsenstein, 1988.
35. Gogarten et al., 1989.
36. Miyata et al., 1991.
37. Penny et al., p. 218, 1994.
38. Woese, p. 7, 1991.
39. Cavalier-Smith, p. 275, 1991.
40. Cavalier-Smith, 1991; Song & Fambrough, 1994.
41. Cedergren, 1995.
42. Raff, p. 59, 1996.
43. Raff, p. 55, 1996.
44. Raff, p. 51, 1996.
45. See Weston, 1994.
46. Lai, 1995.
47. Throughout the book, "bacteriophage" and "phage" are used interchangeably.

48. Baltimore, 1971.
49. Nomenclature utilizing + and – signs to designate a type of nucleic acid are extremely hazardous. (–)strand DNA is the sense strand transcribed into mRNA, whereas (+)strand RNA is the sense or mRNA strand. Here, this nomenclature is restricted to RNA in an effort to avoid confusion.
50. Polymerases are identified as "directed" or "dependent" according to the template nucleic acid. RNA directed DNA polymerases are also identified as RdRp.
51. Goldbach & de Haan, p. 108, 1994.
52. Cann, 1993; Morse, 1994; Gibbs et al., 1995.
53. Chao, 1994.
54. Gorbelanya, 1995.
55. Viruses are grouped in families which can be identified by a capital first letter and the "-viridae" suffix. Members of families are more often identified by the family's best known member or an abbreviated version of its name and a "-like" ending.
56. The 5′ terminal nucleoside (usually a purine [adenosine or guanosine]) of eukaryotic, cytoplasmic mRNA (but not the mRNA of mitochondria or chloroplasts) is enzymatically "capped" with a methylated guanosine connected to the transcript by a 5′–5′ triphosphate linkage. The terminal purines beneath the cap can also be methylated at several positions.
57. Mononegrivales is the only superfamily of viruses approved by the International Committee on Taxonomy of Viruses.
58. McGeoch & Davison, 1995.
59. See Erickbush, 1994 for discussion and citations.
60. Retroelements, coined by Howard Temin, are DNA insertion sequences that encode a reverse transcriptase.
61. Erickbush, p. 126, 1994.
62. Erickbush, 1994.
63. Fox et al., 1977; Woese & Fox, 1977.
64. Woese et al., p. 4577, 1990.
65. Some efforts have been made to reconcile viruses with life forms, see, e.g., Haldane, 1994.
66. Woese et al. (1990), give several reasons why "domain" should be adopted as "a new rank at the top of the existing hierarchy" (i.e., above kingdom), to which I would like to add one more: "domain" moves nomenclature away from an undesirable overlay of social values. I would not like to see "Empire," Linnaeus's term for the very highest taxonomic category (adopted by Cavalier-Smith [1991]), return to nomenclature, for the same reason. "Domain" is an unfortunate choice, however, because the term is already defined in molecular biology (where it generally means a sequence of bases in DNA encoding a portion of a polypeptide chain). A neutral term, such as "Division,"

would be better if it were not already preempted by plant systematists (as an equivalent of an animal phylum). Perhaps a neologism, such as "Phylon" might replace "domain," especially if it could be adapted without the implication of monophylety currently attached to domain.

67. Hori et al., 1991.
68. I would much prefer avoiding the hassle over how many superkingdoms should be recognized and what they should be called. Here, I have adopted the nomenclature of Woese et al., 1990 not because I think it is the best or for reasons of priority but because it is the briefest. I go along with Lake and Rivera (p. 901, 1996), "In the end, when the necessary genomes are available, and when methods are developed to properly test these theories at the genome level, we will know the answer."
69. Sogin, p. 189, 1994. Also see Stetter, 1994.
70. Lazcano, 1994, p. 67.
71. Gogarten et al., 1989; Iwabe et al., 1989.
72. Woese & Fox (1977) called the common ancestor of the three domains the "progenote" and the ancestor of the eukaryan branch the urkaryote. Woese et al. (1990) resists giving names to other common branches.
73. See Runnegar, 1994.
74. Runnegar, 1994.
75. Sogin, 1994.
76. Bult et al., 1996. See discussion by Morell, 1996b; Edgell & Doolittle, 1997.
77. Woese et al. (1990) were not unaware of a problem with the monophylety of their domains, since they are at pains to explain that "The position of the root was determined by comparing (the few known) sequences of pairs of paralogous genes that diverged from each other before the three primary lineages emerged from their common ancestral condition."
78. Sogin, 1994.
79. Raff, p. 105, 1996.
80. Clayton et al., p. 462, 1997.
81. Maddison & Maddison, 1992.
82. See Fox et al., 1977.
83. Nei, p. 12, 1987.
84. See Hori et al., 1991.
85. Doolittle et al., p. 470, 1996a.
86. Hasegawa & Fitch, 1996.
87. Gogarten et al., 1996.
88. Doolittle et al., p. 1753, 1996b, respond that none of the corrections suggested by critics "lead to prokaryote–eukaryote divergence times greater than 2500 Ma." and the effects of horizontal gene transfer would be offset by the inclusion of paralogous genes and vice versa.
89. Gu, 1997.

90. See Woese & Fox, 1977; Woese et al., 1990; Woese & Pace, 1993.
91. Gogarten et al., p. 6664, 1989.
92. The TψCG sequence identifies one of the arms of tRNA. T = thymidine; ψ = pseudouracil, a substituted base; C = cytosine; G = guanine.
93. See Woese & Pace, 1993.
94. Cavalier-Smith, p. 288, 1991.
95. Woese & Pace, p. 111, 1993; also see Woese, 1987.
96. Murein contains the amino sugar N-acetyl muramic acid which forms an alternating copolymer with N-acetylglucosamine (GlcNAc) and is cross-linked into a two- or three-dimensional network by short peptides (made by enzymes and usually containing both L- and D-amino acids) attached to the muramic acid.
97. According to Cavalier-Smith, p. 288, 1991. "all eubacteria except mycoplasmas, Gram-positive bacteria and *Thermotoga*, have a second lipoprotein membrane, located outside the plasma membrane [lipoproteins], and attached to it by adhesion zones known as Bayer's patches."
98. According to Cavalier-Smith, 1991, Gram-positive bacteria + mycoplasmas = phylum Posibacteria. An extra-thick peptidoglycan layer is found in Gram-positive bacteria while reduced in Gram-negative bacteria and wall-less mycoplasmas (thought to have evolved through the loss of murein).
99. Fraser et al., p. 397, 1995.
100. See Pierson, 1994.
101. See Golubic, 1994.
102. See Cerff, 1995.
103. Saimi et al., 1994.
104. Guy & Durell, 1994.
105. See Cerff, 1995; Martin & Müller, 1998.
106. Stetter, p. 150, 1994.
107. See Woese et al., p. 4577, 1990, for references to data.
108. Kandler, p. 159, 1994.
109. Cavalier-Smith, pp. 287–288, 1991.
110. Woese & Pace, 1993.
111. Bult et al., pp. 1069–1070, 1996.
112. Bult et al., p. 1072, 1996.
113. Pierson, p. 162, 1994.
114. Cavalier-Smith, p. 288, 1991.
115. Borst et al., 1984.
116. See Birky, 1991; Clegg et al., 1991.
117. See Clegg et al., 1991; Sugita et al., 1995.
118. Sugiura, 1989.

119. Clegg et al., 1991.
120. Hori et al., 1991.
121. Pierson, 1994.
122. Palmer, 1991; Gillham, 1994.
123. Miklos & Campbell, p. 504, 1994.
124. Gray, 1989.
125. Reviewed in Gillham, 1994.
126. See Birky, 1991.
127. Attardi & Schatz, 1988.
128. Forsburg & Guarente, 1989.
129. See Saimi et al., 1994.
130. Woese et al., p. 4577, 1990.
131. Sogin, 1994.
132. Sogin, p. 497, 1989.
133. Cavalier-Smith, 1991.
134. Vossbrinck et al., 1987; Sogin et al., 1989; Sogin, 1989.
135. Viscogliosi et al., 1993.
136. Klenk et al., 1995.
137. Brown & Doolittle, 1995.
138. Kamaishi et al., 1996; Yamamoto et al., 1997.
139. Sogin, 1991.
140. Keeling, p. 88, 1998.
141. Keeling, p. 89, 1998.
142. The absence of mitochondria, peroxisomes and golgi apparatus is not restricted to these organisms but is characteristic of other intracellular, eukaryan parasites as well.
143. Sogin et al., 1989.
144. Upcroft & Upcroft, 1998.
145. Keeling, p. 91, 1998.
146. Keeling, p. 91, 1998.
147. Kamaishi et al., 1996.
148. Keeling & Doolittle, 1996; Edlind et al., 1996.
149. Taylor, 1994.
150. Elwood et al., 1985; Sogin & Elwood, 1986; Greenwood et al., 1991.
151. Siddall et al., 1995.
152. Sogin et al., 1986.
153. Christen et al., 1991; Wainright et al., 1993; Philippe et al., 1994; Pawlowski et al., 1996.
154. Smothers et al., 1994; Schlegel et al., 1996.

155. Turbeville et al., 1992. Also see Patterson, 1989.
156. Christen et al., 1991; Christen, 1994.
157. Sogin, p. 497, 1989.
158. Raff, p. 490, 1994. See tree drawn by Tuberville et al. 1992.
159. Christen et al., 1991; Christen, 1994.
160. Doolittle et al., p. 473, 1996a.
161. Raff, p. 491, 1994.
162. Doolittle et al., p. 472, 1996a.
163. Bult et al., p. 1070, 1996.
164. See Wilkie & Yokoyama, 1994; Runnegar, 1994.
165. Runnegar, p. 288, 1994.
166. Zillig et al., p. 185, 1992.
167. Bult et al., p. 1072, 1996.
168. Kandler, 1994.
169. Doolittle, 1998.
170. Bult et al., p. 1070, 1996.
171. Iwabe et al., 1989; Gogarten et al., 1989; Zillig et al., 1989; Brown & Doolittle, 1995. Also see discussion by Morell, 1996a.
172. Bult et al., p. 1068, 1996.
173. Bult et al., p. 1069, 1996.
174. Bult et al., p. 1069, 1996.
175. Miklos & Campbell, 1994.
176. Oshima, 1991. Also see Fabbri et al., 1998; Li et al., 1998.
177. Woese, p. 13, 1991. Also see Bult et al., 1996.
178. Woese, p. 14, 1991.
179. Cavalier-Smith, 1991.
180. Keeling & Doolittle, 1997.
181. Kandler, p. 154, 1994. Also see Chapter 2.
182. See Zillig et al., 1992. Also see Benner et al 1993 for other examples of enzymes incongruously shared by Archaea and Bacteria.
183. Donald Williamson (1992) argues that such a process is continuous.
184. Zillig et al., 1989; Martin & Müller, 1998.

CHAPTER FOUR

1. Apologies to Crick, 1988.
2. Raff, p. 52, 1996.

3. The quotation is adapted from Burke et al., p. 142, 1998, but could have been taken from any number of molecular tracts on molecular evolution.
4. Deleuze & Guattari, 1987.
5. Devolution is a palindrome of the new field of evo–devo, studying the evolution of development.
6. See motto at head of chapter; Deleuze, p. xix, 1994.
7. Of course, a predisposition toward singularity and bifurcational branching are understandable, inasmuch as they are consistent with religious and cultural bias.
8. Davies et al., p. 187, 1993.
9. For example, Joyce, p. 391, (1991) 1994.
10. Using American notation, 10^9 is a billion (i.e., a thousand million); 10^8 is a hundred million; 10^7 is ten million, and 10^6 is a million. A Ga thus is 10^9 years or 1000 My, while a My or Ma is 10^6 years or 0.001 Ga. A period of 10^7 years is 0.001 Ga or 10 My, while a period of 10^8 years is 0.01 Ga or 100 My. "Ga" and "Ma" do not imply "ago" but can be used to signify time's passage before the present when followed by "ago."
11. Kasting & Chang, 1992; Chang, 1994; Lowe, 1994.
12. Mojzsis et al., 1996; Hayes, 1996.
13. Prior to Oparin, T. C. Chamberlin and R. T. Chamberlin (1908) considered conditions that would have favored organic synthesis in the early Earth, and, contemporaneously but independently of Oparin, J. B. S. Haldane (1929) speculated on an the origin of life, but both these efforts were also largely ignored.
14. Oparin, 1964. Also see Oparin, 1924, in Deamer and Fleischaker, 1994.
15. Urey, 1952.
16. Miller, 1953; also Miller & Urey, 1959.
17. Oró, 1994, for citations.
18. In addition, spectroscopic data show the presence of linear $HC_{11}N$ and C_3 in the circumstellar envelope of a cool carbon star.
19. Chang, 1994; Deamer et al., 1994.
20. Of course, massively destructive impacts also constrained the timing of the origin of life. Chang, 1994.
21. Oró, 1994, p. 51.
22. Contemporary eucaryal bio-membranes are composed chiefly of phospholipids consisting of fatty acids linked to glycerol, phosphate, and some amine, such as choline. Fluidity in these membranes is a function of "kinks" in the fatty acid chains due to unsaturated bonds and the presence of cholesterol.
23. Deamer et al., 1994. The Murchison meteorite also contained octanoic (C8) and aromatic carboxylic acids.
24. Crick credits J. B. S. Haldane with having had a similar idea.

25. Crick, 1981.
26. Behe, p. 249, 1996.
27. Hoyle & Wickramasinghe, p. 204, 1984.
28. Hoyle & Wickramasinghe, 1984.
29. Hoyle, 1984, p. 105. Hoyle's argument seems a turnabout on the persistent criticism of the Gaia hypothesis: "If Earth is alive, how come it has not reproduced?" In the context of Hoyle's microbial grain model, one would hardly describe life in the Universe as sterile and isolated. Earthlife might be an excellent example of reproduction (to say nothing of growth and evolution) in the galactic biosphere.
30. Hoyle, 1984, p. 63.
31. Hoyle, 1984, p. 67.
32. Typically, the IVF patient is treated with a hormonal regimen designed to induce the maturation of several eggs (i.e., induced polyovulation); a clutch of "ripe" eggs is aspirated from the patient's ovary; fertilization and culture are carried out *in vitro* and a few days are allowed for cell division to take place, for blastocysts (so-called preembryos) to form (consisting of a cyst or ball of somewhere between 50 and 100 cells) and for the uterus to open its "window of opportunity" during which it may accept blastocysts; blastocysts are then transferred to the patient's uterus; if implantation is successful, each blastocyst develops placental connections at its surface to maternal tissue and extraembryonic membranes and an embryo within; the embryo becomes a fetus, and, following birth, a neonate or baby. Most attempts at IVF are presently (1998) unsuccessful, but a high proportion of successful attempts are multiple births of nonidentical siblings, resulting from the transfer of several preembryos to the patient's uterus.
33. In early August, 1996, release of the McKay et al., 1996 analysis of a meteorite allegedly from Mars bearing microfossils was circulated in news media around the world and along with other material from the American space agency NASA was promptly placed on the world wide web. The President of the United States, Bill Clinton, was widely quoted as having declared the news as "arguably the biggest discovery in the history of science." Also Kerr, 1996a,b.
34. As proudly displayed on NASA's "Meteorites from Mars!" page at its Web cite (curator Anita Dodson); update 9 August, 1996.
35. McKay et al., 1996. The difficulty facing exobiologists is abundantly documented by McKay et al.: on the one hand, evidence of life is based on similarity with terrestrial life (i.e., terrestrial biomarkers); on the other hand, evidence of indigenous life depends on differences with terrestrial biomarkers. What is more, indigenous biomarkers might be totally different from terrestrial biomarkers and thus ignored by the terrestrial exobiologist.
36. A nanometer (nm) is one billionth (10^{-9}) meters.
37. Kerr, 1996a,b.
38. Bada et al. 1998; Jull et al., 1998.

39. McKay et al., 1996.
40. Davies, 1995, p. 21.
41. Davies, 1995, p. 37.
42. Davies, 1995, p. 32.
43. Oró, 1994.
44. Cairns-Smith, 1985.
45. Gedulin & Arrhenius, 1994.
46. Wächtershäuser, 1994.
47. One such pigment might have been the complex ion, ferrocyanide which absorbs near-UV light (300–400 nm) in a photochemical reaction that releases cyanide radicals and absorbs a proton. The bacterium, *Rhodospirillum rubrum*, illuminated in the presence of ferrocyanide at high concentration, generates a pH gradient across the chromatophore membrane forcing protons outward through proton-ATPase and proton-PPase ports and fueling the synthesis of adenosine triphosphate (ATP) or pyrophosphate (PPi). Baltscheffsky & Baltscheffsky, 1994. Another possible pigment is polycyclic aromatic hydrocarbons (PAH) of the sort delivered by carbonaceous meteorites. Floating as a thin film, PAH, absorbing light in the near UV range, might have promoted the production of amphiphiles as well as donating electrons to molecular condensation and generating proton gradients. Deamer et al., 1994;
48. Baltscheffsky & Baltscheffsky, 1994.
49. Orgel, 1986.
50. Wächtershäuser, 1994.
51. Senapathy, 1994
52. Gilbert, 1986.
53. Sharp, 1985.
54. The ordinary, textbook definition of "intron" is a sequence of DNA which is transcribed but the complementary RNA is removed from the nascent transcript before translation. In the debates over the RNA world and early- versus late-introns hypotheses, "introns" are often considered the portion of the RNA transcript removed during transcript processing. In what follows, "transcripts of introns" and "RNA introns" will refer to transcribed RNA sequences, while "introns" (unmodified) will refer to DNA sequences. "Intronic RNA" will be used for RNA in the RNA world that was not presumably transcribed from DNA.
55. The exception is the presence of an internal coding region (for a mutase, endonuclease or even reverse transcriptase) which may be translated prior to excision from the intron.
56. Joyce, (1991) 1994.
57. "Ribozyme" was coined by Kruger et al., p. 154, 1982, for the IVS intron of *Tetrahymena*: "Because the IVS RNA is not an enzyme but has some enzyme-like characteristics, we call it a ribozyme, an RNA molecule that has the intrinsic ability to break and form covalent bonds.... If the RNA moiety of the

small nuclear RNP bound a nucleotide cofactor or participated in catalysis in any way, it would be a ribozyme." Ribozymes are thus a loosely defined group of RNA particles (folded RNA molecules) with catalytic activity.

58. Gold et al., p. 497, 1993.
59. "Autocatalysis" originally referred to RNA's ability to cleave its diester bonds *in vitro* in the absence of protein. Today, "autocatalysis" is more likely to refer to the rearrangement of sequences in RNA presumably mediated by RNA.
60. Group I introns were originally identified in fungal mitochondrial genes, but are now known to be more widespread. Today, group I introns are characterized by *in vitro*, self-extracting introns of rRNA genes found in the ciliate, *Tetrahymena thermophila*, which are studied more easily through their mutants cloned in *Escherichia coli* after transfer by bacteriophage. RNA transcripts of group I introns are identified by a common mechanism of splicing via transesterification and an internal secondary structure built around "conserved" or "consensus" sequences (P–Q and S–R). Group II introns are characterized by secondary structure and "consensus" sequences near the boundary (splicing junction) sequences. In addition to mitochondrial genes, group II introns occur in the genes of plastids of land plants and resemble introns in nuclear, pre-mRNA introns.
61. Naked group I and II intronic RNA are credited with these activities *in vitro*. The same activity is generally conceded to utilize protein *in vivo*.
62. Ribonuclease P is an endoribonuclease found in *Escherichia coli*. Its own RNA generates the 5′ end of tRNAs by cleaving transcripts behind leaders or between consecutive tRNA molecules.
63. Pan et al., 1993.
64. Spliceosome complexes arc assembled by the accumulation of small nuclear RNA particles, snRNA, which, when complexed with protein in ribonucleoprotein particles are snRPs or "snurps," each with a single RNA and about 10 proteins.
65. In forming the complex, a U2 snurp complexes with a transcribed intron, followed by the U5 and U4/U6 snurps. When the left exon in the same transcript is cleaved and the intron forms a loop (known as lariat formation), the U4 snurp is released and the right exon is cleaved and joined to the left one. The spliceosomal complex then disperses, freeing the snurps to cut and splice again. The snurp components of the spliceosome are required for nuclear splicing. U1, which may not ordinarily be part of a spliceosome complex, exhibit complementarity with a few junctional base pairs, and removal of the 5′ terminal nucleotides of U1, U2, U4 and U6 effectively blocks splicing, but the role of base-pairing between snurp and intron RNA remains unclear.
66. Maizels & Weiner, 1993.
67. Noller et al., 1992.
68. Lazcano, 1994, p. 76.
69. Turner & Bevilacqua, 1993.

70. Szostak & Ellington, p. 530, 1993.
71. L-19 RNA would seem to be an exception, since, following intramolecular rearrangement, its polypyrimidine sequence can bind and add cytosine oligonucleotides to its 5′ end.
72. Cech, 1993.
73. Konarska & Sharp, 1989.
74. Here, the initial 5′ part of an infectious mRNA is translated to a polymerase which replicates a full-length negative (–) template strand that is, in turn, differentially transcribed into a variety of mRNAs.
75. Joyce, (1991) 1994 and Joyce & Orgel, 1993 for a critique and additional citations.
76. Even proteinaceous RNA replicases are error-prone, since they lack proofreading functions. Catalytic RNA, on the other hand, may be able to act as a 3′-exoribonuclease, and thus act as a proofreading enzyme although lacking RNA replicase activity.
77. Joyce, p. 394, (1991) 1994.
78. Moore et al. (1993) allow protein into their RNA world.
79. Davies et al., 1993.
80. The use of "intronic" as an adjective for RNA was suggested to me by Paula Grabowski (personnel communication), second author of the paper introducing the term "ribozyme," (above), in order to avoid confusion with introns in DNA (as usually defined) and transcribed introns (or RNA introns, the transcribed but untranslated form of introns). In what follows, "exonic" will also be used as an adjective for RNA in the RNA world made without transcription.
81. Gilbert, 1987, Gilbert et al., 1995.
82. Gō, 1981, Gō & Noguti, 1995.
83. Moore et al., 1993.
84. Cerff, p. 205, 1995.
85. Sugita et al., 1995.
86. Clegg et al., p. 139, 1991
87. Doolittle, 1989, for citations.
88. Cerff, p. 218, 1995. Also Moore et al., 1993; Baserga & Steitz, 1993; Tani et al., 1995.
89. Eigen et al., 1989.
90. "Domains" here refer to the Archaea, Bacteria and Eucarya.
91. Joyce, (1991) 1994.
92. The use of triplets in the genetic code has long been a subject of speculation. See Weiss & Cherry, 1993 for a discussion of the evolution of the code and the ability of ribosome to operate in the translation of codons.
93. Crick, 1968.
94. Woese et al., 1966.

95. Yarus, p. 1757, 1988.
96. RajBhandary & Söll, 1995.
97. For discussion of modification of tRNA in the course of evolution see Söll, 1993; Cusack, 1995.
98. The structure of tRNAs is discussed by Cedergren, 1995.
99. TψC is a unique triplet of nitrogenous bases identifying the TψC arm of tRNA when drawn two dimensionally as a cloverleaf-shaped molecule. T is thymidine; ψ is pseudouridine; C is cytosine
100. Maizels and Weiner, 1993; Schimmel, 1995. The portion of tRNA containing the TψC stem, acceptor stem, and the 3′-terminal CCA may represent an ancient, RNA-genomic tag required for the initiation of RNA replication and present long before the anticodon-containing domain was added to the tRNA molecule and before the advent of protein synthesis.
101. Saks et al., p. 1665, 1998.
102. McClain, 1995.
103. Nureki & Yokoyama, 1995.
104. Maizels & Weiner, 1993.
105. Shiba, 1995.
106. Martinis & Schimmel, 1995.
107. Noller, 1993. Specifically, the acceptor-TψC domain of tRNA would have coevolved with 23S rRNA while the anticodon-containing domain of tRNA coevolved with 16S rRNA.
108. Eigen et al., 1989.
109. Hanyu et al., 1986.
110. King et al., 1991.
111. RajBhandary & Söll, p. 1, 1995.
112. In the standard nomenclature for RNA, U = uridine, G = guanine, C = cytosine, and A = adenine.
113. Watanabe & Osawa, 1995.
114. Yarus, 1993.
115. Eigen, p. 43, 1992.
116. Lazcano, p. 63, 1994.
117. Weiss & Cherry, p. 74, 1993.
118. Moore et al., p. 303, 1993.
119. Noller et al., 1992.
120. Weiss & Cherry, p. 82, 1993.
121. Noller, p. 148, 1993.
122. Gray et al., 1995. The same sort of ligation might also have taken place without ribozymal intervention.

123. Ikawa et al., 1995. Copy choice, like DNA recombination, covalently joins nucleic acid from two parental molecules but requires replication.
124. Cerff, 1995.
125. Oshima, 1991.
126. Atkins, p. 544, 1993. Replicase associated recombination is also known as copy-choice.
127. Atkins, pp. 544–545, 1993, goes on to suggest that RNAs may have been preserved as the genetic material in some viruses because, "in the tight virions of some viruses, it is likely that water is excluded, so that the hydrolytic instability of RNA is not relevant."
128. "Reverse transcriptases" would seem to be misnamed for the extant enzyme used *in vitro* by molecular biologists to make cDNA from RNA if the original direction of information flow was from RNA to DNA. The flow from DNA to RNA catalyzed by present-day DNA-dependent RNA polymerases would then represent the "reverse" reaction. Atkins, 1993.
129. Fraser et al., 1995.
130. Gilbert et al., 1995.
131. Gojobori et al., 1995.
132. "Domain" here refers to a sequence in DNA encoding a particular sequence of amino acid residues in a polypeptide chain or protein.
133. Yura & Gō, p. 192, 1996: "The finding of HTH modules in a variety of proteins supports the notion that HTH modules were shuffled during protein evolution. The shuffling was plausibly carried out by exon shuffling ... because correspondence of intron positions and module boundaries suggests that a module was encoded by a single exon."
134. Helfman, p. 106, 1994: "Most genes contain multiple exons, e.g., collagen has more than 50 exons, and the dystrophin gene has more than 70 exons. ... [T]he dystrophin gene which is over two million base pairs long ... contains a number of introns which are larger than 200,000 nucleotides."
135. Gilbert et al., p. 244, 1995.
136. Long & Langley, 1993; Helfman, 1994.
137. Cerff, 1995.
138. Benner et al., p. 44, 1993. Also see Benner et al., 1989.
139. Wilkie & Yokoyama, 1994.
140. Hibbert, 1996.
141. "[A] protein can be dissected within a region of a secondary structure without losing its activity.... [G]enetic units do not necessarily correspond to structural units." Shiba, p. 19, 1995.
142. Stoltzfus et al., 1994.
143. Wilkie & Yokoyama, p. 261, 1994.
144. Doolittle, 1989.
145. Lawrence & Roth, 1996.

146. Lawrence & Roth, p. 1844, 1996.
147. Lawrence & Roth, p. 1845, 1996. The example given is from lambdoid and T4 family bacteriophages in which genes encoding proteins that function together as a logical groups (head or tail proteins) are clustered.
148. Lawrence & Roth, p. 1849, 1996.
149. Also called lateral gene transfer. Lawrence & Hartl, 1992; Kidwell, 1993.
150. Smaller, 1992, p. 21. Also Campos et al., 1992.
151. Syvanen, 1994.
152. Palmer, 1995.
153. Gillham, 1994.
154. Charlesworth & Langley, 1991; Houck et al., 1991; Cummings, 1994; Syvanen, 1994.
155. Hibbett, 1996 for citations.
156. Lawrence & Ochman, p. 385, 1997.
157. Lawrence & Ochman, p. 394, 1997.
158. Gojobori et al., 1995.
159. Sogin, p. 181, 1994.
160. Woese et al., 1990.
161. Sogin, p. 181, 1994.
162. Crawford & Milkman, 1991, for a view of reticulate evolution among duplicated genes.
163. Kidwell, 1993.
164. Gibbs et al., 1995.
165. Another class of transposons, simple transposons, resembles the genome of bacteriophage (Mu). They are found in prokaryotes but do not undergo reverse transcription.
166. Spradling, p. 77, 1994.
167. Spradling, 1994. Also see for nonLTR retrotransposons in insect 28S rRNA gene.
168. Bult et al., 1996.
169. Woese et al., p. 4577, 1990.
170. Tani et al., 1995 for critical discussion of the spliceosomal intron family in small nuclear RNA particles (snRNA).
171. Iwabe et al., 1989; Gogarten et al., 1989.
172. As of 1993, the list of other ribosomal proteins represented in the three domains of life included 15 separate proteins with one to four representatives each. Benner et al., 1993.
173. Benner et al., 1993.
174. Koonin, 1997 commenting on Bult et al., 1996.
175. Saimi et al., p. 183, 1994.
176. Song & Fambrough, 1994; Tang et al., 1996.

177. Oshima, 1991.
178. Gogarten et al., 1989.
179. Two of the glycolytic enzymes, triosphosphate isomerase (TPI) and glyceraldehyde-3-phosphate dehydrogenase (GAPDH) have provided evidence on both sides of the exon shuffling theory of genetic evolution. Cerff, 1995.
180. Martin, 1996.
181. Applebury, p. 246, 1994.
182. Wilkie & Yokoyama, p. 255, 1994.
183. Wilkie & Yokoyama, p. 256, 1994.
184. Gorbalenya, 1995.
185. Morse, 1994.
186. Falkow, 1975.
187. Heringa & Argos, p. 88, 1994.
188. McGeoch & Davidson, p. 71, 1995.
189. McGeoch & Davidson, p. 74, 1995.
190. See Heringa & Argos, p. 89, 1994 for citations.
191. Gorbalenya, p. 61, 1995.
192. Gorbalenya, 1995.
193. Manfred Eigen's term, "quasispecies," introduced to help solve the mysteries of life's origins, seems merely to beg the question of dynamics in early life forms rather than provide a useful model. "Quasispecies" nevertheless may be used for a distribution of related but nonidentical genomes (a replicon population) in a population at equilibrium. It offers a convenient theoretical model for examining effects of environmental changes and random-sampling on RNA viral populations (as well as avoiding all the ambiguity over "viral species").
194. Gorbalenya, p. 56, 1995.
195. Erickbush, p. 144, 1994.
196. Erickbush, 1994.
197. Goldbach & de Hann, p. 117, 1994.
198. Falkow, p. 134, 1975.
199. Hori et al., 1991.
200. Lake et al., 1984.
201. Rivera & Lake, 1992.
202. Lake et al., 1984, Lake, 1987a–c, 1988, Rivera & Lake, 1992.
203. Reviewed by Miyata et al., 1991.
204. Runnegar, 1994.
205. See Woese, 1991.
206. Miyata et al., p. 347, 1991.
207. Cavalier-Smith, 1987; Cavalier-Smith & Chao, 1996.

208. Zillig et al., 1989, Zillig et al., 1992; Gupa & Golding, 1996.
209. Dyer & Obar, 1994.
210. Cavalier-Smith, 1991.
211. Yang et al., 1985; Clegg et al., 1991, 1994.
212. Margulis, p. 322, 1991a.
213. Margulis, 1993.
214. Gray, p. 40, 1989.
215. Sapp, 1985; Hinkle, 1991.
216. Margulis, 1990c; Sapp, 1994.
217. Trench, 1991.
218. Margulis, p. 306, 1991a; Sapp, 1991.
219. Margulis & Cohen, p. 327, 1994.
220. Southward & Southward, 1996.
221. See reviews in Margulis & Fester, 1991.
222. Reviewed by Margulis et al., 1996. Also see Muscatine & McAuley, 1982.
223. Schwemmler, 1984a.
224. Schwemmler, 1984b, 1989, 1991; Goodwin, 1994.
225. Muscatine, 1974.
226. McMenamin & McMenamin, 1990.
227. Palmer, 1992.
228. Reviewed by Gillham, 1994.
229. Upcroft & Upcroft, p. 262, 1998. Upcroft & Upcroft, 1998, credit Gupa & Golding, 1996, with the idea of bacterial/archaean fusion.
230. Lake & Rivera, 1994.
231. Cavalier-Smith, p. 298, 1991.
232. Quoted by Vogel, p. 1604, 1997. Also Clark & Roger, 1995; Edlind et al., 1996; Yamamoto et al., 1997; Roger et al., 1998.
233. Clark & Roger, 1995; Rosenthal et al., 1997.
234. Roger et al., 1998.
235. Ironically, some archezoans once thought to have never acquired mitochondria are now regarded as having lost them. Keeling, 1998.
236. Martin & Müller, 1998.
237. Martin & Müller, p. 40, 1998.
238. Rosenthal et al., 1997; Keeling & Doolittle, 1997.
239. Brul & Stumm, 1994.
240. Doolittle, 1998.
241. Kidwell, p. 239, 1993.
242. Clegg et al., 1991.
243. Sugita et al., 1995.

244. Sugita et al., pp. 82–83, 1995.
245. Clegg et al., p. 139, 1991.
246. Clegg et al., 1991.
247. Sugiura, 1989.
248. Smith et al., 1992.
249. Attardi & Schatz, 1988.
250. Kuroiwa, 1998.
251. Sociobiologists advocating theories of "selfish genes" will have a hard time explaining these "selfish," gene-less organelles.
252. Gillham, p. 43, 1994.
253. Cavalier-Smith, p. 298, 1991.
254. Taylor, 1994. Chromophytes include Chrysophyta, Haptophyta, Cryptophyta, Xanthophyta, Eustigmatophyta, Bacillariophyta, and Phaeophyta.
255. Cerff, 1995.
256. Golubic, p. 335, 1994.
257. Taylor, 1994.
258. Birky, p. 123, 1991.
259. This freshwater heterotrophic, jakobid flagellate resembling retortamonad, was earlier thought to lack mitochondria.
260. Lang et al., 1997.
261. Cavalier-Smith, 1991.
262. Margulis, p. 311, 1991a, traces the idea back to B. M. Kozo-Polyanski (1890–1957) who claimed "that cell motility was derived symbiotically from "flagellated cytodes" by which he meant primitive flagellated bacteria." Also see Hinkle, 1991.
263. Hall et al., 1989.
264. Taylor, p. 316, 1994.
265. Köhler et al., 1997.
266. Buss & Seilacher, 1994; Lom, 1990; Shostak, 1993a,b.
267. Shostak & Kolluri, 1995.
268. Tarter, 1961; Nanney, 1985.
269. Cavalier-Smith, p. 299, 1991.

CHAPTER FIVE

1. Popper, 1972, 1976; also see Sober, 1994.
2. Kuhn, 1970.

3. E.g., Gross & Levitt, 1994.
4. de Duve, pp. 2–3, 1995.
5. See Ruse, 1996. As a historian, Ruse's ideas of how paradigms are built are simplistic, if not vulgar. For example, he suggests (Ruse, p. 277, 1996) that, in order to build the 20th century DNA paradigm, "The first thing that Jim Watson did after discovering the double helix was to write a huge textbook (Watson 1965)." But Ruse can be forgiven for such misdemeanors, since he makes no attempt to represent himself as a philosopher of science.
6. These are the titles of Adrian Desmond's monumental biography of Huxley (1994, 1997).
7. Ruse agrees with Desmond on this point. See Desmond, 1994, 1997.
8. Ruse, p. 277, 1996.
9. Ruse, p. 278, 1996.
10. One might contemplate, however, the appearance of adaptive mutability among excess revertants generated under conditions of severe constraint. Benson, 1997. Also see Jablonka & Lamb, 1995.
11. For the most persistent believer in evolutionary molecules see Dawkins, 1965, 1976, 1995, 1997.
12. Zuckerkandl & Pauling, 1965.
13. Genera, families, orders, classes, phyla [or divisions], kingdoms and domains, defined by biochemical, morphological, and behavioral characteristic sets.
14. Confusion over taxa is most obvious at the level of species. The standard criterion for the "biospecies" is only relevant for sexually reproducing species and then only hypothetically to the degree that the members of the species are potentially capable of interbreeding. A rational criterion for identifying taxa in a nested hierarchy might employ some function of the number of generations since a last common ancestor, although such a criterion would introduce its own ambiguity given uncertainty in the fossil record and the rate of change in "DNA clocks."
15. If there are 30 to 40 million extant species, and these represent 1% of all the species that have even lived, and species have existed on Earth for somewhere between 3000 and 4000 million years, and assuming further that, over that period of time, species originate and become extinct at a constant average rate, then the average species exists for about a million years. This period would presumably be longer for organisms with longer generation times and shorter for organisms with shorter generation times, but no one can sensibly claim that a million years can be equated with "stability."
16. Deleuze, p. 41, 1994.
17. Deleuze, p. 76, 1994.
18. Leigh Brown, 1994.
19. Domingo & Holland, p. 163, 1994, contrast the mutation rate or rate of misincorporation events during nucleic acid synthesis, which is inevitably higher in RNA viruses compared to other life forms, with the rate of fixation,

or accumulation of changed nucleic acid, which, natural selection would presumably dampen. The mutation (or mutant) frequency is a measure of the proportion of mutants in a population.

20. Gibbs & Keese, 1995.
21. See discussion of COGs by Tatusov et al., 1997.
22. Taylor, p. 319, 1994.
23. Chothia, p. 27, 1994.
24. Song & Fambrough, 1994.
25. Tatusov et al., p. 634, 1997.
26. Tatusov et al., p. 631, 1997.
27. Tatusov et al., p. 633, 1997.
28. Domingo & Holland, p. 171, 1994.
29. Milos & Campbell, p. 514, 1994.
30. Watson & Crick, 1953.
31. "Paralogy" and "orthology" were coined by Fitch & Margoliash, 1970.
32. Crawford & Milkman, 1991.
33. Bennett et al., p. 223, 1994.
34. An exception is the bacterial initiator triplet for fMet–tRNA which recognize codons of AUG and GUG differing in the first base of the codon.
35. Watanabe & Osawa, 1995.
36. UGA is also the codon for tryptophan (in addition to UGG) in Micrococcus capricolum. and " a good number of UGA codons occur at the sites that are Trp in the corresponding *Escherichia coli* proteins, and no UGA codon has been identified at the termination site" (Watanabe & Osawa, p. 236, 1995).
37. Raff, p. 152, 1996.
38. Kreitman, 1996; Ohta, 1996.
39. Wu & Hammer, 1991.
40. Miklos & Campbell, p. 516, 1994. Also see Atkins, 1993 for a discussion of rates and frequencies of genetic change.
41. See Charlesworth & Langley, 1991.
42. Spradling, p. 69, 1994.
43. Cann, p. 74, 1993.
44. Lewin, p. 147, 1990.
45. See Gilbert et al., 1995.
46. According to Noguti and Gō, p. 162, 1995: "The correlation between modules and exons suggests that the former were encoded by corresponding exons in genes for ancestral proteins and functioned as evolutionary units; new proteins might be produced by combining various modules through exon shuffling or fusion."
47. Benner et al., 1993.

48. See Cerff, 1995.
49. Engel et al., 1994.
50. See Helfman, 1994.
51. Patthy, 1991a,b, 1994; Takahashi et al., 1995; also see Stoltzfus et al., 1994.
52. Schimke, 1982.
53. Gibbs & Keese, p. 77, 1995.
54. Lai, 1995.
55. Myers et al., 1995.
56. Gibbs & Keese, p. 88, 1995.
57. Heringa & Argos, 1994. Exceptions are, of course, inevitable, such as the similarities found in African swine fever virus and T4 phage.
58. Heringa & Argos, p. 95, 1994.
59. Heringa & Argos, p. 98, 1994.
60. Goldbach & De Haan, 1994.
61. Horizontal gene transfer, or the movement of genes through the environment, contrasts with vertical gene transmission, or the movement of genes through reproductive lineages.
62. Heringa & Argos, 1994.
63. Chao, 1994.
64. Fenner, 1995.
65. Garnett & Antia, p. 67. 1994.
66. Any number of examples could have been chosen. Sex in viruses was selected for its concreteness. Lin Chao explains (p. 247, 1994): "If one follows the definition that sex constitutes genetic exchange among individuals..., segmentation in a multicomponent virus is sex." Elsewhere, Chao explains (p. 235, 1994): "A lone single-component virus infecting a host cell reproduces asexually."
67. See discussion of neoLamarckism in Shostak, 1998.
68. Bennett et al., p. 230, 1994.
69. Tatusov et al., p. 634, 1997. Other instances are readily found, for example, Crawford & Milkman, 1991.
70. Martin, p. 526, 1996.
71. Raff, p. 154, 1996.
72. Lawrence & Oshman, 1997.
73. Houck et al., 1991.
74. Hall, 1997.
75. Brian Goodwin, pp. 226–227, 1984. Also see King, 1996.
76. Arnold, p. 140, 1997.
77. Williamson, p. 197, 1992.
78. Wilson, 1896; Berrill, 1961; Saunders, 1982.

79. Peterson et al., 1997.
80. Gerhart & Kirschner, 1997; Gellon & McGinnis, 1998.
81. See Maddison, p. 49, 1996.
82. Jablonka & Lamb, 1995.
83. Mayr, 1994.
84. Behe, p. 40, 1996.
85. Behe, p. 112, 1996.
86. See Behe, pp. 216 ff., 1996. Elevating design to a scientific concept would seem desirable, but Behe does no more than wish for such a situation. Instead of laying out the testable principles of design, he resorts to the same loose analogies for which he criticizes those who ridicule design as an easy way out of any argument. The issue is not whether design is testable but whether it violates the principles of chemistry, physics and whatever other pretenders to universal laws (e.g., chaos theory) ordinarily expected to constrain theories of life. Behe's technique and weakness is especially apparent in his confusing a qualitative combination lock with a quantitative measure of performance.
87. Bell, p. 227, 1997.
88. Sober, 1994.
89. See Song & Fambrough, 1994.
90. "Exaptation" was coined by Gould & Vrba, 1982. Bell, p. 230, 1997.
91. Darwin (1959) 1968; (1872) 1958.
92. See Eldredge, 1985, Gould, 1994; also see Dawkins, 1997.
93. Miklos & Campbell, p. 514, 1994.
94. See Nei, 1987.
95. Lewontin & Hubby, 1966.
96. Flavell et al., 1986.
97. Nei & Hughes, p. 236, 1991.
98. See Wilkie & Yokoyama, 1994. Their estimates of the rate of amino acid replacement per site per year in *Drosophila*, frog and mammal, shows that "Among the Gα-II group, G_q has the slowest evolutionary rate. For comparison, the most highly conserved Gα genes have evolved somewhat faster than histones and trisphosphate isomerase but more slowly than creatine kinase.... Substitution rates in the most rapidly evolving mammalian Gα genes fall between the hemoglobin α and β genes" (Wilkie & Yokoyama, p. 257, 1994).
99. "Rapid methods for DNA sequencing at low cost are revealing examples of sequence variation that are not consistent with species phylogenies." Kidwell, p. 236, 1993.
100. Engel et al., 1994.
101. The exception is hybridization linked to autopolyploidization events in the case of plants.
102. Burke et al., 1998.

103. Leo Buss, 1994, suggests that "Understanding the origin of life may be profitably explored by decoupling the origins of different features of life."
104. Gojobori et al., p. 259, 1995.
105. See Shostak, 1998.
106. See Hofbauer & Sigmund, 1988; Prigogine & Stengers, 1984; Stewart, 1989.
107. See Holland, 1995; Kauffman, 1993.
108. See Coveney & Highfield, 1995.
109. Avnir et al., 1998; Tsonis, 1998; Mandelbrot et al., 1998. Also see Stewart, 1989.

BIBLIOGRAPHY

Adelmann, H. B. (1942). *The Embryological Treatises of Hieronymus Fabricius of Aquapendente*. Ithaca: Cornell University Press.

Adelmann, H. B. (1966). *Marcello Malpighi and the Evolution of Embryology*. Ithaca: Cornell University Press.

Afzelius, B. A. (1972). Sperm morphology and fertilization biology. In: *Edinburgh Symposium on the Genetics of the Spermatozoon, 1971* (R. A. Beatty and S. Gluecksohn-Waelsch, eds.), pp. 131–143. Edinburgh: University of Edinburgh.

Althusser, L. (1990). *Philosophy and the Spontaneous Philosophy of the Scientists and Other Essays*. London: Verso.

Ameisen, J. C. (1996). The origin of programmed cell death: perspectives. *Science* **272** (31 May), 1278–1279.

Andersson, S. G. E., Zomorodipour, A., Andersson, J. O., Sicheritz-Pontén, T., Alsmark, U. C. M., Podowski, R. M., Näslund, A. K., Eriksson, A.-S., Winkler, H. H., and Kurland, C. G. (1998). The genome sequence of *Rickettsia prowazekii* and the origin of mitochondria. *Nature*, **396**, 133–140.

Anonymous (1998). Biology versus physics? *Nature* **391**, 107.

Appel, T. A. (1987). *The Cuvier–Geoffroy Debate: French Biology in the Decades Before Darwin*. New York: Oxford University Press.

Applebury, M. L. (1994). Relationships of G-protein–coupled receptors: a survey with the photoreceptor Opsin subfamily. In: *Molecular Evolution of Physiological Processes*, 47th Annual Symposium Society of Physiologists, Vol. 49 (D. M. Fambrough, ed.), pp. 235–248. New York: Rockefeller University Press.

Aristotle (1984). *Generation of Animals*. In: *The complete works of Aristotle*, revised Oxford translation. Vol. 1, Book 1 (J. Barnes, ed.), Bollingen series LXXI, 2. Princeton: Princeton University Press.

Arnold, M. L. (1997). *Natural Hybridization and Evolution*. New York: Oxford University Press.

Aronowitz, S. (1988). *Science As Power: Discourse and Ideology in Modern Society*. Minneapolis: University of Minnesota Press.

Arrhenius, G., Sales, B., Mojzsis, S., and Lee, T. (1997). Entropy and charge in molecular evolution — the case of phosphate. *J. Theor. Biol.* **187**, 503–522.

Atkins, J. F. (1993). Contemporary RNA genomes. In: *The RNA World: The Nature of Modern RNA Suggests a Prebiotic RNA World* (R. F. Gesteland and J. F. Atkins, eds.), pp. 535–556. Cold Spring Harbor: Cold Spring Harbor Laboratory Press.

Attardi, G., and Schatz, G. (1988). Biogenesis of mitochondria. *Ann Rev. Cell Biol.* **4**, 289–333.

Avnir, D., Biham, O., Lidar D., and Malcai. O. (1998). Is the geometry of nature fractal? *Science* **279**, 39–40.

Bada, J. L., Glavin, D. P., McDonald G. D., and Becker. L. (1998). A search for endogenous amino acids in Martian Meteorite ALH84001. *Science* **279**, 362–365.

Baer, K. E. v. (1966). *Epistola de ovi mammalium et hominis genesi.* Lipsiae; 1827. Facsimile reproduction. Bruxelles: Culture et Civilisation.

Baer, K. E. V. (1967). *Uper Entwickelungsgeschichte der Thiere*. Königsberg; 1828. Facsimile reproduction. Bruxelles: Culture et Civilisation.

Bajaj, J. E. (1988). Francis Bacon, the first philosopher of modern science. In: *Science, Hegemony and Violence: A Requiem for Modernity* (A. Nandy, ed.), pp. 24–67. Bombay: Oxford University Press.

Baker, H. F., and Ridley, R. M., eds. (1996). *Prion Diseases*. Totowa, NJ: Humana.

Baltimore, D. (1971). Expression of animal virus genomes. *Bacteriol. Rev.* **35**, 235–241.

Baltscheffsky, H., and Baltscheffsky, M. (1994). Molecular origin and evolution of early biological energy conversion. In: *Early Life on Earth, Nobel Symposium* No. 84. (S. Bengtson, ed.), pp. 81–90. New York: Columbia University Press.

Baltscheffsky, H. (1997). Major "anastrophes" in the origin and early evolution of biological energy conversion. *J. Theor. Biol.* **87**, 495–501.

Bandman, E., Moore, L. A., Arrizubieta, M. J., Tidyman, W. E., Herman L., and Wick. M. (1994). The evolution of the chicken sarcomeric myosin heavy chain multigene family. In: *Molecular Evolution of Physiological Processes*, 47th Annual Symposium Society of Physiologists, Vol. 49. (D. M. Fambrough, ed.), pp. 129–139. New York: Rockefeller University Press.

Barinaga, M. (1991). "Is homosexuality biological?" News and Comment. *Science* (30 August), **253**, 956–957.

Barinaga, M. (1996). "A shared strategy for virulence," Research News. *Science* (31 May), **272**, 1261–1263.

Barnes, R. D. (1985). Current perspectives on the origins and relationships of lower invertebrates. In: *The Origins Vertebrates,* The Systematics Association Special Volume 28 (S. Conway Morris, R. Gibson and H. M. Platt, eds.), pp. 360–367. Oxford: Clarendon Press.

Barthélemy-Madaule, M. (1982). *Lamarck the Mythical Precursor: A Study of the Relations between Science and Ideology.* Cambridge: MIT Press.

Baserga, S. J., and Steitz, J. A. (1993). The diverse world of small ribonucleoproteins. In: *The RNA World: The Nature of Modern RNA Suggests a Prebiotic RNA World* (R. F. Gesteland and J. F. Atkins, eds.), pp. 383–381. Cold Spring Harbor: Cold Spring Harbor Laboratory Press.

Bass, B. L. (1993). RNA editing: new uses for old players in the RNA World. In: *The RNA World: The Nature of Modern RNA Suggests a Prebiotic RNA World* (R. F. Gesteland and J. F. Atkins, eds.), pp. 239–381. Cold Spring Harbor: Cold Spring Harbor Laboratory Press.

Bauld, J., D'Amelio E., and Farmer, J. D. (1992). Modern microbial mats. In: *The Proterozoic Biosphere: A Multidisciplinary Study* (J. W. Schopf and C. Klein, eds.), pp. 261–269. Cambridge: Cambridge University Press.

Baylies, M. K., and Bate, M. (1996). *Twist*: a myogenic switch in *Drosophila. Science* **272**, 1481–1484.

Behe, M. J. (1996). *Darwin's Black Box: the Biochemical Challenge to Evolution.* New York: The Free Press.

Bell, G. (1997). *Selection: The Mechanism of Evolution.* New York: Chapman & Hall.

Bengtson, S. (1994). The advent of animal skeletons. In: *Early Life on Earth*, Nobel Symposium No. 84. (S. Bengtson, ed.), pp. 412–425. New York: Columbia University Press.

Benner, S. A., Ellington A. D., and Tauer, A. (1989). Modern metabolism as a palimpsest of the RNA world. *Proc. Natl. Acad. Sci. U.S.A.* **86**, 7054–7058.

Benner, S. A., Cohen, M. A., Gonnet, G. H., Berkowitz D. B., and Johnsson. K. P. (1993). Reading the palimpsest: contemporary biochemical data and the RNA world. In: *The RNA World: The Nature of Modern RNA Suggests a Prebiotic RNA World* (R. F. Gesteland and J. F. Atkins, eds.), pp. 27–70. Cold Spring Harbor: Cold Spring Harbor Laboratory Press.

Bennett, M. V. L., Zheng X., and Sogin, M. (1994). The connexins and their family tree. In: *Molecular Evolution of Physiological Processes*, 47th Annual Symposium Society of Physiologists, Vol. 49 (D. M. Fambrough, ed.), pp. 223–233. New York: Rockefeller University Press.

Benson, S. (1997). Adaptive mutation: a general phenomenon or special case? *BioEssays* **19**, 9–11.

Bergerat, A., de Massy, B., Gadelle, D., Varoutas, P.-C., Nicolas A., and Forterre, P. (1997). An atypical topoisomerase II from archaea with implications for meiotic recombination. *Nature* **326**, 414–417.

Bergström, J. (1994). Ideas on early animal evolution. In: *Early Life on Earth*, Nobel Symposium No. 84 (S. Bengtson, ed.), pp. 460–466. New York: Columbia University Press.

Bermudes, D., and Back. R. C. (1991). Symbiosis inferred from the fossil record. In: *Symbiosis As a Source of Evolutionary Innovation* (L. Margulis and R. Fester, eds.), pp. 72–91. Cambridge: MIT Press.

Bernard, C. (1957/1865). *An Introduction to the Study of Experimental Medicine.* Translated by H. C. Green. New York: Dover Publications.

Bernardi, G. (1996). The distribution of genes in the human genome. In: *Gene Families: Structure, Function, Genetics and Evolution*, Proceedings of the VIII International Congress on Isozymes. (R. S. Holmes and H. A. Lim eds.), pp. 27–41. Isingapore: World Scientific.

Berrill, N. J. (1961). *Growth, Development, and Pattern.* San Francisco: W. H. Freeman.

Bertin, L. (1952). Oviparité, ovoviviparité, viviparité. *Soc. Zool. Fr. Bull.* **77**, 84–88.

Beukes, N. J., and Klein, C. (1992). Models for iron-formation deposition. In: *The Proterozoic Biosphere: A Multidisciplinary Study* (J. W. Schopf and C. Klein, eds.), pp. 147–151. Cambridge: Cambridge University Press.

Birky Jr., C. W. (1991). Evolution and population genetics of organelle genes: mechanisms and models. In: *Evolution at the Molecular Level* (R. K. Selander, A. G. Clark and T. S. Whittam, eds.), pp. 112–134. Sunderland, MA: Sinauer Associates.

Blackburn, E. H. (1993). Telomerase. In: *The RNA World: The Nature of Modern RNA Suggests a Prebiotic RNA World* (R. F. Gesteland and J. F. Atkins, eds.), pp. 557–576. Cold Spring Harbor: Cold Spring Harbor Laboratory Press.

Blackmore, S. (1994). Models, modules, and molecules in morphogenesis. In: *Models in Phylogeny Reconstruction*, Systematics Association Special Volume 52 (R. W. Scotland, D. J. Siebert and D. M. Williams, eds.), pp. 51–63. Oxford: Clarendon Press.

Blattner, F. R., et al. (1997). The complete genome sequence of *Escherichia coli* K-12. *Science* **277**, 1453–1462.

Boiker, J. A., and Raff, R. A. (1996). Developmental genetics and traditional homology. *BioEssays* **18**, 489–49.

Bonner, J. T. (1988). *The Evolution of Complexity by Means of Natural Selection*. Princeton: Princeton University Press.

Bonner, J. T. (1998). The origins of multicellularity. *Integrative Biol.* **1**, 27–36.

Bonnet, C. (1985/1762). *Considérationsds sur les corps organisés*, Texte revu par Francine Markovits et Sophie Bienaymé. Paris: Librairie Arthème Fayard.

Borst, P., Grivell, L. A., and Groot, G. S. P. (1984). Organelle DNA. *Trends Biochem. Sci.* **9**, 128–130.

Bosch, T. C. G., Unger, T. F., Fisher, D. A., and Steele, R. E. (1989). Structure and expression of STK, an src-related gene in the simple metazoan Hydra attenuata. *Mol. Cell. Biol.* **9**, 4141–4151.

Botstein, D., Chervitz, S. A., and Cherry, J. M. (1997). Yeast as a model organism. *Science* **277**, 1259–1260.

Boveri, T. (1964/1902). On multipolar mitosis as a means of analysis of the cell nucleus [Translated by S. Glucksohn-Waelsch]. In: *Foundations of Experimental Embryology* (B. H. Willier and M. J. Oppenheimer, eds.), pp. 74–97. Englewood Cliffs, NJ: Prentice-Hall.

Bowie, J. U., Lüthy, R., and Eisenbert, D. (1995). Three-dimensional profiled: assessing the compatibility of an amino acid sequence with a three-dimensional structure. In: *Tracing Biological Evolution in Protein and Gene Structure* (M. Gō and P. Schimmel, eds.), pp. 297–309. Amsterdam: Elsevier.

Brndsted, H. V. (1969). *Planarian Regeneration*. New York: Pergamon Press.

Brown J. R., and Doolittle, W. F. (1995). Root of the universal tree of life based on ancient aminoacyl-tRNA synthetase gene duplications. *Proc. Natl. Acad. Sci. U.S.A.* **92**, 2441–2445.

Brul, S., and Stumm, C. K. (1994). Symbionts and organelles in anaerobic protozoa and fungi. *Trends Ecol. Evol.* **9**, 319–324.

Buick, R. (1992). The antiquity of oxygenic photosynthesis: Evidence from stromatolites in sulphate-deficient Archean lakes. *Science* **255**, 74–77.

Bult, C. J., White, O., Olsen, G. J., Zhou, L., et al. (1996). Complete genome sequence of the methanogenic archaeon, *Methanococcus jannaschii*. *Science* **273**, 1058–1073.

Burke, W. D., Malik, H. S., Lathe III, W. C., and Eickbush, T. H. (1998). Are retrotransposons long-term hitchhikers? Scientific correspondence. *Nature* (12 March), **392**, 141–142.

Buss, L. W. (1987). *The Evolution of Individuality*. Princeton: Princeton University Press.

Buss, L. W. (1994). Protocell life cycles. In: *Early Life on Earth*, Nobel Symposium No. 84 (S. Bengtson, ed.), pp. 133–134. New York: Columbia University Press.

Buss, L. W., and Seilacher. A. (1994). The phylum Vendobionta: a system group of the Eumetazoa? *Paleobiology* **20**, 1–4.

Cairns-Smith, A. G. (1985). *Seven Clues to the Origin of Life: A Scientific Detective Story*. Cambridge: Cambridge University Press.

Campbell, R. D. (1979). Development of Hydra lacking interstitial and nerve cells ("epithelial Hydra"). In: *Determinants of Spatial Organization*, Thirty-Seventh Symp. Soc. Devel. Biol. (S. Subtelny and I. R. Konigsberg, eds.), pp. 267–293. New York: Academic Press.

Campos, J., Fusté, M. C., Vazquez, J., Saez-Nieto J. A., and Viñas, M. (1992). Origin and spread of penicillin-resistance in *Neisseria meningitidis*. In: *Gene Transfers and Environment*, Proceedings of the Third European Meeting on Bacterial Genetics and Ecology (M. J. Gauthier, ed.), pp. 175–180. Berlin: Springer-Verlag.

Cann, A. J. (1993). *Principles of Molecular Virology*. London: Academic Press.

Cardon, L. R., Smith, S. D., Fulker, D. W., Kimberling, W. J., Pennington B. R., and Defries, J. C. (1994). Quantitative trait locus for reading disability on chromosome 5. *Science* **266**, 276–279.

Carter, G. S. (1954). On Hadzi's interpretation of animal physiology. *Systemat. Zool.* **3**, 163–167.

Caskey, C. T. (1993). Presymptomatic diagnosis: A first step toward genetic health care. *Science* **262**, 48–49.

Cavalier-Smith, T. (1975). The origin of nuclei and of eukaryotic cells. *Nature* **256**, 463–468.

Cavalier-Smith, T. (1987a). The origin of cells: a symbiosis between genes, catalysts, and membranes. *Cold Spring Harbor Symposia on Quant. Biol.* **LII**, 805–824.

Cavalier-Smith, T. (1987b). The origin of eukaryote and archaebacterial cells. *Ann. N.Y. Acad. Sci.* **503**, 17–54.

Cavalier-Smith, T. (1991). The evolution of cells. In: *Evolution of Life: Fossils, Molecules, and Culture* (S. Osawa and T. Honjo, eds.), pp. 271–304. Tokyo: Springer-Verlag.

Cavalier-Smith, T. (1992). The number of symbiotic origins of organelles. *BioSystems* **28**, 91–106.

Cavalier-Smith, T., and Chao, E. F. (1996). Molecular phylogeny of the free-living archaezoan *Trepomonas agilis* and the nature of the first eukaryote. *J. Mol. Evol.* **43**, 551–562.

Cech, T. R. (1993). Structure and mechanism of the large catalytic RNAs: Group I and Group II introns and ribonuclease P. In: *The RNA World: The Nature of Modern RNA Suggests a Prebiotic RNA World* (R. F. Gesteland and J. F. Atkins, eds.), pp. 239–269. Cold Spring Harbor: Cold Spring Harbor Laboratory Press.

Cedergren, R. (1995). Using sequence variation to model RNA three-dimensional structures. In: *Tracing Biological Evolution in Protein and Gene Structure* (M. Gō and P. Schimmel, eds.), pp. 149–160. Amsterdam: Elsevier.

C. elegans Sequencing Consortium, The (1998). Genome sequence of the nematode *C. elegans*: A platform for investigating biology. *Science*, **282**, 2012–2018.

Cerff, R. (1995). The chimeric nature of nuclear genomes and the antiquity of introns as demonstrated by the GAPDH gene system. In: *Tracing Biological Evolution in Protein and Gene Structure* (M. Gō and P. Schimmel, eds.), pp. 205–223. Amsterdam: Elsevier.

Chait, B. T., Wang, R., Beavis, R. C., and Kent, S. B. H. (1993). Protein ladder sequencing. *Science* **262**, 89–92.

Chambers, R. (1994/1844). *Vestiges of the Natural History of Creation and Other Evolutionary Writings*. Edited with a new Introduction by James A. Secord. Chicago: The University of Chicago Press.

Chamberlin, T. C., and Chamberlin, R. T. (1994/1908). Early terrestrial conditions that may have favored organic synthesis. *Science* **28**, 897–911, 1908. Reprinted in D. W. Deamer and G. R. Fleischaker, eds., *Origins of Life: The Central Concepts*, pp. 73–814. Boston: Jones & Bartlett.

Chang, S. (1994). The planetary setting of prebiotic evolution. In: *Early Life on Earth*, Nobel Symposium No. 84 (S. Bengtson, ed.), pp. 10–23. New York: Columbia University Press.

Chao, L. (1994). Evolution of genetic exchange in RNA viruses. In: *The Evolutionary Biology of Viruses* (S. S. Morse, ed.), pp. 233–250. New York: Raven Press.

Charlesworth, B., and Langley, C. H. (1991). Population genetics of transposable elements in Drosophila. In: *Evolution at the Molecular Level* (R. K. Selander, A. G. Clark and T. S. Whittam, eds.), pp. 150–176. Sunderland, MA: Sinauer Associates.

Chatton, E. (1938). *Titres et Travoux Scientifiques (1906–1937) de Edouard Chatton*, E. Sottano, ed. France: Sète.

Chay, T. R., and Lee, Y. S. (1990). Bursting, beating, and chaos by two functionally distinct inward current inactivations in excitable cells. In: *New York Academy of Sciences*, Vol. 591: *Mathematical Approaches to Cardiac Arrhythmias*, pp. 328–350.

Chenuil, P. H., Chenuil, A., and Adoutte. A. (1994). Can the Cambrian explosion be inferred through molecular phylogeny? *Development* (1994 Suppl., *The Evolution of Developmental Mechanisms*), 15–25.

Chothia, C. (1994). Protein families in the metazoan genome. *Development* (1994 Suppl., *The Evolution of Developmental Mechanisms*), 27–33.

Christen, R. (1994). Molecular phylogeny and the origin of Metazoa. In: *Early Life on Earth*, Nobel Symposium No. 84 (S. Bengtson, ed.), pp. 466–474. New York: Columbia University Press.

Christen, R., Ratto, A., Baroin, A., Perasso, R., Grell, K. G., and Adoutte. A. (1991). Origin of metazoans, a phylogeny deduced from sequences of the 28S ribosomal RNA. In: *The Early Evolution of Metazoa and the Significance of Problematic Taxa* (A. M. Simonetta and S. Conway Morris, eds.), pp. 1–9. Cambridge: Cambridge University Press.

Clark, C. G., and Roger. A. J. (1995). Direct evidence for secondary loss of mitochondria in *Entamoeba histolytica*. *Proc. Natl. Acad. Sci. U.S.A.* **92**, 6518–6521.

Clayton, R. A., White, O., Ketchum, K. A., and Venter. C. (1997). The first genome from the third domain of life. *Nature* **387**, 459–462.

Clegg, M. T., Learn, G. H., and Golenberg, E. M. (1991). Molecular evolution of chloroplast DNA. In: *Evolution at the Molecular Level* (R. K. Selander, A. G. Clark and T. S. Whittam, eds.), pp. 135–149. Sunderland, MA: Sinauer Associates.

Clegg, M. T., Gaut, B. S., Learn Jr., G. H., and Morton, B. R. (1994). Rates and patterns of chloroplast DNA evolution. *Proc. Natl. Acad. Sci. U.S.A.* **91**, 2795–2801.

Cole, S. T., et al. (1998). Deciphering the biology of *Mycobacterium tuberculosis* from the complete genome sequence. *Nature* **393**, 537–544.

Collins, F., and Galas, D. (1993). A new five-year plan for the U.S. Human Genome Project. *Science* **262**, 43–46.

Conway Morris, S. (1989). Burgess Shale faunas and the Cambrian explosion. *Science* **246**, 339–346.

Conway Morris, S. (1994a). Why molecular biology needs palaeontology. *Development* (1994 Suppl., *The Evolution of Developmental Mechanisms*), 1–13.

Conway Morris, S. (1994b). Early metazoan evolution: first steps to an integration of molecular and morphological data. In: *Early Life on Earth*, Nobel Symposium No. 84 (S. Bengtson, ed.), pp. 450–459. New York: Columbia University Press.

Cook, R. (1987). *Outbreak*. New York: Putnam.

Cooperative Human Linkage Center (CHLC), Généthon, Yale University and Centre d'Etude du Polymorphisme Humain (CRPH) (1994). A comprehensive human linkage map with centimorgan density. *Science* **265**, 2064.

Copeland, H. F. (1956). *Classification of the Lower Organisms*. Palo Alto: Pacific Books.

Copeland, N. G., Jenkins, N. A., Gilbert, D. J., Eppig, J. T., Maltais, L. J., Miller, J. C., Dietrich, W. F., Weaver, A., Lincoln, S. E., Steen, R. G., Stein, L. D., Nadeau, J. H., and Lander, E. S. (1993). A genetic linkage map of the mouse: current applications and future prospects. *Science* **262**, 57–66.

Corliss, J. O. (1987). Protistan phylogeny and eukaryogenesis. *Int. Rev. Cytol.* **100**, 319–370.

Coveney, P., and Highfield, R. (1995). *Frontiers of Complexity: The Search for Order in a Chaotic World*. New York: Fawcett Columbine.

Cox, L. S., and Lane. D. P. (1995). Tumour suppressors, kinases and clamps: how p53 regulates the cell cycle in response to DNA damage. *BioEssay* **17**, 501–508.

Crawford, I. P., and Milkman, R. (1991). Orthologous and paralogous divergence, reticulate evolution, and lateral gene transfer in bacterial trp genes. In: *Evolution at the Molecular Level* (R. K. Selander, A. G. Clark and T. S. Whittam, ed.), pp. 77–95. Sunderland, MA: Sinauer Associates.

Crick, F. H. C. (1958). On protein synthesis. *Symp. Soc. Exp. Biol.* **12**, 138–163.

Crick, F. H. C. (1968). The origin of the genetic code. *J. Mol. Biol.* **38**, 367–379.

Crick, F. (1981). *Life Itself: Its Origin and Nature*. New York: Simon & Schuster.

Crick, F. (1988). *What Mad Pursuit: A Personal View of Scientific Discovery*. New York: Basic Books.

Cui, X., Wise, R. P., and Schnable, P. S. (1996). The rf2 nuclear restorer gene of male-sterile T-cytoplasm maize. *Science* **272**, 1334–1336.

Cummings, M. P. (1994). Transmission patterns of eukaryotic transposable elements: arguments for and against horizontal transfer. *Trends Ecol. Evol.* **9**, 141–145.

Cusack, S. (1995). Structure, function and evolution of amionoacyl-tRNA synthetases. In: *Tracing Biological Evolution in Protein and Gene Structure* (M. Gō and P. Schimmel, eds.), pp. 37–52. Amsterdam: Elsevier.

Cuticchia, A. J., Chipperfield, M. A., Porter, C. J., Kearns, W., and Pearson, P. L. (1993). Managing all those bytes: the Human Genome Project. *Science* **262**, 47–48.

Darwin, C. R. (1958/1872). *On the Origin of Species by Means of Natural Selection or the Preservation of Favoured Races in the Struggle for Life.* Edited J. W. Burrow. Harmondsworth: Penguin Classics. (1968/1859). 6th ed., The Mentor Edition. New York: Mentor Book.

Darwin, F. (1995/1902). *The Life of Charles Darwin*. London: Senate Studio Editions Ltd. [London: John Murray, 1902].

Davies, P. (1995). *Are We Alone? Philosophical Implications of the Discovery of Extraterrestrial Life*. New York: Basic Books.

Davies, J., von Ahsen, U., and Schroeder, R. (1993). Antibiotics and the RNA world: a role for low-molecular weight effectors in biochemical evolution? In: *The RNA World: The Nature of Modern RNA Suggests a Prebiotic RNA World* (R. F. Gesteland and J. F. Atkins, eds.), pp. pp. 185–204. Cold Spring Harbor: Cold Spring Harbor Laboratory Press.

Dawe, C. (1969). Phylogeny and oncogeny. In: *Neoplasm and Related Disorders of Invertebrate and Lower Vertebrate Animals*, National Cancer Institute Monograph, No. 31 (C. J. Dawe and J. C. Harshbarger, eds.), pp. 1–39.

Dawkins, R. (1965). *The Blind Watchmaker*. New York: W. W. Norton.

Dawkins, R. (1976). *The Selfish Gene*. Oxford: Oxford University Press.

Dawkins, R. (1995). *River out of Eden: A Darwinian View of Life*. New York: Basic Books.

Dawkins, R. (1997). Human chauvinism. *Evolution* **51**, 1015–1020.

Deamer, D. W., and Fleischaker, G. R. (1994). *Origins of Life: The Central Concept*. Boston: Jones & Bartlett.

Deamer, D. W., Mahon, E. H., and Bosco, G. (1994). Self-assembly and function of primitive membrane structures. In: *Early Life on Earth*, Nobel Symposium No. 84 (S. Bengtson, ed.), pp. 107–123. New York: Columbia University Press.

Deckert, G., Warren, P. V., Gaasterland, T., Young, W. G., Lenox, A. L., Graham, D. E., Overbeek, R., Snead, M. A., Keller, M., Aujay, M., Huber, R., Feldman, R. A., Short, J. M., Olsen, G. J., and Swanson, R. V. (1998). The complete genome of the hyperthermophilic bacterium *Aquifex aeolicus*. *Nature* **392**, 353–358.

de Duve, C. (1995). *Vital Dust: Life as a Cosmic Imperative*. New York: Basic Books.

Deleuze, G. (1994). *Difference and Repetition*. Translated by Paul Patton. New York: Columbia University Press.

Deleuze, G., and Guattari, F. (1987). *A Thousand Plateaus: Capitalism and Schizophrenia*. Translation and foreword by B. Massumi. Minneapolis: University of Minnesota Press.

Desmond, A. (1994). *Huxley: The Devil's Disciple,* Vol. 1. London: Michael Joseph.

Desmond, A. (1997). *Huxley: Evolution's High Priest*, Vol. 2. London: Michael Joseph.

Desmond, A., and Moore, J. (1991). *Darwin*. London: Michael Joseph.

Devereux, J., Haeberli, P., and Smithies, O. (1984). A comprehensive set of sequence analysis programs for the VAX. *Nucleic Acids Res.* **12**, 387–395.

DiBerardino, M. A. (1997). *Genomic Potential of Differentiated Cells*. New York: Columbia University Press.

Dobner, T., Horikoshi, N., Rubenwolf, S., and Shenk. T. (1996). Blockage by adenovirus E4orf6 of transcriptional activation by the p53 tumor suppressor. *Science* **272**, 1470–1473.

Domingo, E., and Holland, J. J. (1994). Mutation rates and rapid evolution of RNA viruses. In: *The Evolutionary Biology of Viruses* (S. S. Morse, ed.), pp. 161–184. New York: Raven Press.

Doolittle, R. F. (1985). The geneology of some recently evolved vertebrate proteins. *Trends Biochem. Sci.* **10**, 233–2575.

Doolittle, R. F. (1986). *Of URFS and ORFS: A Primer on How to Analyze Derived Amino Acids Sequences*. Mill Valley, CA: University Science Books.

Doolittle, R. F., Feng, D.-F., Tsang, S., Cho, G., and Little, E. (1996a). Determining divergence times of the major kingdoms of living organisms with a protein clock. *Science* **271**, 470–477.

Doolittle, R. F., Feng, D.-F., Tsang, S., Cho G., and Little, E. (1996b). "Dating the cenancester of organisms," Response. *Science* **274**, 1751–1753.

Doolittle, W. F. (1989). Whatever happened to the progenote? In: *The Hierarchy of Life: Molecules and Morphology in Phylogenetic Analysis*. Proceedings from Nobel Symposium 70 (B. Fernholm, K. Bremer and H. Jörnvall, eds.), pp. 63–72. Amsterdam: Excerpta Medica.

Doolittle, W. F. (1998). "A paradigm gets shifty," News and Views. *Nature* (25 March), **392**, 15–16.

Doolittle, W. F., and Brown, J. R. (1994). Tempo, mode, the progenote, and the universal root. *Proc. Natl. Acad. Sci. U.S.A.* **91**, 6721–6728.

Dover, G. (1997). "There's more to life than selection and neutrality," Correspondence. *BioEssays* (January). **19**, 91–93.

Duboule, D. (1994). Temporal colinearity and the phylotypic progression: a basis for the stability of a vertebrate Bauplan and the evolution of morphologies through heterochrony. *Development* (1994 Suppl., *The Evolution of Developmental Mechanisms*), 135–142.

Driesch, H. (1964/1892). Entwicklungsmechanische. Studien. I. *Z. Wiss. Zool.* **53**, 160–178, 183–184. (1892). The potency of the first two cleavage cells in echinoderm development: experimental production of partial and double formations. Abridged and translated by L. Matzger, M. Hamburger, V. Hamburger and T. W. Hall with references, plate VII, and explanation of figures. In: *Foundations of Experimental Embryology* (B. H. Willier and J. M. Oppenheimer, eds.), pp. 40–50. Englewood Cliffs, NJ: Prentice-Hall.

Dyer, B. D., and Obar, R. A. (1994). *Tracing the History of Eukaryotic Cells: The Enigmatic Smile*. New York: Columbia University Press.

Edgell, D. R., and Doolittle, W. F. (1997). Archaebacterial genomics: the complete genome sequence of *Methanococcus jannaschii*. *BioEssays* **19**, 1–4.

Edlind, T. D., Li, J., Visvesvara, G. S., Vodkin, M. H., McLaughlin, G. L., and Katiyar, S. K. (1996). Phylogenetic analysis of β-tubulin sequences from amitochondrial protozoa. *Mol. Phylogenet Evol.* **5**, 359–367.

Eigen, M. (1992). *Steps Toward Life: A Perspective on Evolution*. With R. Winkler-Oswatitsch. Translation by Paul Woolley. Oxford: Oxford University Press.

Eigen, M., Lindemann, B. F., Tietze, M., Winkler-Oswatitsch, R., Dress, A., and Von Haeseler, A. (1989). How old is the genetic code? Statistical geometry of tRNA provides an answer. *Science* **244**, 673–679.

Eldredge, N. (1985). *Time Frames: The Evolution of Punctuated Equilibria*. Princeton: Princeton Science Library.

Elwood, H. J., Olsen, G. J., and Sogin, M. L. (1985). The small-subunit ribosomal RNA gene sequences from the hypotrichous ciliates *Oxytricha nova* and *Stylonychia pustulata*. *Mol. Biol. Evol.* **2**, 399–410.

Engel, J., Efimov, V. P., and Mauer, P. (1994). Domain organizations of extracellular matrix proteins and their evolution. *Development* (1994 Suppl., *The Evolution of Developmental Mechanisms*), 35–42.

Erickbush, T. H. (1994). Origin and evolutionary relationships of retroelements. In: *The Evolutionary Biology of Viruses* (S. S. Morse, ed.), pp. 121–157. New York: Raven Press.

Erickson, H. P., and Bourdon, M. A. (1989). Tenascin: an extracellular matrix protein prominent in specialized embryonic tissues and tumors. *Ann. Rev. Cell Biol.* **5**, 71–92.

(The) EU Arabidopsis Genome Project (Bevan, M., et al.) (1998). Analysis of 1.9 Mb of contiguous sequence from chromosome 4 of *Arabidopsis thaliana*. *Nature* **391**, 485–488.

Fabbri, S., Fruscoloni, P., Bufardeci, E., Di Nocola Negri, E., Baldi, M. I., Gandini Attardi, D., Mattoccia, E., and Tocchini-Valentini, G. P. (1998). Conservation of substrate recognition mechanisms by tRNA splicing endonucleases. *Science* **280**, 284–292.

Falkow, S. (1975). *Infectious Multiple Antibiotic Resistance*. London: Pion.

Fawcett, D. W. (1975). A comparative view of sperm ultrastructure. *Biol. Reprod., Suppl.* **2**, 90–127.

Felsenstein, J. (1984). The statistical approach to inferring evolutionary trees and what it tells us about parsimony and compatibility. In: *Cladistics: Perspectives in the Reconstruction of Evolutionary History* (T. Ducan and T. F. Stuessy, eds.), pp. 169–191. New York: Columbia University Press.

Felsenstein, J. (1988). *Phylogeny Inference Package*, Version 3.1. University of Washington, Seattle.

Fenner, F. (1995). Classical studies of virus evolution. In: *Molecular Basis of Virus Evolution* (A. J. Gibbs, C. H. Calisher and F. García Arenal, eds.), pp. 13–30. Cambridge: Cambridge University Press.

Fietz, M. J., Concordet, J.-P., Barbosa, R., Johnson, R., Krauss, S., McMahon, A. P., Tabin, C., and Ingham, P. W. (1994). The hedgehog gene family in Drosophila and vertebrate development. *Development* (1994 Suppl., *The Evolution of Developmental Mechanisms*), 43–51.

Fitch, W. M., and Margoliash, E. (1970). The usefulness of amino acid and nucleotide sequences in evolutionary studies. In: *Evolutionary Biology*, Vol. 4. (Th. Dobzhansky, M. K. Hecht and W. C. Streere eds.), pp. 67–109. New York: Plenum.

Flavell, R. A., Allen, H., Burkly, L. C., Sherman, D. H., Waneck, G. L., and Widera, G. (1986). Molecular biology of the H-2 histocompatibility complex. *Science* **233**, 437–443.

Fleischmann, R. D., et al. (1995). Whole-genome random sequencing and assembly of *Haemophilus influenzae* Rd. *Science* **269**, 496–512.

Forman, P. (1997). "Assailing the seasons," Book Reviews. *Science* (2 May), **276**, 750–752.

Forney, J. (1997). DNA rearrangements and mating-type determination in *Paramecium tetraurelia*. *BioEssays* **19**, 5–8.

Forsburgh, S. L., and Guarente, L. (1989). Communication between mitochondria and the nucleus in regulation of cytochrome gene in the yeast *Saccharomyces cerevisiae*. *Ann. Rev. Cell Biol.* **5**, 153–180.

Foucault, M. (1973). *The Birth of the Clinic: An Archaeology of Medical Perception.* Translated by A. M. Sheridan Smith. New York: Pantheon.

Fox, G. E., Magrum, L. J., Balch, W. E., Wolfe, R. S., and Woese, C. R. (1977). Classification of methanogenic bacteria by 16S ribosomal RNA characterization. *Proc. Natl. Acad. Sci. U.S.A.* **74**, 4537–4541.

Fraser, C. M., et al. (1995). The minimal gene complement of *Mycoplasma genitalium*. *Science* **270**, 397–403.

Fraser, C. M., et al. (1997). Genomic sequence of a Lyme disease spirochaete, *Borellia burgdorferi*. *Nature* **390**, 580–586.

Fraser, C. M., et al. (1998). Complete genome sequence of *Treponema pallidum*, the Syphilis spirochete. *Science* **281**, 375–388.

Fryberg, C., Ryan, L., McNally, L., Kenton, M., and Fryberg, E. (1994). The actin protein superfamily. In: *Molecular Evolution of Physiological Processes*, 47th Annual Symposium Society of Physiologists, Vol. 49 (D. M. Fambrough, ed.), pp. 173–178. New York: Rockefeller University Press.

Frydman, J., and Hartl, F. U. (1996). Principles of chaperone-assisted protein folding: differences between in vitro and in vivo mechanisms. *Science* **272**, 1497–1502.

Fukami-Kobayashi, K., Mizutani, M., and Gō, M. (1995). Correlation between module boundaries and intron positions in hemoglobins from various taxa. In: *Tracing Biological Evolution in Protein and Gene Structure* (M. Gō and P. Schimmel, eds.), pp. 271–282. Amsterdam: Elsevier.

Gans, C. (1989). Stages in the origin of vertebrates: analysis by means of scenarios. *Biol. Rev.* **64**, 221–268.

Garnett, G. P., and Antia, R. (1994). Population biology of virus–host interactions. In: *The Evolutionary Biology of Viruses* (S. S. Morse, ed.), pp. 51–73. New York: Raven Press.

Garrod, A. E. (1923/1909). *Inborn Errors of Metabolism*, 2nd ed. London: Oxford University Press.

Gedulin, B., and Arrhenius, G. (1994). Sources and geochemical evolution of RNA precursor molecules: the role of phosphate. In: *Early Life on Earth*, Nobel Symposium No. 84 (S. Bengtson, ed.), pp. 91–106. New York: Columbia University Press.

Gehling, J. G. (1991). The case for Ediacaran fossil roots to the metazoan tree. *Memoirs Geological Society India* No. 20, pp. 181–223.

Geison, G. L. (1995). *The Private Science of Louis Pasteur*. Princeton: Princeton University Press.

Gellon, G., and McGinnis, W. (1998). Shaping animal body plans in development and evolution by modulation of *Hox* expression patterns. *BioEssays* **20**, 116–125.

Gerhart, J., and Kirschner, M. (1997). *Cells Embryos, and Evolution: Toward a Cellular and Developmental Understanding of Phenotypic Variation and Evolutionary Adaptability*. Oxford: Blackwell Science.

Gerhold, D., and Caskey, C. T. (1996). Its the genes! EST access to human genome content. *BioEssays* **18**, 973–981.

Gibbs, A., and Keese. P. K. (1995). In search of the origins of viral genes. In: *Molecular Basis of Virus Evolution* (A. J. Gibbs, C. H. Calisher and F. García Arenal, eds.), pp. 76–90. Cambridge: Cambridge University Press.

Gibbs, A. J., Calisher, C. H., and García Arenal, F. (1995). Introduction. In: *Molecular Basis of Virus Evolution* (A. J. Gibbs, C. H. Calisher and F. García Arenal, eds.), pp. 1–10. Cambridge: Cambridge University Press.

Gilbert, W. (1985). Genes-in-pieces revisited. *Science* **228**, 823–824.

Gilbert, W. (1986). The RNA world. *Nature* **319**, 618.

Gilbert, W. (1987). The exon theory of genes. *Cold Spring Harbor Symposia on Quantitative Biology* **52**, 901–905.

Gilbert, W. (1992). A vision of the grail. In: *The Code of Codes: Scientific and Social Issues in the Human Genome Project* (D. J. Kevles and L. Hood, eds.), pp. 83–97. Cambridge: Harvard University Press.

Gilbert, W., Long, M., Rosenberg, C., and Glynias, M. (1995). Tests of the exon theory of genes. In: *Tracing Biological Evolution in Protein and Gene Structure* (M. Gō and P. Schimmel, eds.), pp. 237–247. Amsterdam: Elsevier.

Gillham, N. W. (1994). *Organelle Genes and Genomes*. New York: Oxford University Press.

Ginzburg, C. (1986). *Clues, Myths, and the Historical Method*. Translated by John and Anne C. Tedeschi. Baltimore: The Johns Hopkins University Press.

Glaessner, M. F. (1984). *The Dawn of Animal Life: A Biohistorical Study.* Cambridge: Cambridge University Press.

Gō, M. (1981). Correlation of DNA exonic regions with protein structural units in haemoglobin. *Nature* **291**, 90–92.

Gō, M., and Noguti, T. (1995). Putative origin of introns deduced from protein anatomy. In: *Tracing Biological Evolution in Protein and Gene Structure* (M. Gō and P. Schimmel, eds.), pp. 229–235. Amsterdam: Elsevier.

Goffeau, A., Barrell, B. G., Bussey, H., Davis, R. W., Dujon, B., Feldmann, H., Galibert, F., Hoheisel, J. D., Jacq, C., Johnston, M., Louis, E. J., Mewes, H. W., Murakami, Y., Philippsen, P., Tettelin, H., and Oliver. S. G. (1996). Life with 6000 genes. *Science* **274**, 546–567.

Goffeau, A., et al. (1997). The yeast genome directory. *Nature* **387** (Suppl.), 1–105.

Gogarten, J. P., Kibak, H., Dittrich, P., Taiz, L., Bowman, E. J., Bowman, B. J., Manolson, M. F., Poole, R. J., Date, T., Oshima, T., Konishi, J., Denda, K., and Yoshida, M. (1989). Evolution of the vacuolar H^+-ATPase: implications for the origin of eukaryotes. *Proc. Natl. Acad. Sci. U.S.A.* **85**, 6661–666.

Gogarten J. P., Olendzenski, L., Hilario, E., Simon, C., and Holsinger. K. E. (1996). Dating the Cenancester of organisms. *Science* **274**, 1750–1751.

Gojobori, T., Endo, T., and Ikeo, K. (1995). Domain evolution of serine protease and its inhibitor genes. In: *Tracing Biological Evolution in Protein and Gene Structure* (M. Gō and P. Schimmel, eds.), pp. 249–260. Amsterdam: Elsevier.

Gold, L., Allen, P., Binkley, J., Brown, D., Schneider, D., Eddy, S. R., Tuerk, C., Green, L., MacDougal, S., and Tasset, D. (1993). RNA: the shape of things to come. In: *The RNA World: The Nature of Modern RNA Suggests a Prebiotic RNA World* (R. F. Gesteland and J. F. Atkins, eds.), pp. 497–509. Cold Spring Harbor: Cold Spring Harbor Laboratory Press.

Goldbach, R., and de Haan, P. (1994). RNA viral supergroups and the evolution of RNA viruses. In: *The Evolutionary Biology of Viruses* (S. S. Morse, ed.), pp. 105–119. New York: Raven Press.

Goldmann, L. (1977). *Lukács and Heidegger: Towards a New Philosophy.* Translated by W. Q. Boelhower. London: Routledge & Kegan Paul.

Goldman, M. A., Holmquist, G. P., Gray, N. C., Caston, L. A., and Nag. A. (1984). Replication timing of genes and middle repetitive sequences. *Science* **224**, 684–692.

Goldschmidt, R. B. (1956). *Portraits from Memory: Recollections of a Zoologist.* Seattle: University of Washington Press.

Goldschmidt, R. B. (1960). *In and Out of the Ivory Tower: The Autobiography of Richard B. Goldschmidt*. Seattle: University of Washington Press.

Golubic, S. (1994). The continuing importance of cyanobacteria. In: *Early Life on Earth* (S. Bengston, ed.), pp. 335–340, New York: Columbia University Press.

Goodman, M., Czelusniak, J., Koop, B. F., Tagle, D. A., and Slightom, J. L. (1987). Globins: A case study of molecular phylogeny. *Cold Spring Harbor Symposia on Quantitative Biology* **52**, 875–890.

Goodson, H. V. (1994). Molecular evolution of the myosin superfamily: application of phylogenetic techniques to cell biological questions. In: *Molecular Evolution of Physiological Processes*, 47th Annual Symposium Society of Physiologists, Vol. 49 (D. M. Fambrough, ed.), pp. 141–157. New York: Rockefeller University Press.

Goodwin, B. C. (1984). A relational or field theory of reproduction and its evolutionary implications. In: *Beyond Neo-Darwinism: An Introduction to the New Evolutionary Paradigm* (M.-W. Ho and P. T. Saunders, eds.), pp. 219–241. Orlando: Academic Press.

Goodwin, B. (1994). *How the Leopard Changed Its Spots: The Evolution of Complexity*. New York: Touchstone Book.

Gorbalenya, A. E. (1995). Origin of RNA viral genomes: approaching the problem by comparative sequence analysis. In: *Molecular Basis of Virus Evolution* (A. J. Gibbs, C. H. Calisher and F. García Arenal, eds.), pp. 49–66. Cambridge: Cambridge University Press.

Goreau, T. F., Goreau, N. I., and Goreau, T. J. (1979). Corals and coral reefs. *Sci. Am.* **241**, 124–137.

Gould, S. J. (1977). *Ontogeny and Phylogeny*. Cambridge: Belknap Press.

Gould, S. J. (1989). *Wonderful Life: The Burgess Shale and the Nature of History*. New York: W. W. Norton.

Gould, S. J. (1994). Tempo and mode in the macroevolutionary reconstruction of Darwinism. *Proc. Natl. Acad. Sci. U.S.A.* **91**, 6764–6771.

Gould, S. J., and Lewontin. R. C. (1979). The spandrels of San Marco and the Panglossian paradigm: a critique of the adaptionist programme. *Proc. Roy. Soc. London B.* **205**, 581–598.

Gould, S. J., and Vrba, E. S. (1982). Exaptation — a missing term in the science of form. *Paleobiology* **8**, 4–15.

Gray, M. W. (1989). Origin and evolution of mitochondrial DNA. *Ann. Rev. Cell Biol.* **5**, 25–50.

Gray, M. W., Greenwood, S. J., Smallman, D. S., Spencer, D. F., and Schnare, M. N. (1995). Ribosomal RNA in pieces: a modern paradigm of the primordial ribosome. In: *Tracing Biological Evolution in Protein and Gene Structure* (M. Gō and P. Schimmel, eds.), pp. 65–76. Amsterdam: Elsevier.

Gray, P. (1970). *The Encyclopedia of the Biological Sciences*, 2nd ed. New York: Van Nostrand Reinhold.

Green, P., Lipman, D., Hiller, L., Waterston, R., States, D., and Claverie, J.-M. (1993). Ancient conserved regions in new gene sequences and the protein databases. *Science* **259**, 1711–1716.

Greenberg, J. M. (1959). Ancestors, embryos, and symmetry. *Systemat. Zool.* **8**, 212–221.

Greenwood, S. J., Sogin, M. L., and Lynn, D. H. (1991). Phylogenetic relationships within the Class Oligohymenophorea, Phylum Ciliophora, inferred from the complete small subunit rRNA gene sequences of *Colpidium campylum, Glaucoma chattoni*, and *Opisthonecta henneguyi. J. Mol. Evol.* **33**, 163–174.

Grimstone, A. V. (1959). Cytology, homology, and phylogeny — a note on "organic design." *Am. Nat.* **93**, 273–282.

Grobstein, C. (1988). *Science and the Unborn: Choosing Human Futures*. Basic Books, New York.

Grosberg, R. K., Levitan, D. R., and Cameron, B. B. (1996). Characterization of genetic structure and genealogies using RAPD-PCR markers: a random primer for the novice and nervous. In: *Molecular Zoology: Advances, Strategies, and Protocols* (J. D. Ferrais and S. R. Palumbi, eds.), pp. 67–100. New York: Wiley-Liss.

Gross, P. R., and Levitt, N. (1994). *Higher Superstition: The Academic Left and its Quarrels with Science*. Baltimore: The Johns Hopkins University Press.

Grotzinger, J. P. (1994). Trends in Precambrian carbonate sediments and their implication for understanding evolution. In: *Early Life on Earth* (S. Bengston, ed.), pp. 245–258. New York: Columbia University Press.

Gu, X. (1997). The age of the common ancestor of eukaryotes and prokaryotes: statistical inferences. *Mol. Biol. Evol.* **14**, 861–866.

Gupa, R. S., and Golding, G. B. (1996). The origin of the eukaryotic cell. *Trends Biochem. Sci.* **21**, 166–171.

Guy, H. R., and Durell, S. R. (1994). Using sequence homology to analyze the structure and function of voltage-gated ion channels. In: *Molecular Evolution of Physiological Processes*, 47th Annual Symposium Society of Physiologists, Vol. 49 (D. M. Fambrough, ed.), pp. 197–212. New York: Rockefeller University Press.

Hadzi, J. (1963). *The Evolution of the Metazoa*. New York: Macmillan.

Haeckel, E. (1892). *The History of Creation*, 2 vols. Translated by E. R. Lankester from the 8th edition of Natürliche Schöpfungsgeschichter. London: Kegan Paul, Trench, Trubner & Co.

Haeckel, E. H. P. A. (1914). *The Evolution of Man: A popular Exposition of the Principal points of Human Ontogney and Phylogeny*. New York: D. Appleton (pref. 1876).

Haeckel, E. H. P. (1901). *The Riddle of the Universe at the Close of the Nineteenth Century*. Translated by Joseph McCabe. New York: Harper & Brothers.

Haeckel, E. (1905). *The Wonders of Life: A Popular Study of Biological Philosophy*. Supplementary Volume to "The Riddle of the Universe." Translated by Joseph McCabe. New York: Harper & Brothers.

Haldane, J. B. S. (1994). *The Origin of Life*. From *The Rationalist Annual*, 1929. Reprinted in J. D. Bernal, *Origin of Life*, pp. 242–249. London: Weidenfeld & Nicolson; 1967. In: *Origins of Life: The Central Concepts* (D. W. Deamer and G. R. Fleischaker, eds.), pp. 73–81. Boston: Jones & Bartlett.

Hall, J. L., Ramanis, Z., and Luck, D. J. L. (1989). Basal body/centriolar DNA: molecular genetic studies in *Chlamydomonas*. *Cell* **59**, 121–132.

Hall, R. M. (1997). Mobile gene cassettes and integrons: moving antibiotic resistance genes in Gram-negative bacteria. In: *Antibiotic Resistance: Origins, Evolution, Selection and Spread*, Ciba Foundation Symposium No. 207 (D. J. Chadwick and J. Goode, eds.), pp. 192–202.

Hamer, D. H., Hu, S., Magnuson, V. L., Hu, N., and Pattatucci. A. M. L. (1993). A linkage between DNA markers on the X chromosome and male sexual orientation. *Science* **261**, 321–327.

Hanelt, B., Van Schyndel, D., Adema, C. M., Lewis, L. A., and Loker, E. S. (1996). The hypogenetic position of *Phopalura ophicomae* (Orthonectida) based on 18S ribosomal DNA sequence analysis. *Mol. Biol. Evol.* **13**, 1187–1191.

Hanks, S. K., and Hunter, T. (1995). The eukaryotic protein kinase superfamily: kinase (catalytic) domain structure and classification. *FASEB J.* **9**, 576–596.

Hanson, E. D. (1977). *The Origin and Early Evolution of Animals*. Middletown: Wesleyan University Press.

Hanyu, N., Kuchino, Y., Nishimur, S., and Beier, H. (1986). Dramatic events in ciliate evolution: alteration of UAA and UAG termination codons to glutamine codons due to anticodon mutations in *Tetrahymena* $tRNA^{Gln}$. *EMBO J.* **5**, 1307–1311.

Haraway, D. J. (1997). *Modest_Witness @ Second_Millenium. FemaleMan©_Meets_OncoMouse™: Feminism and Technoscience*. New York: Routledge.

Hardison, R. C. (1991). Evolution of globin gene families. In: *Evolution at the Molecular Level* (R. K. Selander, A. G. Clark and T. S. Whittam, eds.), pp. 272–289. Sunderland, MA: Sinauer Associates.

Hartl, D. L. (1996). EST! EST!! EST!!! *BioEssays* **18**, 1021–1023.

Harvey, W. (1952/1652). *Generatione Animalium: Anatomical Exercises on the Generation of Animals*. Translated by R. Willis. Reprinted, Chicago: Encyclopedia Brittanica, Inc., Great Books of the Western World.

Hasegawa, M., and Fitch, W. M. (1996). Dating the cenancester of organisms. *Science* **274**, 1750.

Hayes, J. M. (1994). Global methanotrophy at the Archean–Proterozoic transition. In: *Early Life on Earth* (S. Bengston, ed.), pp. 220–236. New York: Columbia University Press.

Hayes, J. M. (1996). "The earliest memories of life on Earth," News and Views. *Nature* (7 November), **384**, 21–22.

Hayes, J. M., Des Marais, D. J., Lambert, I. B., H. Strauss, and Summons, R. E. (1992). Unsolved problems and conclusions. In: *The Proterozoic Biosphere: A Multidisciplinary Study* (J. W. Schopf and C. Klein, eds.), pp. 133–134. Cambridge: Cambridge University Press.

Hedrick, P. W., Klitz, W., Robinson, W. P., Kuhner, M. K., and Thomson, G. (1991). Population genetics of HLA. In: *Evolution at the Molecular Level* (R. K. Selander, A. G. Clark, and T. S. Whittam, eds.), pp. 248–271. Sunderland, MA: Sinauer Associates.

Helfman, D. M. (1994). The generation of protein isoform diversity by alternative RNA splicing. In: *Molecular Evolution of Physiological Processes*, 47th Annual Symposium Society of Physiologists, Vol. 49 (D. M. Fambrough, ed.), pp. 105–115. New York: Rockefeller University Press.

Heringa, J., and Argos, P. (1994). Evolution of viruses as recorded by their polymerase sequences. In: *The Evolutionary Biology of Viruses* (S. S. Morse, ed.), pp. 87–103. New York: Raven Press.

Hershey, A. D. (1966). The injection of DNA into cells by phage. In: *Phage and the Origins of Molecular Biology* (J. Cairns, G. S. Stent and J. D. Watson, eds.), pp. 100–108. Cold Spring Harbor: Cold Spring Harbor Laboratory.

Hetzer, M., Wurzer, G., Schweyen, R. J., and Mueller, M. W. (1997). *Trans*-activation of group II intron splicing by nuclear U5 snRNA. *Nature* **386**, 417–420.

Hibberd, D. (1975). Observations on the ultrastructure of the choanoflagellate *Codosiga botrytis* (Ehr.) Saville-Kent, with special reference to the flagellar apparatus. *J. Cell Sci.* **17**, 191–219.

Hibbett, D. S. (1996). Phylogenetic evidence for horizontal transmission of group I introns in the nuclear ribosomal DNA of mushroom-forming fungi. *Mol. Biol. Evol.* **13**, 903–917.

Hillis, D. M. (1996). "Evolutionary virology," Book Reviews. *Science* (30 August), **273**, 1179–1180.

Himmelreich, R., Hilbert, H., Plagens, H., Pirkl, E., Li, B.-C., and Herrmann, R. (1996). Complete sequence analysis of the genome of the bacterium *Mycoplasma penumoniae. Nucleic Acids Res.* **24**, 4420–4449.

Hinkle, G. (1991). Status of the theory of the symbiotic origin of undulipodia (cilia). In: *Symbiosis as a Source of Evolutionary Innovation: Speciation and Morphogenesis* (L. Margulis and R. Fester, eds.), pp. 135–142. Cambridge: MIT Press.

Hofbauer, J., and Sigmund, K. (1988). *The Theory of Evolution and Dynamical Systems*. Cambridge: Cambridge University Press.

Hogan, B. L. M., Blessing, M., Winnier, G. E., Suzuki, N., and Jones, C. M. (1994). Growth factors in development: the role of the TGF-β related polypeptide signaling molecules in embryogenesis. *Development* (1994 Suppl., *The Evolution of Developmental Mechanisms*), 53–60.

Hofmann, H. J. (1994). Proterozoic carbonaceous compressions ("metaphytes" and "worms"). In: *Early Life on Earth* (S. Bengston, ed.), pp. 342–357. New York: Columbia University Press.

Holland, H. D. (1992). Distribution and paleoenvironmental interpretation of Proterozoic paleosols. In: *The Proterozoic Biosphere: A Multidisciplinary Study* (J. W. Schopf and C. Klein, eds.), pp. 152–155. Cambridge: Cambridge University Press.

Holland, J. H. (1995). *Hidden Order: How Adaptation Builds Complexity*. Reading, MA: Helix Books.

Hollande, A. (1952). Ordre des Choanoflagellés ou Craspédomonadines. In: P. P. Grassóe, ed., *Traitè de Zool.* **1**(1), 579–598.

Holley, R. W., Apgar, J., Everett, G. A., Madison, J. T., Marquisee, M., Merrill, S. H., Penswick, J. R., and Zamir, A. (1965). Structure of a ribonucleic acid. *Science* **147**, 1462–1465.

Hori, H., Satow, Y., Inoue, I., and Chihara, M. (1991). Archaebacteria vs Metabacteria: phylogenetic tree of organisms, indicated by comparison of 5S ribosomal RNA sequences. In: *Evolution of Life: Fossils, Molecules, and Culture* (S. Osawa and T. Honjo, eds.), pp. 325–336. Tokyo: Springer-Verlag.

Hortsch, M., and Goodman, C. S. (1991). Cell and substrate adhesion molecules in *Drosophila*. *Ann. Rev. Cell Biol.* **7**, 505–557.

Houck, M. A., Clark, J. B., Peterson, K. R., and Kidwell, M. G. (1991). Possible horizontal transfer of *Drosophila* genes by the mite *Protolaelaps regalis*. *Science* **253**, 1125–1129.

Hoyle, F. (1984/1983). *The Intelligent Universe*. New York: Holt, Rinehart & Winston.

Hoyle, F., and Wickramasinghe, C. (1984). *From Grains to Bacteria*. Cardiff: University College Cardiff Press.

Hubbard, R. (1990). *The Politics of Women's Biology*. New Brunswick: Rutgers University Press.

Hue, K., and Shimura, Y. (1995). Mechanisms of alternative RNA splicing. In: *Tracing Biological Evolution in Protein and Gene Structure* (M. Gō and P. Schimmel, eds.), pp. 87–95. Amsterdam: Elsevier.

Huyhn, T. V., Young, R. A., and Davis, R. A. (1984). Constructing and screening cDNA libraries in gt10 and gt11. In: *DNA Cloning: A Practical Approach*, Vol. 1 (D. M. Glover, ed.), pp. 49–78. Oxford: IRL Press.

Ikawa, Y., Shiraishi, H., and Inoue, T. (1995). The *Tetrahymena* ribozyme tolerates diverse activator domains that replace P5abc. In: *Tracing Biological Evolution in Protein and Gene Structure* (M. Gō and P. Schimmel, eds.), pp. 115–123. Amsterdam: Elsevier.

Ivker, F. B. (1972). A hierarchy of histo-incompatibility in *Hydractinia echinata*. *Biol. Bull.* **143**, 162–174.

Iwabe, N., Kuma, K.-I., Hasegawa, M., Osawa, S., and Miyata, T. (1989). Evolutionary relationship of archaebacteria, eubacteria, and eukaryotes inferred from phylogenetic trees of duplicated genes. *Proc. Natl. Acad. Sci. U.S.A.* **85**, 9355–9359.

Jablonka, E., and Lamb, M. J. (1995). *Epigenetic Inheritance and Evolution: The Lamarckian Dimension*. Oxford: Oxford University Press.

Jegla, T., and Salkoff, L. (1994). Molecular evolution of K+ channels in primitive eukaryotes. In: *Molecular Evolution of Physiological Processes*, 47th Annual Symposium Society of Physiologists, Vol. 49 (D. M. Fambrough, ed.), pp. 213–222. New York: Rockefeller University Press.

Joyce, G. F. (1994/1991). The rise and fall of the RNA world. *New Biologist* **3**, 399–407; 1991. Reprinted in D. W. Deamer and G. R. Fleischaker, eds., *Origins of Life: The Central Concepts*, pp. 391–399. Boston: Jones & Bartlett.

Joyce, G. F., and Orgel, L. E. (1993). Prospects for understanding the origin of the RNA world. In: *The RNA World: The Nature of Modern RNA Suggests a Prebiotic RNA World* (R. F. Gesteland and J. F. Atkins, eds.), pp. 1–25. Cold Spring Harbor: Cold Spring Harbor Laboratory Press.

Judson, H. R. (1979). *The Eighth Day of Creation: Makers of the Revolution in Biology*. New York: Simon & Schuster.

Jull, A. J. T., Courtney, C., Jeffrey, D. A., and Beck, J. W. (1998). Isotopic evidence for a terrestrial source of organic compounds found in Martian meteorites Allan Hills 84001 and Elephant Moraine 79001. *Science* **279**, 366–369.

Kable, M. L., Seiwert, S. D., Heidmann, S., and Stuart, K. (1996). RNA editing: a mechanism for gRNA-specified uridylate insertion into precursor mRNA. *Science* **273**, 1189–1195.

Kamaishi, T., Hashimoto, T., Nakamura, Y., Nakamura, F., Murata, S., Okada, N., Okamoto, K.-I., Shimzu, M., and Hasegawa, M. (1996). Protein phylogeny of translation elongation factor EF-1α suggests Microsporidians are extremely ancient eukaryotes. *J. Mol. Evol.* **42**, 257–263.

Kandler, O. (1994). The early diversification of life. In: *Early Life on Earth* (S. Bengston, ed.), pp. 152–160. New York: Columbia University Press.

Kasting, J. F., and Chang, S. (1992). Formation of the Earth and the origin of life. In: *The Proterozoic Biosphere: A Multidisciplinary Study* (J. W. Schopf and C. Klein, eds.), pp. 9–12. Cambridge: Cambridge University Press.

Kauffman, S. A. (1993). *The Origins of Order: Self-Organization and Selection in Evolution*. New York: Oxford University Press.

Keeling, P. J. (1998). A kingdom's progress: Archezoa and the origin of eukaryotes. *BioEssays* **20**, 87–95.

Keeling, P. J., and Doolittle, W. F. (1996). Alpha-tubulin from early-diverging eukaryotic lineages and the evolution of the tubulin family. *Mol. Biol. Evol.* **13**, 1297–1305.

Keeling, P. J., and Doolittle, W. F. (1997). Evidence that eukaryotic triosphosphate isomerase is of alpha-proteobacterial origin. *Proc. Natl. Acad. Sci. U.S.A.* **94**, 1270–1275.

Kerr, R. A. (1996a). "Ancient life on Mars?" News. *Science* (16 August), **273**, 864–866.

Kerr, R. A. (1996b). "Martian rocks tell divergent stories," News. *Science* (8 November), **274**, 918.

Kerr, R. A. (1998) "Requiem for life on Mars? Support for microbes fades," News Focus. *Science* (20 November) **282**, 1398–1400.

Kevles, D. J. (1992). Out of eugenics: the historical politics of the human genome. In: *The Code of Codes: Scientific and Social Issues in the Human Genome Project* (D. J. Kevles and L. Hood, eds.), pp. 3–36. Cambridge: Harvard University Press.

Khakhina, L. N. (1993). *Concepts of Symbiogenesis: A Historical and Critical Study of the Research of Russian Botanists*. Translated by S. Merkel and R. Coalson. New Haven: Yale University Press.

Khorana, H. G. (1995). Transfer RNA: discovery, early work, and total synthesis of a tRNA gene. In: *tRNA: Structure, Biosynthesis, and Function* (D. Söll and U. L. RajBhandary, eds.), pp. 5–16. Washington, DC: ASM Press.

Kidwell, M. G. (1993). Lateral transfer in natural populations of eukaryotes. *Ann. Rev. Genet.* **27**, 235–256.

Kimura, M. (1983). *The Neutral Theory of Molecular Evolution*. Cambridge: Cambridge University Press.

King, J. (1996). Could transgenic supercrops one day breed superweeds? Research News. *Science* **274**, 180–191.

King, J. A., Maley, M. E., Ling, K.-Y., Kanbrocki, J. A., and Kung, C. (1991). Efficient expression of the *Paramecium* calmodulin gene in *Escherichia coli* after four TAA-to-CAA changes through a series of polymerase chain reactions. *J. Protozool.* **38**, 441–447.

Klein, C. (1992). Introduction. In: *The Proterozoic Biosphere: A Multidisciplinary Study* (J. W. Schopf and C. Klein, eds.), pp. 137–138. Cambridge: Cambridge University Press.

Klein, C., and Beukes, N. J. (1992). Time distribution, stratigraphy, and sedimentologic setting, and geochemistry of Precambrian iron-formations. In: *The Proterozoic Biosphere: A Multidisciplinary Study* (J. W. Schopf and C. Klein, eds.), pp. 139–146. Cambridge: Cambridge University Press.

Klenk, H.-P. Zillig, W., Lanzendörfer, M., Grampp, B., and Palm, P. (1995). Location of protist lineages in phylogenetic tree inferred from sequences of DNA-dependent RNA polymerases. *Arch. Protistenkd.* **134**, 221–230.

Klenk, H.-P., et al. (1997). The complete genome sequence of the hyperthermophilic, sulphate-reducing archaeon *Archaeoglobus fulgidus*. *Nature* **390**, 364–370.

Knoll, A. H. (1992). The early evolution of eukaryotes: a geological perspective. *Science* **256**, 622–627.

Köhler, C., Delwiche, F., Denny, P. W., Tilney, L. G., Webster, P., Wilson, R. J. M., Palmer, J. D., and Roos, D. S. (1997). A plastid of probably green algal origin in apicomplexan parasites. *Science* **275**, 1485–1489.

Konarska, M. M., and Sharp, P. A. (1989). Replication of RNA by the DNA-dependent RNA polymerase of phage T7. *Cell* **57**, 423–431.

Koonin, E. V. (1997). "Evidence for a family of archaeal ATPases," Technical Comments. *Science* (7 March), **275**, 1489–1490.

Koonin, S. E. (1998). An independent perspective on the Human Genome Project. *Science* **279**, 36–37.

Kornberg, A. (1982). *Supplement to DNA Replication.* San Francisco: W. H. Freeman.

Kreitman, M. (1991). Detecting selection at the level of DNA. In: *Evolution at the Molecular Level* (R. K. Selander, A. G. Clark and T. S. Whittam, eds.), pp. 204–221. Sunderland, MA: Sinauer Associates.

Kreitman, M. (1996). The neutral theory is dead. Long live the neutral theory. *BioEssays* **18**, 678–683.

Kruger, K., Grabowski, P. J., Zaug, A. J., Sands, J., Gottschling, D. E., and Cech, T. R. (1982). Self-splicing RNA: autoexcision and autocyclization of the ribosomal RNA intervening sequence of Tetrahymena. *Cell* **31**, 147–157.

Kuhn, T. S. (1970). *The Structure of Scientific Revolutions*, 2nd ed., Enlarged. Chicago: University of Chicago Press.

Kunst, F., et al. (1997). The complete genome sequence of the Gram-positive bacterium *Bacillus subtilis. Nature* **390**, 249–256.

Kuroiwa, T. (1998). The primitive red algae *Cyanidium caldarium* and *Cyanidioschyzon merolae* as model system for investigating the dividing apparatus of mitochondria and plastids. *BioEssays* **20**, 344–355.

Lackey, J. B. (1959). Morphology and biology of a species of Protospongia. *Trans. Am. Microsc. Soc.* **78**, 202–206.

Lai, M. M. C. (1995). Recombination and its evolutionary effect on viruses with RNA genomes. In: *Molecular Basis of Virus Evolution* (A. J. Gibbs, C. H. Calisher and F. García Arenal, eds.), pp. 119–132. Cambridge: Cambridge University Press.

Lajtha, L. G. (1979). Stem cell concepts. *Differentiation* **14**, 23–34.

Lake, J. A. (1987a). Determining evolutionary distances from highly diverged nucleic acid sequences: Operator metrics. *J. Mol. Evol.* **26**, 79–73.

Lake, J. A. (1987b). A rate-independent technique for analysis of nucleic acid sequences: evolutionary parsimony. *Mol. Biol. Evol.* **4**, 167–197.

Lake, J. A. (1987c). Prokaryotes and Archaebacteria are not monophyletic: rate invariant analysis of rRNA genes indicates that eukaryotes and eocytes form a monophyletic taxon. *Cold Spring Harbor Symposia on Quantitative Biology* **52**, 839–846.

Lake, J. A. (1988). Origin of the eukaryotic nucleus determined by rate-invariant analysis of rRNA sequences. *Nature* **331**, 184–196.

Lake, J. A., and Rivera, M. (1994). Was the nucleus the first endosymbiont? *Proc. Natl. Acad. Sci. U.S.A.* **91**, 2880–2881.

Lake, J. A., and Rivera, M. (1996). "Methanococcus genome," Letters. *Science* (8 November), **274**, 901.

Lake, J. A., Henderson, E., Oakes, M., and Clark, M. W. (1984). Eocytes: a new ribosome structure indicates a kingdom with a close relationship to eukaryotes. *Proc. Natl. Acad. Sci. U.S.A.* **81**, 3786–3790.

Lamarck, J.-B. (1984b/1809). *Zoological Philosophy: An Exposition with Regard to the Natural History of Animals*. Translated by Hugh Elliot. Chicago: University of Chicago Press.

Lander, E. S. (1996). The new genomics: global views of biology. *Science* **274**, 536–539.

Lang, B. F., Burger, G., O'Kelly, C. J., Cedergren, R., Golding, G. B., Lemieux, C., Sankoff, D., Turmel, M., and Gray. M. W. (1997). An ancestral mitochondrial DNA resembling a eubacterial genome in miniature. *Nature* **387**, 493–497.

Laval, M. (1971). Ultrastructure et mode de nutrition du choanoflagellé *Salpinoeca pelagica* sp. nov. *Protistologica* **7**, 324–336.

Lawrence, J. G., and Hartl, D. L. (1992). Inference of horizontal genetic transfer from molecular data: an approach using the bootstrap. *Genetics* **131**, 753–760.

Lawrence, J. G., and Roth, J. R. (1996). Selfish operons: horizontal transfer may drive the evolution of gene clusters. *Genetics* **143**, 1843–1860.

Lawrence, J. G., and Ochman, H. (1997). Amelioration of bacterial genomes: rates of change and exchange. *J. Mol. Evol.* **44**, 383–397.

Lazcano, A. (1994). The transition from nonliving to living; The RNA world, its predecessors, and its descendants. In: *Early Life on Earth*, Nobel Symposium No. 84 (S. Bengtson, ed.), pp. 60–69, 70–80. New York: Columbia University Press.

Leblond, C. P., and Walker. B. E. (1956). Renewal of cell populations. *Physiol. Rev.* **36**, 255–279.

Leigh Brown, A. J. (1994). Methods of evolutionary analysis of viral sequences. In: *The Evolutionary Biology of Viruses* (S. S. Morse, ed.), pp. 75–84. New York: Raven Press.

LeVay, S. (1991). A difference in hypothalamic structure between heterosexual and homosexual men. *Science* **253**, 1034–1037.

Levings III, C. S. (1996). "Infertility treatment: a nuclear restorer gene in maize," Perspectives. *Science* (31 May), **272**, 1279–1280.

Levy, S. (1992). *Artificial Life: The Quest for a New Creation*. London: Jonathan Cape.

Lewin, B. (1990). *Genes, IV*. Oxford: Oxford University Press.

Lewontin, R. C., and Hubby, J. L. (1966). A molecular approach to the study of genic heterozygosity in natural populations, II: Amount of variation and degree of heterozygosity in natural populations of *Drosophila pseudoobscura*. *Genetics* **54**, 595–609.

Li, H., Trotta, C. R., and Abelson, J. (1998). Crystal structure and evolution of a transfer RNA splicing enzyme. *Science* **280**, 279–286.

Livnah, O., Stura, E. A., Johnson, D. L., Middleton, S. A., Mulcahy, L. S., Wrighton, N. C., Dower, W. J., Joliffe, L. K., and Wilson, I. A. (1996). Functional mimicry of a protein hormone by a peptide agonist: the EPO receptor complex at 2.8 Å. *Science* **273**, 464–471.

Lobry, J. R. (1996). Origin of replication of *Mycoplasm genitalium*. *Science* **272**, 745–746.

Lom, J. (1990). Phylum Myxozoa. In: *Handbook of Protoctista* (L. Margulis, J. O. Corliss, M. Melkonian, and D. J. Chapman, eds.), pp. 36–52. Boston: Jones and Bartlett.

Long, M., and Langley, C. (1993). Natural selection and the origin of Ijingwei, I: a chimeric processed functional gene in *Drosophila*. *Science* **260**, 91–95.

Lovejoy, A. O. (1964/1936). *The Great Chain of Being: A Study of the History of an Idea*. Cambridge: Harvard University Press.

Lovelock, J. (1990). *The Ages of Gaia*. New York: Bantam Books.

Lovelock, J. (1996). The Gaia hypothesis. In: *Gaia in Action* (P. Bunyard, ed.), pp. 13–31. Edinburgh: Flores.

Lowe, D. R. (1992). Major events in the geological development of the Precambrian Earth. In: *The Proterozoic Biosphere: A Multidisciplinary Study* (J. W. Schopf and C. Klein, eds.), pp. 67–75. Cambridge: Cambridge University Press.

Lowe, D. R. (1994). Early environments: constraints and opportunities for early evolution. In: *Early Life on Earth*, Nobel Symposium No. 84 (S. Bengtson, ed.), pp. 24–35. New York: Columbia University Press.

Lurie, E. (1988). *Louis Agassiz: A Life in Science*. Baltimore: The Johns Hopkins University Press.

Maddison, W. P., and Maddison, D. R. (1992). *MacClade: Analysis of Phylogeny and Character Evolution*, Version 3. Sunderland, MA: Sinauer Associates.

Maddison, W. P. (1996). Molecular approaches and the growth of phylogenetic biology. In: *Molecular Zoology: Advances, Strategies, and Protocols* (J. D. Ferrais and S. R. Palumbi, eds.), pp. 47–63. New York: Wiley-Liss.

Maizels, N., and Weiner, A. M. (1993). The genomic tag hypothesis: modern viruses as molecular fossils of ancient strategies for genomic replication. In: *The RNA World: The Nature of Modern RNA Suggests a Prebiotic RNA World* (R. F. Gesteland and J. F. Atkins, eds.), pp. 577–602. Cold Spring Harbor: Cold Spring Harbor Laboratory Press.

Malpighii, M. (1968/1675). *Anatome Plantarum. Appendix, Ovo Incubato: Observationes continens*. Londini: Regiae Societati; 1675. Facsimile reproduction. Bruxelles: Culture et Civilisation.

Manak, J. R., and Scott, M. P. (1994). The evolution of cell lineage in nematodes. *Development* (1994 Suppl., *The Evolution of Developmental Mechanisms*), 61–77.

Mandelbrot, B. B. (1998). "Is nature fractal?" Letters. *Science* **279**, 783–784.

Margoliash, E., and Smith. E. L. (1965). Structural and functional aspects of cytochrome *c* in relation to evolution. In: *Evolving Genes and Proteins* (V. Bryson and H. J. Vogel, eds.), pp. 221–242. New York: Academic Press.

Margulis, L. (1990a). Introduction. In: *Handbook of Protoctista* (L. Margulis, J. O. Corliss, M. Melkonian and D. J. Chapman, eds.), pp. xi–xxiii. Boston: Jones & Bartlett.

Margulis, L. (1990b). Big trouble in biology: physiological autopoiesis versus mechanistic new-Darwinism. In: *Speculations* (J. Brockman, ed.), pp. 211–235. Englewood Cliffs, NJ: Prentice-Hall.

Margulis, L. (1990c). Words as battle cries — symbiogenesis and the new field of endocytobiology. *BioScience* **40**, 673–677.

Margulis, L. (1991a). Symbiosis in evolution: origins of cell motility. In: *Evolution of Life: Fossils, Molecules, and Culture* (S. Osawa and T. Honjo, eds.), pp. 305–324. Tokyo: Springer-Verlag.

Margulis, L. (1991b). Symbiogenesis and Symbionticism. In: *Symbiosis as a Source of Evolutionary Innovation: Speciation and Morphogenesis* (L. Margulis and R. Fester, eds.), pp. 1–24. Cambridge: MIT Press.

Margulis, L. (1991c). *Symbiosis in Cell Evolution: Microbial Communities in the Archean and Proterozoic Eons*, 2nd ed. New York: Freeman.

Margulis, L. (1996). Jim Lovelock's Gaia. In: *Gaia in Action* (P. Bunyard, ed.), pp. 45–57. Edinburgh: Flores.

Margulis, L., and Cohen, J. E. (1994). Combinatorial generation of taxonomic diversity: implication of symbiogenesis for the Proterozoic fossil record. In: *Early Life on Earth*, Nobel Symposium No. 84 (S. Bengtson, ed.), pp. 327–334. New York: Columbia University Press.

Margulis, L., and Fester, R. (1991). *Symbiosis as a Source of Evolutionary Innovation: Speciation and Morphogenesis*. Cambridge: MIT Press.

Margulis, L., and Sagan, D. (1986). *Microcosmos: Four Billion Years of Evolution from our Microbial Ancestors*. New York: Summit.

Margulis, L., and Schwartz, K. V. (1982). *Five Kingdoms*. San Francisco: W. H. Freeman.

Margulis, L., Guerrero, R., and Bunyard, P. (1996). We are all symbionts. In: *Gaia in Action* (P. Bunyard, ed.), pp. 160–177. Edinburgh: Flores.

Markert, C. L. (1996). Isozymes: model systems for analyzing the origin, evolution, regulation, and function of gene families. In: *Gene Families: Structure, Function, Genetics and Evolution*, Proceedings of the VIIIth International Congress on Isozymes (R. S. Holmes and H. A. Lim, eds.), pp. 4–7. Isingapore: World Scientific.

Marshall, E. (1995a). "Dispute splits schizophrenia study," News and Comment. *Science* (12 May), **268**, 792–4.

Marshall, E. (1995b). "NIH's 'Gay gene' study questioned," News and Comment. *Science* (30 June), **268**, 1841.

Marshall, E. (1998). "Physicists urge technology push to reach 2005 target," News and Comment, *Science* (2 January), **279**, 23.

Martin, W. F. (1996). Is something wrong with the tree of life? *BioEssays* **18**, 523–527.

Martin, W., and Müller, M. (1998). The hydrogen hypothesis for the first eukaryote. *Nature* **392**, 37–41.

Martinis, S. A., and Schimmel, P. (1995). Small RNA oligonucleotide substrates for specific aminoacylations. In: *tRNA: Structure, Biosynthesis, and Function* (D. Söll and U. L. RajBhandary, eds.), pp. 349–370. Washington, DC: ASM Press.

Maynard Smith, J. (1970). Natural selection and the concept of a protein space. *Nature* **225**, 563–564.

Maynard Smith, J. (1991). A Darwinian view of symbiosis. In: *Symbiosis as a Source of Evolutionary Innovation* (L. Margulis and R. Rester, eds.), pp. 26–39. Cambridge: MIT Press.

Mayr, E. (1982). *The Growth of Biological Thought*. Cambridge: Harvard University Press.

Mayr, E. (1988). *Toward a New Philosophy of Biology: Observations of an Evolutionist*. Cambridge: Harvard University Press.

Mayr, E. (1994). Driving forces in evolution: an analysis of natural selection. In: *The Evolutionary Biology of Viruses* (S. S. Morse, ed.), pp. 29–48. New York: Raven Press.

McClain, W. H. (1995). The tRNA identity problem: past, present, and future. In: *tRNA: Structure, Biosynthesis, and Function* (D. Söll and U. L. RajBhandary, eds.), pp. 335–347. Washington, DC: ASM Press.

McGeoch, D. J., and Davison, A. J. (1995). Origins of DNA viruses. In: *Molecular Basis of Virus Evolution* (A. J. Gibbs, C. H. Calisher and F. García Arenal, eds.), pp. 67–75. Cambridge: Cambridge University Press.

McKay, D. S., Gibson Jr., E. K., Thomas-Keprta, K. L., Vali, H., Romanek, C. S., Clemett, S. J., Chillier, X. D. F., Maechling, C. R., and Zare, R. N. (1996). Search for past life on Mars: possible relic biogenic activity in Martian meteorite ALH84001 *Science* **273**, 924–930.

McMenamin, M. A. S., and Schulte McMenamin, D. L. (1990). *The Emergence of Animals: The Cambrian Breakthrough.* New York: Columbia University Press.

Mehra, J. (1994). *The Beat of a Different Drum: The Life and Science of Richard Feynman.* Oxford: Clarendon Press.

Mestel, R. (1996). "Putting prions to the test," Special News Report: Prions. *Science* (12 July), **273**, 184–189.

Mewes, H. W., Albermann, K., Bähr, M., Frishman, D., Gleissner, A., Hani, J., Heumann, K., Kleine, K., Maierl, A., Oliver, S. G., Pfeiffer F., and Zollner, A. (1997). Overview of the yeast genome: The Yeast Genome Directory. *Nature* **387** (Suppl.), 7–8.

Miklos, G., and Campbell, L. G. (1994). From protein domains to extinct phyla: reverse-engineering approaches to the evolution of biological complexities. In: *Early Life on Earth*, Nobel Symposium No. 84 (S. Bengtson, ed.), pp. 501–516. New York: Columbia University Press.

Miller, S. L. (1953). A production of amino acids under possible primitive earth conditions. *Science* **117**, 528–529.

Miller, S. L., and Urey, Y. C. (1959). Organic compound synthesis on the primitive earth. *Science* **130**, 245–251.

Milkman, R. (1997). Recombination and population structure in *Escherichia coli. Genetics* **146**, 745–750.

Mitman, G., and Fausto-Sterling, A. (1992). Whatever happened to *Planaria*? C. M. Child and the physiology of inheritance. In: *The Right Tools for the Job: At Work in Twentieth-Century Life Sciences* (A. E. Clarke and J. H. Fugimura, eds.), pp. 172–197. Princeton: Princeton University Press.

Miyata, T., Iwabe, K., Kuma, K.-I., Kawahishi, Y.-I., Hasagawa, M., Kishino, H., Mukohata, Y., Ihara, K., and Osawa, S. (1991). Evolution of Archaebacteria: Phylogenetic relationships among Archaebacteria, Eubacteria, and Eukaryotes. In: *Evolution of Life: Fossils, Molecules, and Culture* (S. Osawa and T. Honjo, eds.), pp. 337–351. Tokyo: Springer-Verlag.

Mojzsis, S. J., Arrhenius, G., McKeegan, K. D., Harrison, T. M., Nutman, A. P., and Friend, C. R. L. (1996). Evidence for life on Earth before 3,800 million years ago. *Nature* **384**, 55–59.

Monod, J. (1971). *Chance and Necessity: An Essay on the Natural Philosophy of Modern Biology*. Translated from the French by A. Wainhouse. New York: Knopf.

Moore, J. A. (1983). Thomas Hunt Morgan — the geneticist. *Am. Zool.* **23**, 855–865.

Moore, J. A. (1986). Science as a way of knowing — genetics. *Am. Zool.* **26**, 583–747.

Moore, J. A. (1987). Science as a way of knowing — developmental biology. *Am. Zool.* **27**, 415–573.

Moore, M. J., Query, C. C., and Sharp, P. A. (1993). Splicing of precursors to mRNA by the spliceosome. In: *The RNA World: The Nature of Modern RNA Suggests a Prebiotic RNA World* (R. F. Gesteland and J. F. Atkins, eds.), pp. 303–357. Cold Spring Harbor: Cold Spring Harbor Laboratory Press.

Moore, P. B. (1993). Ribosomes and the RNA world. In: *The RNA World: The Nature of Modern RNA Suggests a Prebiotic RNA World* (R. F. Gesteland and J. F. Atkins, eds.), pp. 119–135. Cold Spring Harbor: Cold Spring Harbor Laboratory Press.

Morell, V. (1996a). Manic-depression findings spark polarized debate. Research News. *Science* (5 April), **272**, 31–32.

Morell, V. (1996b). "Life's last domain," Research News. *Science* (23 August), **273**, 1043–1045.

Morell, V. (1997). "Microbiology's scarred revolutionary," Research News. *Science* (2 May), **276**, 699–702.

Morgan, T. H. (1926). *The Theory of the Gene*. New Haven: Yale University Press.

Morse, S. S. (1994). Toward an evolutionary biology of viruses. In: *The Evolutionary Biology of Viruses* (S. S. Morse, ed.), pp. 1–28. New York: Raven Press.

Murra, K., Selleck, P., Hooper, P., Hyatt, A., Gould, A., Gleeson, L., Westbury, H., Hiley, L., Selvey, L., Rodwell, B., and Ketterer, P. (1995). A morbillivirus that caused fatal disease in horses and humans. *Science* **268**, 94–97.

Muscatine, L. (1974). Endosymbiosis of cnidarians and algae. In: *Coelenterate Biology: Reviews and New Perspectives* (L. Muscatine and H. M. Lenhoff, eds.), pp. 359–395. New York: Academic Press.

Muscatine, L., and McAuley, P. J. (1982). Transmission of symbiotic algae to eggs of green hydra. *Cytobios* **33**, 111–124.

Myers, G., Tautz, N., and Theil, H.-J. (1995). Cellular sequences in viral genomes. In: *Molecular Basis of Virus Evolution* (A. J. Gibbs, C. H. Calisher and F. García Arenal, eds.), pp. 91–102. Cambridge: Cambridge University Press.

Needham, J. (1959). *A History of Embryology*. Cambridge: University Press.

Nei, M. (1987). *Molecular Evolutionary Genetics*. New York: Columbia University Press.

Nei, M., and Hughes, A. L. (1991). Polymorphism and evolution of the major histocompatibility complex loci in mammals. In: *Evolution at the Molecular Level* (R. K. Selander, A. G. Clark and T. S. Whittam, eds.), pp. 222–247. Sunderland, MA: Sinauer Associates.

Nigg, E. A. (1995). Cyclin-dependent protein kinases: key regulators of the eukaryotic cell cycle. *BioEssay* **17**, 471–480.

Noguti, T., and Gō, M. (1995). Modules of barnase: the physicochemical basis for their structures. In: *Tracing Biological Evolution in Protein and Gene Structure* (M. Gō and P. Schimmel, eds.), pp. 161–174. Amsterdam: Elsevier.

Noller, H., Hoffarth, F. V., and Zimniak. L. (1992). Unusual resistance of peptidyl transferase to protein extraction procedures. *Science* **256**, 1416–1419.

Noller, H. R. (1993). On the origin of the ribosome: coevolution of subdomains of tRNA and rRNA. In: *The RNA World: The Nature of Modern RNA Suggests a Prebiotic RNA World* (R. F. Gesteland and J. F. Atkins, eds.), pp. 137–156. Cold Spring Harbor: Cold Spring Harbor Laboratory Press.

Novelli, M. R., Williamson, J. A., Tomlinson, I. P. M., Elia, G., Hodgson, S. V., Talbot, I. C., Bodmer, W. F., and Wright. N. A. (1996). Polyclonal origin of colonic adenomas in an XO/SY patient with FAP. *Science* **272**, 1187–1190.

Nureki, O., and Yokoyama, S. (1995). Architectures of class I aminoacyl-tRNA synthetases. In: *Tracing Biological Evolution in Protein and Gene Structure* (M. Gō and P. Schimmel, eds.), pp. 23–35. Amsterdam: Elsevier.

Ohta, T. (1994). Early evolution of genes and genomes. In: *Early Life on Earth*, Nobel Symposium No. 84 (S. Bengtson, ed.), pp. 134–142. New York: Columbia University Press.

Ohta, T. (1996). The current significance and standing of neutral and nearly neutral theories. *BioEssays* **18**, 673–677.

Olby, R. (1985). *Origins of Mendelism*, 2nd ed. Chicago: University of Chicago Press.

Olby, R. (1994). *The Path to the Double Helix: The Discovery of DNA.* Unabridged, corrected and enlarged republication. New York: Dover Publications.

Olive, P. J. W. (1985). In: *Symbiosis as a Source of The Origins and Relationships of Lower Invertebrates.* The Systematics Association Special Volume 28 (S. Conway Morris, R. Gibson and H. M. Platt, eds.), pp. 42–59. Oxford: Clarendon Press.

Olsen, G. J. (1987). Earliest phylogenetic branchings: Comparing rRNA-based evolutionary trees inferred with various techniques. *Cold Spring Harbor Symposia on Quantitative Biology* **52**, 825–837.

Oparin, A. I. (1964). *Life: Its Nature, Origin and Development.* Translated from the Russian by Ann Synge. New York: Academic Press.

Oppenheim, R. W. (1982). Preformation and epigenesis in the origins of the nervous system and behaviour: issues, concepts, and their history. In: *Perspectives in Ethnology*, Vol. 5 (P. P. G. Bateson and P. H. Klopfer, eds.), pp. 1–10. New York: Plenum Press.

Oppenheimer, J. M. (1967). *Essays in the History of Embryology and Biology.* Cambridge: MIT Press.

Orgel, L. E. (1986). RNA catalysis and the origins of life. *J. Theoretical Biol.* **123**, 127–149.

Oró, J. (1994). Early chemical stages in the origin of life. In: *Early Life on Earth*, Nobel Symposium No. 84 (S. Bengtson, ed.), pp. 48–59. New York: Columbia University Press.

Oshima, T. (1991). Early biochemical evolution: speculations on the biochemistry of primitive life. In: *Evolution of Life: Fossils, Molecules, and Culture* (S. Osawa and T. Honjo, eds.), pp. 353–359. Tokyo: Springer-Verlag.

Oshima, T., Yaoi, T., and Gō, M. (1995). Module replacement converted coenzyme specificity of isocitrate dehydrogenase. In: *Tracing Biological Evolution in Protein and Gene Structure* (M. Gō and P. Schimmel, eds.), pp. 197–203. Amsterdam: Elsevier.

Ourisson, G. (1994). Biomarkers in the Proterozoic record. In: *Early Life on Earth*, Nobel Symposium No. 84 (S. Bengtson, ed.), pp. 259–269. New York: Columbia University Press.

Pace, N. R. (1997). A molecular view of microbial diversity and the biosphere. *Science* **276**, 734–740.

Palmer, J. D. (1991). Plastid chromosomes: structure and evolution. In: *The Molecular Biology of Plastids* (L. Bogorad and I. Vasil, eds.), pp. 5–53. San Diego, Academic Press.

Palmer, J. D. (1992). Comparison of chloroplast and mitochondrial genome evolution in plants. In: *Plant Gene Research: Cell Organelles* (R. G. Hermann, ed.), pp. 99–133. Wien: Springer-Verlag.

Palmer, J. D. (1995). Rubisco rules fall: gene transfer triumphs. *BioEssay* **17**, 1005–1008.

Pan, T., Long, D. M., and Uhlenbeck, O. C. (1993). Divalent metal ions in RNA folding and catalysis. In: *The RNA World: The Nature of Modern RNA Suggests a Prebiotic RNA World* (R. F. Gesteland and J. F. Atkins, eds.), pp. 271–302. Cold Spring Harbor: Cold Spring Harbor Laboratory Press.

Parham, P. (1994). Functional polymorphism in Class I Major Histocompatibility Complex genes. In: *Molecular Evolution of Physiological Processes*, 47th Annual Symposium Society of Physiologists, Vol. 49 (D. M. Fambrough, ed.), pp. 93–103. New York: Rockefeller University Press.

Patino, M. M., Liu, J.-J., Glover, J. R., and Linkquist, S. (1996). Support for the prion hypothesis for inheritance of a phenotypic trait in yeast. *Science* **273**, 622–626.

Patterson, D. J. (1989). Phylogenetic relations of major groups: Conclusions and prospects. In: *The Hierarchy of Life* (B. Fernholm, K. Bremer and W. L. Diver, eds.), pp. 471–488. Amsterdam: Excerpta Medica.

Patthy, L. (1991a). Exons: original building blocks of protein? *BioEssay* **13**, 187–192.

Patthy, L. (1991b). Modular exchange principles in protein. *Curr. Opinion Struct. Biol.* **1**, 351–361.

Patthy, L. (1994). Introns and Exons. *Curr. Opinion Struct. Biol.* **4**, 383–392.

Pawlowski, J., Montoya-Burgos, J.-I., Fahrni, J. F., Wüest, J., and Zininetti, L. (1996). Origin of Mesozoa inferred from 18S rRNA gene sequences. *Mol. Biol. Evol.* **13**, 1128–11232.

Pennisi, E., and Roush, W. (1997). "Developing a new view of evolution," Special News Report. *Science* (4 July) **277**, 34–37.

Penny, D., Lockhart, P. J., Steel, M. A., and Hendy, M. D. (1994). The role of models in reconstructing evolutionary trees. In: *Models in Phylogeny Reconstruction*, Systematics Association Special Volume 52 (R. W. Scotland, D. J. Siebert and D. M. Williams, eds.), pp. 211–230; Oxford: Clarendon Press.

Peterson, K. J., Andrew Cameron, R., and Davidson, E. H. (1997). Set-aside cells in maximal indirect development: evolutionary and developmental significance. *BioEssays* **19**, 623–631.

Pflug, H. D. (1970). Zur Fauna der Nama-Schichten in Südwest-Afrika, I: Pteridinia, Bau und systematishche Zugehörigkeit. *Palaeontographica Abt. A*. **137**, 226–262.

Philippe, H., Chenuil, A., and Adoutte, A. (1994). Can the Cambrian explosion be inferred through molecular phylogeny? *Development* (1994 Suppl., *The Evolution of Developmental Mechanisms*), pp. 15–25.

Pierson, B. K. (1994). The emergence, diversification, and role of photosynthetic eubacteria. In: *Early Life on Earth*, Nobel Symposium No. 84 (S. Bengtson, ed.), pp. 161–180. New York: Columbia University Press.

Pollard, T. D., and Quirk, S. (1994). Prolifins, ancient actin binding proteins with highly divergent primary structures. In: *Molecular Evolution of Physiological Processes*, 47th Annual Symposium Society of Physiologists, Vol. 49 (D. M. Fambrough, ed.), pp. 117–128. New York: Rockefeller University Press.

Pool, R. (1993). "Evidence for homosexuality gene," Research News. *Science* (16 July), **261**, 291–292.

Popper, K. R. (1972). *Objective Knowledge: An Evolutionary Approach*. Oxford: Clarendon Press.

Popper, K. (1976). *Unended Quest*. La Salle, Ill: Open Court Press.

Preston, R. (1994). *The Hot Zone*. New York: Random House.

Prigogine, I., and Stengers, I. (1984). *Order Out of Chaos: Man's New Dialogue with Nature*. Toronto: Bantam Books.

Rabinow, P. (1996). *Making PCR: A Story of Biotechnology*. Chicago: University of Chicago Press.

Raff, R. A. (1994). Developmental mechanisms in the evolution of animal form: origins and evolvability of body plans. In: *Early Life on Earth*, Nobel Symposium No. 84 (S. Bengtson, ed.), pp. 489–500. New York: Columbia University Press.

Raff, R. A. (1996). *The Shape of Life: Genes, Development and the Evolution of Animal Form*. Chicago: University of Chicago Press.

RajBhandary, U. L., and Söll, D. (1995). Transfer RNA in its fourth decade. In: *tRNA: Structure, Biosynthesis, and Function* (D. Söll and U. L. RajBhandary, eds.), pp. 1–4. Washington, DC: ASM Press.

Ramón y Cajal, Santiago. (1989). *Recollections of My Life*. Translated by E. Horne Craigie with the Assistance of Juan Cano. Cambridge: The MIT Press.

Restak, R. (1991). *The Brain Has a Mind of Its Own: Insights from a Practicing Neurologist*. New York: Harmony Books.

Retallack, G. J. (1994). Were the ediacaran fossils lichens? *Paleobiology* **20**, 523–544.

Rieger, R. M. (1994). Evolution of the "lower" Metazoa. In: *Early Life on Earth*, Nobel Symposium No. 84 (S. Bengtson, ed.), pp. 475–488. New York: Columbia University Press.

Rieppel, O. (1994). Species and history. In: *Models in Phylogeny Reconstruction*, Systematics Association Special Volume 52 (R. W. Scotland, D. J. Siebert and D. M. Williams, eds.), pp. 31–50. Oxford: Clarendon Press.

Riley, M. (1993). Functions of gene products in *E. coli*. *Microbiol. Rev.* **57**, 862–952.

Ringertz, N. R., and Savage, R. E. (1976). *Cell Hybrids*. New York: Academic Press.

Rivera, M., and Lake, J. (1992). Evidence that eukaryotes and eocyte prokaryotes are immediate relatives. *Science* **257**, 74–76.

Roger, A. J., Svärd, S. G., Tovar, J., Clark, C. G., Smith, M. W., Gillin, F. D., and Sogin, M. L. (1998). A mitochondrial-like chaperonin 60 gene in *Giardia lamblia*: evidence that diplomonads once harbored an endosymbiont related to the progenitor of mitochondria. *Proc. Natl. Acad. Sci. U.S.A.* **95**, 229–234.

Rogers, J. (1996). *The Matter of Revolution: Science, Poetry, and Politics in the Age of Milton*. Ithaca: Cornell University Press.

Rosenthal, B., Mai, Z., Caplivski, D., Ghosh, S., De La Vega, H., Graf, T., and Samuelson, J. (1997). Evidence for the bacterial origins of genes encoding fermentation enzymes of the amitochondriate protozoan parasite *Entamoeba histolytica*. *J. Bacteriol.* **179**, 3736–3745.

Roughgarden, J. (1975). Evolution of marine symbiosis — a simple cost-benefit model. *Ecology* **56**, 1208.

Rowen, L., Mahairas, G., and Hood, L. (1997). Sequencing the human genome. *Science* **278**, 605–607.

Runnegar, B. (1994). Proterozoic eukaryotes: evidence from biology and geology. In: *Early Life on Earth*, Nobel Symposium No. 84 (S. Bengtson, ed.), pp. 287–297. New York: Columbia University Press.

Ruse, M. (1996). *Monad to Man: The Concept of Progress in Evolutionary Biology*. Cambridge: Harvard University Press.

Ruse, M. (1997). (). origin of life: philosophical perspectives. *J. Theor. Biol.* **187**, 473–482.

Saccharomyces Genome Database (SGD) at http://genome-www.stanford.edu/Saccharomyces/.

Sagan, D. (1990). *Biosphere*. New York: McGraw-Hill.

Saiki, R. K., Scharf, S., Faloona, F., Mullis, K. B., Horn, G. T., Erlich, H. A., and Arhneim, N. (1985). Enzymatic amplification of β-globin genomic sequences and restriction site analysis for diagnosis of sickle cell anemia. *Science* **230**, 1350–1354.

Saimi, Y., Martinac, B., Preston, R. R., Zhou, X.-L., Sukharev, S., Blount, P., and Kung, C. (1994). Ion Channels of microbes. In: *Molecular Evolution of Physiological Processes*, 47th Annual Symposium Society of Physiologists, Vol. 49 (D. M. Fambrough, ed.), pp. 179–195. New York: Rockefeller University Press.

Saks, M. E., Sampson, J. R., and Abelson, J. (1998). Evolution of a transfer RNA gene through a point mutation in the anticodon. *Science* **279**, 1665–1670.

Salvani-Plawen, L. V. (1978). On the origin and evolution of the lower Metazoa. *Wiss. zool. Syst. Evolut.-Forsch.* **16**, 40–88.

Sapp, J. (1985). Concepts of organization and leverage of ciliate protozoa. In: *A Conceptual History of Modern Embryology* (S. F. Gilbert, ed.), *Developmental Biology: A Comprehensive Synthesis*, Vol. 7., pp. 229–258. New York: Plenum Press.

Sapp, J. (1991). Living together: symbiosis and cytoplasmic inheritance. In: *Symbiosis as a Source of Evolutionary Innovation: Speciation and Morphogenesis* (L. Margulis and R. Fester, eds.), pp. 15–25. Cambridge: MIT Press.

Sapp, J. (1994). *Evolution by Association: A History of Symbiosis*. New York: Oxford University Press.

Saunders Jr., J. W. (1982). *Developmental Biology: Patterns/ Problems/ Principles*. New York: Macmillan Publishing.

Schiffmann, Y. (1994). Instability of the homogeneous state as the source of localization, epigenesis, differentiation, and morphogenesis. *Int. Rev. Cytol.* **154**, 309–375.

Schimenti, J. C. (1994). Gene conversion and the evolution of gene families in mammals. In: *Molecular Evolution of Physiological Processes*, 47th Annual Symposium Society of Physiologists, Vol. 49 (D. M. Fambrough, ed.), pp. 85–91. New York: Rockefeller University Press.

Schimke, R. T. (1982). Studies on gene duplication and amplifications — an historical perspective. In: *Gene amplification* (R. T. Schimke, ed.), pp. 1–6. Cold Spring Harbor: Spring Harbor Laboratory.

Schimmel, P. (1995). Implications of an operational RNA code for amino acids. In: *Tracing Biological Evolution in Protein and Gene Structure* (M. Gō and P. Schimmel, eds.), pp. 1–10. Amsterdam: Elsevier.

Schlegel, M., Lom, J., Stechmann, A., Bernhard, D., Leipe, D., Dyková, I., and Sogin. M. L. (1996). Phylogenetic analysis of complete small subunit ribosomal RNA coding regions of *Myxidium lieberkuehni*: evidence that Myxozoa are Metazoa and related to the Bilateria. *Arch. Protistenke.* **147**, 1–9.

Schopf, J. W. (1992). Paleobiology of the Archean. In: *The Proterozoic Biosphere: A Multidisciplinary Study* (J. W. Schopf and C. Klein, eds.), pp. 25–39. Cambridge: Cambridge University Press.

Schopf, J. W. (1994). The oldest known records of life: early Archean stromatolites, microfossils, and organic matter. In: *Early Life on Earth*, Nobel Symposium No. 84 (S. Bengtson, ed.), pp. 191–206. New York: Columbia University Press.

Schrödinger, E. (1944). *What is Life? The Physical Aspects of the Living Cell.* Cambridge: Cambridge University Press.

Schuler, G. D., et al. (1996). A gene map of the human genome. *Science* 274:540–546.

Schwemmler, W. (1984a/1979). *Reconstruction of Cell Evolution: A Periodic System.* [Translated from *Mechanismen der Zellevolution: Grundriss einer modernen Zelltheorie.* Berlin, Walter de Gruyter; 1979]. Boca Raton: CRC Press.

Schwemmler, W. (1984b). Analysis of possible gene transfer between an insect host and its bacterial endosymbionts. *Int. Rev. Cytol., Suppl.* **14**, 247–266.

Schwemmler, W. (1989). Insect endocytobiosis as a model system for egg cell differentiation. In: *Insect Endocytobiosis: Morphology, Physiology, Genetics, Evolution* (W. Schwemmler and G. Gassner, eds.), pp. 38–53. Boca Raton: CRC Press.

Schwemmler, W. (1991). Symbiogenesis in insects as a model for morphogenesis, cell differentiation, and speciation. In: *Symbiosis as a Source of Evolutionary Innovation: Speciation and Morphogenesis* (L. Margulis and R. Fester, eds.), pp. 178–204. Cambridge: MIT Press.

Seilacher, A. (1989). Vendozoa: organismic construction in the Proterozoic biosphere. *Lethaia* **22**, 229–239.

Seilacher, A. (1994). Early multicellular life: late Proterozoic fossils and the Cambrian explosion. In: *Early Life on Earth*, Nobel Symposium No. 84 (S. Bengtson, ed.), pp. 389–400. New York: Columbia University Press.

Senapathy, P. (1994). *Independent Birth of Organisms: A New Theory that Distinct Organisms Arose Independently from the Primordial Pond, Showing that Evolutionary Theories Are Fundamentally Incorrect.* Madison: Genome Press.

Sharp, P. A. (1985). On the origin of RNA splicing and introns. *Cell* **42**, 397–400.

Shiba, K. (1995). Dissection of an enzyme into two fragments at intron-exon boundaries. In: *Tracing Biological Evolution in Protein and Gene Structure* (M. Gō and P. Schimmel, eds.), pp. 11–21. Amsterdam: Elsevier.

Shiva,. V. (1990). Reductionist science as epistemological violence. In: *Science, Hegemony and Violence: A Requiem for Modernity* (A. Nandy, ed.), pp. 232–245. Bombay: Oxford University Press.

Short, R. V. (1977). The discovery of the ovaries. In: *The Ovary* (S. Zuckerman and B. J. Weir, eds.), Vol. 1, pp. 1–67. Orlando: Academic Press.

Shostak, S. (1993a). Symbiogenetic origins of Cnidaria: updating an hypothesis. *Invertebrate Repro. Devel.* **23**, 167–168.

Shostak, S. (1993b). A symbiogenetic theory for the origins of cnidocysts in Cnidaria. *BioSystems* **29**, 49–58.

Shostak, S. (1998). *Death of Life: The Legacy of Molecular Biology*. London: Macmillan.

Shostak, S., and Kolluri, V. (1995). Origins of diversity in cnidarian cnidocysts. *Symbiosis* **19**, 1–29.

Siddal, M. E., Stokes, N. A., and Burreson, E. M. (1995). Molecular phylogenetic evidence that the Phylum Haplosporidia has an alveolate ancestry. *Mol. Biol. Evol.* **12**, 573–581.

Singer, C. (1959). *A Short History of Scientific Ideas to 1900*. London: Oxford University Press.

Smalla, K. (1992). Screening of erythromycin resistant *Bacillus* spp. from aerogeological samplings for recombinant DNA. In: *Gene Transfers and Environment: Proceedings of the Third European Meeting on Bacterial Genetics and Ecology* (M. J. Gauthier, ed.), pp. 21–25. Berlin: Springer-Verlag.

Smith, M. C. M., Burns, N., Sayers, J. R., Sorrell, J. A., Casjens, S. R., and Hendrix. R. W. (1998). "Bacteriophage collagen," Letters. *Science* (20 March), **279**, 1834.

Smith, M. W., Feng, D.-F., and Doolittle, R. F. (1992). Evolution by acquisition: the case for horizontal transfer. *Trends Biochem. Sci.* **17**, 489–493.

Smith, P. G., and Cousens, S. N. (1996). "Is the new variant of Creutzfeldt-Jakob disease from mad cows?" Perspectives. *Science* (9 August), **273**, 748.

Smothers, J. F., von Dohlen, C. D., Smith Jr., L. H., and Spall, R. D. (1994). Molecular evidence that the myxozoan protists are metazoans. *Science* **265**, 1719–1721.

Snow, M. H. L., Tam, P. P. L., and McLaren, A. (1981). On the control and regulation of size and morphogenesis in mammalian embryos. In: *Levels of Genetic Control in Development* (S. Subtelny and U. K. Abbot, eds.), pp. 201–217. New York: Alan R. Liss.

Sober, E. (1994). *From a Biological Point of View: Essays in Evolutionary Philosophy*. Cambridge Studies in Philosophy and Biology. Cambridge: Cambridge University Press.

Sogin, M. L. (1989). Evolution of eukaryotic microorganisms and their small subunit ribosomal RNAs. *Am. Zool.* **29**, 487–499.

Sogin, M. L. (1991). Early evolution and the origin of eukaryotes. *Curr. Opin. Genet. Devel.* **1**, 457–463.

Sogin, M. L. (1994). The origin of eukaryotes and evolution into major kingdoms. In: *Early Life on Earth* (S. Bengston, ed.), pp. 181–192. New York: Columbia University Press.

Sogin, M. L., and Elwood, H. J. (1986). Primary structure of the *Paramecium tetraurelia* small-subunit rRNA coding regions: Phylogenetic relationships within the Ciliophora. *J. Mol. Evol.* **23**, 53–60.

Sogin, M. L., Elwood, H. J., and Gunderson, J. H. (1986). Evolutionary diversity of eukaryotic small-subunit rRNA genes. *Proc. Natl. Acad. Sci. U.S.A.* **83**, 1383–1387.

Sogin, M. L., Gunderson, J. H., Elwood, H. J., Alonso, R. A., and Peattie, D. A. (1989). Phylogenetic meaning of the kingdom concept: An unusual ribosomal RNA from *Giardia lamblia*. *Science* **243**, 75–77.

Sollner-Webb, B. (1996). "Trypanosome RNA editing: resolved," Perspectives. *Science* (30 August), **273**, 1182–1183.

Söll, D. (1993). Transfer RNA: an RNA for all seasons. In: *The RNA World: The Nature of Modern RNA Suggests a Prebiotic RNA World* (R. F. Gesteland and J. F. Atkins, eds.), pp. 157–184. Cold Spring Harbor: Cold Spring Harbor Laboratory Press.

Song, Y., and Fambrough, D. M. (1994). Molecular evolution of the Calcium-transporting ATPases analyzed by the maximum parsimony method. In: *Molecular Evolution of Physiological Processes*, 47th Annual Symposium Society of Physiologists, Vol. 49 (D. M. Fambrough, ed.), pp. 271–283. New York: Rockefeller University Press.

Southward, A. J., and Southward, E. C. (1996). Symbiosis and sulphur-fueled animal life in the sea: Gaian implications. In: *Gaia in Action* (P. Bunyard, ed.), pp. 183–199. Edinburgh: Flores.

Spradling, A. (1994). Transposable elements and the evolution of heterochromatin. In: *Molecular Evolution of Physiological Processes*, 47th Annual Symposium Society of Physiologists, Vol. 49 (D. M. Fambrough, ed.), pp. 69–83. New York: Rockefeller University Press.

Steen, R. G. (1996). *DNA and Destiny: Nature and Nurture in Human Behavior*. New York: Plenum Press.

Steitz, T. A. (1993). Similarities and differences between RNA and DNA recognition proteins. In: *The RNA World: The Nature of Modern RNA Suggests a Prebiotic RNA World* (R. F. Gesteland and J. F. Atkins, eds.), pp. 219–237. Cold Spring Harbor: Cold Spring Harbor Laboratory Press.

Steitz, T. A., Silvian, L. F., and Rath. V. L. (1995). Aminoacyl tRNA synthetase and polymerase: modular design and evolution. In: *Tracing Biological Evolution in Protein and Gene Structure* (M. Gō and P. Schimmel, eds.), pp. 125–133. Amsterdam: Elsevier.

Stent, G. (1971). *Molecular Genetics: An Introductory Narrative*. San Francisco: W. H. Freeman.

Stetter, K. O. (1994). The lesson of Archaebacteria. In: *Early Life on Earth*, Nobel Symposium No. 84 (S. Bengtson, ed.), pp. 143–151. New York: Columbia University Press.

Stewart, I. (1989). *Does God Play Dice? The Mathematics of Chaos*. Oxford: Basil Blackwell Ltd.

Stoltzfus, A., Spencer, D. F., Zuker, M., Logsdon, J. M., and Doolittle, W. F. (1994). Testing the exon theory of genes: the evidence from protein structure. *Science* **265**, 202–207.

Strauss, H., and Moore, T. B. (1992). Abundances and isotopic compositions of carbon and sulfur species in whole rock and kerogen samples. In: *The Proterozoic Biosphere: A Multidisciplinary Study* (J. W. Schopf and C. Klein, eds.), pp. 709–798. Cambridge: Cambridge: University Press.

Sugita, M., Ohta, M., and Sigiura, M. (1995). Structure and function of RNA-binding proteins in chloroplasts and cyanobacteria. In: *Tracing Biological Evolution in Protein and Gene Structure* (M. Gō and P. Schimmel, eds.), pp. 77–86. Amsterdam: Elsevier.

Sugiura, M. (1989). The chloroplast chromosomes in land plants. *Ann. Rev. Cell Biol.* **5**, 51–71.

Summons, R. E. (1992). Abundance and composition of extractable organic matter. In: *The Proterozoic Biosphere: A Multidisciplinary Study* (J. W. Schopf and C. Klein, eds.), pp. 101–115. Cambridge: Cambridge University Press.

Suomalainen, E., Saura, A., and Lokki, J. (1987). *Cytology and Evolution in Parthenogenesis*. Boca Raton: CRC Press.

Syvanen, M. (1994). Horizontal gene flow: evidence and possible consequences. *Ann. Rev. Genet.* **28**, 237–261.

Szostak, J. W., and Ellington, A. D. (1993). In vitro selection of functional RNA sequences. In: *The RNA World: The Nature of Modern RNA Suggests a Prebiotic RNA World* (R. F. Gesteland and J. F. Atkins, eds.), pp. 511–533. Cold Spring Harbor: Cold Spring Harbor Laboratory Press.

Takahashi, K.-I., Gō, M., and Noguti, T. (1995). Mechanical stability of protein modules determined by molecular dynamics simulations. In: *Tracing Biological Evolution in Protein and Gene Structure* (M. Gō and P. Schimmel, eds.), pp. 175–185. Amsterdam: Elsevier.

Tang, X., Halleck, M. S., Schlegel, R. A., and Williamson, P. (1996). A subfamily of P-type ATPases with aminophospholipid transporting activity. *Science* **272**, 1495–1497.

Tani, T., Takahashi, Y., Urushiyama, S., and Ohshima, Y. (1995). Spliceosomal introns in the spliceosomal small nuclear RNA genes. In: *Tracing Biological Evolution in Protein and Gene Structure* (M. Gō and P. Schimmel, eds.), pp. 97–114. Amsterdam: Elsevier.

Tateno, M., Mizutani, M., Yura, K., Nureki, O., Yokoyama, S., and Gō, M. (1995). Module structure and function of glutamyl-tRNA synthetase. In: *Tracing Biological Evolution in Protein and Gene Structure* (M. Gō and P. Schimmel, eds.), pp. 53–63. Amsterdam: Elsevier.

Tatusov, R. L., Koonin, E. V., and Lipman, D. J. (1997). A genomic perspective on protein families. *Science* **278**, 631–637.

Taylor, F. J. R. (1994). "Max," The role of phenotypic comparisons in the determination of protist phylogeny. In: *Early Life on Earth*, Nobel Symposium No. 84 (S. Bengtson, ed.), pp. 312–326. New York: Columbia University Press.

Taylor, S. J., and Shalloway, D. (1996). Src and the control of cell division. *BioEssays* **18**, 9–11.

Timmis, J. N., and Scott, N. S. (1984). Promiscuous DNA: sequence homologies between DNA of separate organelles. *Trends Biochem. Sci.* **9**, 271–273.

Tomb, J.-F. (1997). The complete genome sequence of the gastric pathogen *Helicobacter pylori*. *Nature* **388**, 539–547.

Tomizawa, J. (1993). Evolution of functional structures of RNA. In: *The RNA World: The Nature of Modern RNA Suggests a Prebiotic RNA World* (R. F. Gesteland and J. F. Atkins, eds.), pp. 419–445. Cold Spring Harbor: Cold Spring Harbor Laboratory Press.

Towe, K. M. (1994). Earth's early atmosphere: constraints and opportunities for early evolution. In: *Early Life on Earth*, Nobel Symposium No. 84 (S. Bengtson, ed.), pp. 36–47. New York: Columbia University Press.

Trench, R. K. (1991). *Cyanophora paradoxa* Korschikoff and the origins of chloroplasts. In: *Symbiosis as a Source of Evolutionary Innovation: Speciation and Morphogenesis* (L. Margulis and R. Fester, eds.), pp. 143–150. Cambridge: MIT Press.

Tsonis, A. A. (1998). "Fractality in nature," Letters. *Science* (13 March), **279**, 1614–1615.

Tuberville, J. M., Field, K. G., and Raff, R. A. (1992). Phylogenetic position of Phylum Nemertini, inferred from 18S rRNA sequences: molecular data as a test of morphological character homology. *Mol. Biol. Evol.* **9**, 235–249.

Turner, D. H., and Bevilacqua, P. C. (1993). Thermodynamic considerations for evolution by RNA. In: *The RNA World: The Nature of Modern RNA Suggests a Prebiotic RNA World* (R. F. Gesteland and J. F. Atkins, eds.), pp. 447–464. Cold Spring Harbor: Cold Spring Harbor Laboratory Press.

Tuzet, O. (1973). Introduction et place des spongiaires dans la classification. In: P. P. Grassóe, ed., *Traitè de Zool.* **3**:(1), 1–26.

Upcroft, J., and Upcroft, P. (1998). My favorite cell: *Giardia. BioEssays* **20**, 256–263.

Urey, H. C. (1994/1952). On the early chemical history of the Earth and the origin of life. *Proc. Natl. Acad. Sci. U.S.A.* **38**, 351–363; 1952. In: *Origins of Life: The Central Concepts* (D. W. Deamer and G. R. Fleischaker, eds.), pp. 83–95. Boston: Jones & Bartlett.

Valentine, J. W. (1991). Major factors in the rapidity and extent of the metazoan radiation during the Proterzoic Phanerozoic transition. In: *The Early Evolution of Metazoa and the Significance of Problematic Taxa* (A. M. Simonetta and S. Conway Morris, eds.), pp. 11–13. Cambridge: Cambridge University Press.

Vidal, G. (1994). Early ecosystems: limitations imposed by the fossil record. In: *Early Life on Earth*, Nobel Symposium No. 84 (S. Bengtson, ed.), pp. 298–311. New York: Columbia University Press.

Vilà, C., Maldonado, J., Amorim, I. R., Wayne, R. K., Crandall, K. A., and Honeycutt, R. L. (1997). "Man and his dog," Letters. *Science* (10 October 1997), **278**, 206–207.

Viscogliosi, E., Philippe, H., Baroin, A., Perasso, R., and Brugerolle, G. (1993). Phylogeny of trichomonads based on partial sequence of large subunit rRNA and on cladistic analysis of morphological data. *J. Eukaryot. Microbiol.* **40**, 411–421.

Vogel, G. (1997). Searching for living relics of the cell's early days. *Science* **277**, 1604.

Vossbrinck, C. R., Maddox, J. V., Friedman, S., Debrunner-Vossbrinck, B. A., and Woese, C. R. (1987). Ribosomal RNA sequence suggests microsporidia are extremely ancient eukaryotes. *Nature* **326**, 411–414.

Wächtershäuser, G. (1994). Vitalysts and virulysts: a theory of self-expanding reproduction. In: *Early Life on Earth*, Nobel Symposium No. 84 (S. Bengtson, ed.), pp. 124–132. New York: Columbia University Press.

Wächtershäuser, G. (1997). The origin of life and its methodological challenge. *J. Theor. Biol.* **187**, 483–494.

Wainright, P. O., Hinkle, G., Sogin, M. L., and Stickel, S. K. (1993). Monophyletic origins of the metazoa: an evolutionary link with fungi. *Science* **260**, 240–342.

Wakasugi, K., Ishimori, K., and Morishima, I. (1995). Module substitution in globins: preparation and association characteristics of chimeric hemoglobin subunits and myoglobin. In: *Tracing Biological Evolution in Protein and Gene Structure* (M. Gō and P. Schimmel, eds.), pp. 283–295. Amsterdam: Elsevier.

Walter, M. R. (1994). Stromatolites: the main geological source of information on the evolution of the early benthos. In: *Early Life on Earth*, Nobel Symposium No. 84 (S. Bengtson, ed.), pp. 270–286. New York: Columbia University Press.

Warren, R. P., Odell, J. D., Warren, W. L., Burger, R. A., Maciulis, A., Daniels, W. W., and Torres. A. R. (1995). "Reading disability, attention-deficit hyperactivity disorder, and the immune system," Letters. *Science* (12 May), **268**, 786–787.

Watanabe, K., and Osawa, S. (1995). tRNA sequences and variations in the genetic code. In: *tRNA: Structure, Biosynthesis, and Function* (D. Söll and U. L. RajBhandary, eds.), pp. 225–250. Washington, DC: ASM Press.

Watson, J. D. (1969). *The Double Helix: A Personal Account of the Discovery of the Structure of DNA*. New York: Mentor.

Watson, J. D., and Crick, F. H. C. (1953). Molecular structure of nucleic acids: a structure for deoxyribose nucleic acid. *Nature* **171**, 737–738.

Weinstein, J. N., Myers, T. G., O'Connor, P. M., Friend, S. H., Fornace Jr., A. J., Kohn, K. W., Fojo, T., Bates, S. E., Rubinstein, L. V., Anderson, N. L., Buolamwini, J. K., van Osdol, W. W., Monks, A. P., Scudiero, D. A., Sausville, E. A., Zaharevitz, D. W., Bunow, B., Viswanadhan, V. N., Johnson, G. S., Wittes, R. E., and Paull, K. D. (1997). An information-intensive approach to the molecular pharmacology of cancer. *Science* **275**, 343–349.

Weismann, A. (1882). *Ueber die Dauer des Lebens*. Jena: Gustav Fischer Verlag.

Weismann, A. (1892). *Das Keimplasma. Eine Theorie der Vererbung*. Jena: Gustav Fischer Verlag.

Weismann, A. (1912). *The Germ-plasm: A Theory of Heredity*. Translated by W. Newton Parker and Harriet Rionnfeldt. New York: Charles Scribner's Sons; London: W. Scott.

Weiss, A., Mayer, D. C. G., and Leinwand, L. (1994). Diversity of myosin-based motility: multiple genes and functions. In: *Molecular Evolution of Physiological Processes*, 47th Annual Symposium Society of Physiologists, Vol. 49 (D. M. Fambrough, ed.), pp. 159–171. New York: Rockefeller University Press.

Weiss, R., and Cherry, J. (1993). Speculations on the origin of ribosomal translocation. In: *The RNA World: The Nature of Modern RNA Suggests a Prebiotic RNA World* (R. F. Gesteland and J. F. Atkins, eds.), pp. 71–89. Cold Spring Harbor: Cold Spring Harbor Laboratory Press.

Westfall, J. A., Bradbury, P. C., and Townsend, J. W. (1983). Ultrastructure of the dinoflagellate *Polykrikos*. *J. Cell Sci.* **63**, 245–261.

Weston, P. H. (1994). Methods for rooting cladistic trees. In: *Models in Phylogeny Reconstruction*. Systematics Association Special Volume 52 (R. W. Scotland, D. J. Siebert and D. M. Williams, eds.), pp. 125–155. Oxford: Clarendon Press.

Whittaker, R. H. (1959). On the broad classification of organisms. *Quart. Rev. Biol.* **34**, 210–226.

Wilkie, T. M., and Yokoyama, S. (1994). Evolution of the G protein Alpha subunit multigene family. In: *Molecular Evolution of Physiological Processes*, 47th Annual Symposium Society of Physiologists, Vol. 49 (D. M. Fambrough, ed.), pp. 249–270. New York: Rockefeller University Press.

Wilkins, A. S. (1995). The cell cycle in growth and development: a special issue. *BioEssay* **17**, 469–470.

Williamson, D. I. (1992). *Larvae and Evolution: Toward a New Zoology*. New York: Chapman and Hall.

Wilmut, I., Schnieke, A. E., McWhir, J., Kind, A. J., and Campbell, K. H. S. (1997). Viable offspring derived from fetal and adult mammalian cells. *Nature* **385**, 810–813.

Wilson, E. B. (1896). *The Cell in Development and Inheritance*. London: Macmillan.

Woese, C. R. (1987). Bacterial evolution. *Microbiol. Rev.* **51**, 221–227.

Woese, C. R. (1991). The use of ribosomal RNA in reconstructing evolutionary relationships among bacteria. In: *Evolution at the Molecular Level* (R. K. Selander, A. G. Clark and T. S. Whittam, eds.), pp. 1–24. Sunderland, MA: Sinauer Associates.

Woese, C. R., and Fox, G. E. (1977). Phylogenetic structure of the prokaryotic domain: the primary kingdoms. *Proc. Natl. Acad. Sci. U.S.A.* **74**, 5088–5090.

Woese, C. R., and Pace, N. R. (1993). Probing RNA structure, function, and history by comparative analysis. In: *The RNA World: The Nature of Modern RNA Suggests a Prebiotic RNA World* (R. F. Gesteland and J. F. Atkins, eds.), pp. 91–117. Cold Spring Harbor: Cold Spring Harbor Laboratory Press.

Woese, C. R., Dugre, D. H., Dugre, S. A., Kondo, M., and Saxinger, W. C. (1966). On the fundamental nature and evolution of the genetic code. *Cold Spring Harbor Symp. Quant. Biol.* **31**, 723–736.

Woese, C. E., Kandler, O., and Wheelis, M. L. (1990). Towards a natural system of organisms: proposal for the domains Archaea, Bacteria, and Eucarya. *Proc. Natl. Acad. Sci. U.S.A.* **87**, 4576–4579.

Wray, G. A., Levinton, J. S., and Shapiro, L. H. (196573). Molecular evidence for deep Precambrian divergences among metazoan phyla. *Science* **274**, 56.

Wrighton, N. C., Farrell, F. X., Chang, R., Kashyap, A. K., Barbone, F. P., Mulcahy, L. S., Johnson, D. L., Barrett, R. W., Jolliffe, L. K., and Dower, W. J. (1996). Small peptides as potent mimetics of the protein hormone erythropoietin. *Science* **273**, 458–463.

Wu, C.-I., and Hammer, M. F. (1991). Molecular evolution of ultraselfish genes of meiotic drive systems. In: *Evolution at the Molecular Level* (R. K. Selander, A. G. Clark and T. S. Whittam, eds.), pp. 177–203. Sunderland, MA: Sinauer Associates.

Wyatt, J. R., and Tinoco Jr., I. (1993). RNA structural elements and RNA function. In: *The RNA World: The Nature of Modern RNA Suggests a Prebiotic RNA World* (R. F. Gesteland and J. F. Atkins, eds.), pp. 465–496. Cold Spring Harbor: Cold Spring Harbor Laboratory Press.

Yamamoto, A., Hashimoto, T., Asaga, E., Hasegawa, M., and Goto, N. (1997). Phylogenetic position of the mitochondrion-lacking protozoan *Trichomonas tenax*, based on amino acid sequences of elongation factors 1alpha and 2. *J. Mol. Evol.* **44**, 98–105.

Yang, D., Oyaizu, Y., Oyaizu, H., Olsen, G. J., and Woese, C. R. (1985). Mitochondrial origins. *Proc. Natl. Acad. Sci. U.S.A.* **82**, 4443–4447.

Yarus, M. (1988). A specific amino acid binding site composed of RNA. *Science* **240**, 1751–1758.

Yarus, M. (1993). An RNA-amino acid affinity. In: *The RNA World: The Nature of Modern RNA Suggests a Prebiotic RNA World* (R. F. Gesteland and J. F. Atkins, eds.), pp. 205–217. Cold Spring Harbor: Cold Spring Harbor Laboratory Press.

Yarus, M., and Smith, D. (1995). tRNA on the ribosome: a waggle theory. In: *tRNA: Structure, Biosynthesis, and Function* (D. Söll and U. L. RajBhandary, eds.), pp. 443–469. Washington, DC: ASM Press.

Yomo, T., Trakulnaleamsai, S., and Urabe, I. (1995). Sketch of landscapes in the protein sequence space around catalase I from *Bacillus stearothermophilus*. In: *Tracing Biological Evolution in Protein and Gene Structure* (M. Gō and P. Schimmel, eds.), pp. 261–269. Amsterdam: Elsevier.

Yund, P. O., Cunningham, C. W., and Buss, L. W. (1982). Recruitment and post-recruitment interactions in a colonial hydroid. *Ecology* **68**, 971–982.

Yura, K., and Gō, M. (1995). Helix-turn-helix module distribution and module shuffling. In: *Tracing Biological Evolution in Protein and Gene Structure* (M. Gō and P. Schimmel, eds.), pp. 187–195. Amsterdam: Elsevier.

Zillig, W., Klenk, H.-P., Palm, P., Leffers, H., Pühler, G., Gropp, F., and Garrett, R. A. (1989). Did eukaryotes originate by a fusion event? *Endocytobiosis Cell Res.* **6**, 1–2.

Zillig, W., Palm, P., and Klenk, H. P. (1992). The nature of the common ancestor of the three domains of life and the origin of the Eucarya. In: *Frontiers of Life*, Proceedings of the 3rd Renontres de Blois (J. and K. Trân Thanh Vân, J. C. Mounolou, J. Schneider, and C. McKay, eds.), pp. 181–193. France: Editions Frontieres.

Zuckerkandl, E., and Pauling, L. (1965). Molecules as documents of evolutionary history. *J. Theor. Biol.* **8**, 357–366.

INDEX

Achlya ambisexualis, 131
Acrasiomycota, 171, 219
Actin, *See* Gene families
Actinomycin D, 166
Adenosine triphosphatase (ATPase), 79, 82, 116, 118, 130, 132–133, 138, 144, 147, 183–184, 193, 263n
Alcaligenes, 124
Alcoholism, 19
Alignment and sequence assembly software, 64, 69, 70, 72–73, 77, 80, 81, 88, 101, 249n, 251n
Allopolyploidy, 98
Allozymes, 18, 76, 105. *Also see* Isozymes
Alternative splicing (rearrangement) sites, 218
Alveolates, 87t, 140
Ambisense (ambistranded), *See* Double-stranded (ds)RNA virus
Amitochondriate. *See* Eucarya
Amoebamastigote, 140
Amyotrophic lateral sclerosis, 68
Anabaena, 123, 184, 198, 201
Anastrophy, 37
Ancient conserved regions (ACRs), 75, 81–83
Angiosperms, 131, 230
Animalia, 87t, 97–98. *Also see* Metazoa
Anthozoa, 203
Anticodons, 35, 167–171, 218–219, 243n, 266n
Antigenic drift, 226
AntiLamarckism, 31
Apicomplexa (apicomplexans), 87t, 140, 203
Aquifex, 64, 87t
Arabidopsis, 61, 67, 177
Archaea (Archaebacteria), 87t, 116–117, 120, 126–128, 138, 142–145, 147, 151, 172, 181–183, 186, 192–193, 260n
  Crenarchaeota (crenarcheotes or crenotes), 87t, 126, 128–129, 143
  Euryarchaeota (euryotes or euryarchaeotes), 87t, 127–128
  Methanobacter (Methanobacteriales), 127
  Methanococcus (Methanococcales), 127
  Methanogens, 87t, 99, 122, 126, 127–128, 183, 192, 196, 197
  Methanosarcina (Methanomicrobiales), 127
  Methanospirillum, 127
  Thermococcus-Pyrococcus group, 128
  Thermophiles, 126, 128
  Thermoproteus-Pyrodictium cluster, 129
Archaeoglobus fulgidus, 64, 128, 144
Archamoebae, 87t, 140
Archeobacteria, 87t
Archezoa. *See* Eucarya
Aristotle (384–322 BC), 3, 5, 16, 22–24, 26, 29, 241n
Arnold, Michael, 230
Arrhenius, Svante August (1859–1927), 156
Ascaris suum, 132
Asexual reproduction, 227, 274n
Atrazine herbicide, 229
Attention-deficit hyperactivity disorder, 20
Autoimmune disorders, 20
Autocatalysis, 172, 264n
Autonomous replication sequences (ARSs), 66
Autopolyploidization, 275n
Autopomorphies, 105, 146

Autotrophs, 123–125, 128–129, 163, 196–197
Azobacter, 124

Bacillariophyta (bacillariophytes), 271n
Bacillus subtilis, 64, 66–67, 70, 76, 79, 122, 245n, 249n
Bacteria (Eubacteria), 87t, 97–99, 116–117, 120–126, 138, 142–145, 147
  Agrobacteria, 124
  Bacillus/Lactobacillus, 122
  Bacteroides, 125
  Borrelias, 125
  Chlamydiae, 125
  Cyanobacteria (blue-green bacteria), 87t, 120, 121, 123, 124, 184–185, 194, 198, 201
  Deinococci, 125
  Flavobacteria (yellow bacteria), 87t, 125
  Gram-negative, 87t, 121, 123–124, 126, 192, 258n
  Gram-positive, 87t, 121–122, 123, 258n
  Green sulfur bacteria and green bacteria, 87t, 121, 125
  Leptospira, 124, 125
  Mycoplasma, 65, 121, 122, 170, 174, 195, 219, 249n, 258n
  Planctomyces, 125
  Pseudomonads, 124
  Purple bacteria, 87t, 114, 121, 124, 184–185, 194, 197
  Rhizobacteria, 124
  Rickettsiae, 124, 195, 202
  Spirochetes, 125, 202
  Thermotogales, 121, 122t, 125, 126, 143, 145
  Vibrios, 124
Bacterial artificial chromosome (BAC), 61
Bacteriophage (phage), 3, 14, 16, 20–21, 48–50, 106, 187, 226, 255n, 264n
  λ phage, 49, 111, 246n
  φX174, 54
  att-site integration, 111
  Collagen repeats in, 40
  cos-site in, 111, 246n
  Endonuclease, 111
  Hybrid, 246n
  Insertion sequences in, 66
  Lysogenic, 191
  Lysogens, 66
  M13, 111
  MS2, 106
  Mu, 268n
  Qβ, 106
  Small variety, 111
  T3 and T7, 133
  T4, 111, 164, 166, 168, 226, 268n, 274n
  Transducing, 221
Bacteriorhodopsin, 128, 183
Bacteroides, 125
Bada, Jeffrey, 159
Baltimore, David (b. 1938), 106, 246n
Baltscheffsky, Herrick, 37
Barry, Martin (1802–1855), 25
Bastian, H. Charlton (1837–1915), 10
Bateson, William (1861–1926), 7–8, 33, 37, 239n
Baupläne, 4, 9, 46, 232
Beggiatoa, 124
Behe, Michael, 156, 232–233, 275n
Berg, Paul (b. 1926), 50, 246n, 247n
Bernard, Claude (1813–1878), 17, 31, 242n
Bichat, Marie Francois Xavier (1771–1802), 29, 242n
Bioinformatics, 57
Biomarkers, 159–160, 262n
Biospecies, 272n
Blood relatives, 22–23
Bombyx mori, 115
Bonnet, Charles (1720–1793), 27–28, 232, 242n
Bordetella, 124
Borrelia burgdorferi, 64, 66–67, 71
Boveri, Theodore (1862–1915), 7–8
Brenner, Sydney (b. 1927), 51
Brewer's yeast. *See Saccharomyces cerevisiae*
Bridges, Charles (1889–1938), 8
Brown, Robert (1773–1858), 19, 29
Brownian motion, 19
Bryophyta (bryophytes), 171, 219

Buffon, George-Louis Leclerc, Comte de (1707–1788), 96
Burnet, (Sir) Macfarlane (1899–1985), 186

Caenorhabditis elegans (*C. elegans*), 57, 64, 67, 80, 82
Calvin–Benson Cycle, 123, 124
Camerarius (Camerer), Joachim (1534–1598), 6, 24
Carotene, 97
Caulimoviridae, caulimovirus (calimo-variety), 107, 112, 114
Cavalier-Smith, Thomas, 97, 104, 121, 126, 138, 139, 145, 192, 193, 201, 256n, 258n
C. elegans Sequencing Consortium, 65
Cellulose, 90, 97
Central dogma, 32–33, 36–37
Centre d'Étude du Polymorphisme Humain (CEPH), 56, 57, 60, 61
Centrioles, 139, 202
Centromeres, 62, 66, 180, 249n
Chaperones, 79, 145, 183, 223
Chatton, Edouard (1883–1947), 97
Chemolithoautotrophs, 123, 128–129
Child, Charles Manning (1869–1954), 33
Chilomastix, 196
Chi site, 60
Chitin, 90
Chlamydiae, 125
Chlamydomonas, 114, 130–132, 184, 199
Chloroflexaceae, 121
Chlorophyll, 97, 99, 123, 130–131, 199, 215
Chlorophyta (chlorophytes), 130, 140, 141, 194, 201, 202
Chloroplast, 30, 36, 50, 64, 79, 97–99, 121, 123, 130–131, 133, 142, 143, 145, 165, 167, 179, 182, 195, 196, 198–201, 221, 254n
Choanoflagellates, 87t, 140
Chondrite inclusions, 157
Christian creationism, 9
Chromaffin granules, 183
Chromatin, 40, 114, 124, 129, 180–181, 221
Chromista, 87t, 98
Chromophyta (chromophytes), 90, 130, 201, 271n
Chromosomes, 7–8, 16, 20, 37, 40–41, 50, 57–61, 64–67, 72, 80, 98, 111, 117, 129, 138, 151, 168, 176, 178–180, 185, 190, 193, 199, 203, 221, 223, 229, 230
Chrysomonads, 90
Chrysophyta (chrysophytes), 87t, 140, 271n
Chytrids (chytrid fungi), 87t, 139, 140, 196
Ciliates (ciliated protozoans), 87t, 132, 139, 140, 170, 194, 196, 203, 219
Cistrons, 41, 109, 129, 130, 163, 165, 178, 182, 198
Cladistics, 10, 85–89, 94, 100–105, 115, 150, 200, 231
Claus, George, 157
Cloning vectors, 49, 50, 52, 63, 111, 246n
Clostridium, 122
Clusters of orthologous groups (COGs), 81–83, 216
Clustered correlations maps, 42
Clustering methods, 70, 73, 88, 225
Cnidaria, 131, 132, 140, 141, 140, 170, 194, 219, 203, 242n
Coding sequences (CDSs). *See* Open reading frames
Codon adaptation index (CAI), 70, 76, 250n
Codon bias, 70, 71, 101, 170, 224
Codons, 35, 44, 71–72, 91, 103, 168–170, 218, 219, 220, 224, 243n, 244n
  Initiation (start), 35, 70, 170, 250n, 273n
  Mitochondrial, 132, 171, 219
  Positions in, 175, 224
  Stop (signal), termination site (terminator), 71, 115, 171, 219, 220, 243n, 273n
  Tryptophan, 171, 219, 273n
  Universal, 70, 170, 222, 253n
Codon usage, 76, 178, 179, 250n
Coelomates, 140, 141
Coevolution, 170, 225
Coiter, Volcher (1534–1576), 23
Complementarity of base pairs, 14–15, 48, 245n
Complementary or copy (c)DNA, 50–53, 57, 63, 69, 72, 246n, 267n

Consensus sequences and patterns, 66–67, 70, 75, 82, 84–85, 88, 101, 129, 208, 253n, 208, 264n
Conserved sequences and positions, 81–82, 84, 101, 107, 111–112, 115–116, 144, 164, 176–177, 204, 216, 264n, 275n
Contigs, 63, 64–65, 72, 101
Copland, Herbert Faulkner (1902–1968), 97
Copy-choice, 107, 113, 172, 191, 199, 227, 267n
Cosmid, 49, 246n
Crenarchaeota. *See* Archaea
Crick, Francis Harry Compton (b. 1916), 13, 14, 33, 42, 47, 93, 146, 149, 156, 169, 217, 218, 240n, 245n, 261n
Cronquist, Arthur (1919–1992), 99
Cryptophytes, 141
Ctenophora, 141
Cuvier, Georges Léopold Chrétien Frédéric Dagobert, Baron (1769–1832), 2, 9, 95
Cyanobacteria. *See* Bacteria
Cyanophora paradoxa, 198
Cytochromes, 131, 133, 138, 168, 195, 244n
Cytophaga, 125
Cytoskeleton, 36, 79, 98, 133, 197

Darwin, Charles Robert (1809–1882), 6, 9, 10, 33, 83, 96, 207, 232
Darwinian fitness, 171, 212
Darwinism (Darwinian evolution), 8–9, 11, 31, 38, 83, 208, 210, 233. *Also see* Natural selection
Databases, 51–52, 55, 56–57, 60, 61–62, 68–70, 72, 73, 80, 101, 247n, 248n, 249n
Dawkins, Richard (b. 1941), 35, 239n, 272n
Dayhoff, Margaret, 102
De Graaf, Reijnier (1641–1673), 24
Degradosomes, 78
Deleuze, Gilles (1925–1995), 1, 43, 93, 149, 151, 207, 213, 239n
DeLisi, Charles, 56
Deoxyribonucleic acid (DNA). *Also see* Complementary or copy (c)DNA
  Antisense strand (nonreading), 110, 112, 191, 256n
  Clocks, 9, 118–119, 211, 220, 272n
  Cloning, 49–50
  Colony hybridization technique, 53
  Evolution of, 174
  Libraries (genomic), 50–51, 53, 55, 57, 63, 72
  Paradigm, 13–14, 16, 42, 244n, 272n
  Sense (reading) strand, 34, 106, 110, 112
  Sequencing, 69, 72
Descartes, René (1596–1650), 19
Desulfurococcus mobilis, 129
Developmental Therapeutics Program, 42
Devolution, 140, 149, 151–152, 173, 186, 191, 203, 207–209, 214, 222, 224, 228, 233, 237–238, 261n
De Vries, Hugo (1845–1935), 7, 8, 10, 242n
Diatoms, 87t, 90, 140
Dictyosome. *See* Golgi apparatus
Dictyostelium discoideum, 115, 140
Dideoxyribonucleotides, 54–55
Dinoflagellata (dinoflagellates), 87t, 90, 139–140, 179, 201
Dinucleotide bias, 66
Diphtheria toxin, 129
Dislocation mutagenesis, 225
Distance matrix analysis and methods, 88, 90, 102–103, 139, 200
DNA polymerase, 41, 52–53, 71, 78, 113, 122t, 145, 172–174, 188, 225, 226, 256n. *Also see* Reverse transcriptase
  DNA-dependent (directed) DNA polymerase (families A–C [I–III], X), 122t, 142, 226
  DNA/RNA and RNA/DNA switching viruses, 111–114
  Error-prone, 225
  RNA-directed and DNA-directed DNA polymerase, 113
  RNA-directed DNA polymerase (RdDp, reverse transcriptase), 112, 114, 188
DNA viruses, 107, 110–111, 181, 187–189, 226, 227. *Also see* Bacteriophage, Double-stranded (ds)DNA viruses, and Single-stranded (ss)DNA viruses

Dobzhansky, Theodosius (1900–1975), 11, 12
Domain replacement strategy. *See* Site-directed mutagenesis
Domains (in polypeptide coding regions). *See* Modules (domains), motifs, and repeats
Domains of life, 87t, 118, 120, 142, 168, 181, 183, 253n, 268n. *See* Archaea, Bacteria, Eucarya
Doolittle, Russell, 119, 142
Double-stranded (ds)DNA viruses, 107, 110–111, 173, 188, 191, 223. *Also see* Bacteriophage
  Adenoviridae (adenovirus), 106, 107, 111
  African swine fever virus, 226, 274n
  Autographa californica, 223
  Baculoviridae (baculovirus), 223
  Cytomegalovirusus, 106
  Herpesviridae (herpesvirus), 107
  Myxoma, 227
  Papovaviridae (papovavirus), 107, 111
  Polyomavirus, 111
  Poxviridae (poxvirus), 107, 226
Double-stranded (ds)RNA viruses, 107, 109, 191
  Ambisense (ambistranded), 109
  Bipartite and multipartite, 109
  Birnaviridae, 109
  Reoviridae (reoviruses), 107, 109
Driesch, Hans (1867–1941), 27
Drosophila, 8, 51, 80, 82, 114, 115, 132, 144, 177, 179, 181, 223, 229, 243n, 275n
Drosophila-related expressed sequences (DRES), 55, 80
Dujardin, Felix (1801–1862), 29
Duplicate genes, 76, 217, 233
Dulbecco, Renato (b. 1914), 246n
Dutrochet, (René Joachim) Henri (1776–1847), 29

Early-Earth, 20, 152–153, 165, 241n
Earthlife, 152–157, 160, 163, 172, 203, 262n
E. coli. *See Escherichia coli*
Ectodermal Growth Factor (EGF), 81, 82, 175
Editosomes, 68
Edman degradation, 47
Eldredge, Niles, 233
Elongation factors (EFs), 40, 78, 116, 122t, 131, 138, 140, 142, 144, 182, 192, 198
Emboîtement, 26
Embryophyta, 40
Endomembranous system, 98, 133
Endoplasmic reticulum, 98, 139, 249n
Endosymbionts, 98, 124, 167, 194–199, 201–202, 223
Engler, Adolf (1844–1930), 99
Enhancer elements, 129, 250n
Enzymes
  3-methyladenine glycosylase, 66
  Adenosine triphosphate (ATP) synthase, 79, 131
  Adenyl cyclase, 185
  Aminoacyl-tRNA synthetases, 35, 138, 144, 169–170, 183, 185
  Aminocyclopropanecarboxylate synthase, 185
  Aminotransferases, 78, 183
  Asparte proteinase, 113, 115
  Carboxylase/oxidase. *See* Ribulose bisphosphate carboxylase/oxidase (RUBISCO)
  Coupled sensor protein kinases, 79
  Cytochrome *c* oxidase, 132, 133
  Cytochrome *c* reductase, 132
  Dehydrogenase, 131, 178, 184, 197, 269n
  Deoxyribonuclease (DNase), 166
  Endonucleases, 78, 111, 167, 246n, 263n
  Endoribonuclease, 264n
  Exonucleases, 68, 166, 225–226
  Exoribonuclease, 265n
  Glyceraldehyde-3-phosphate dehydrogenase (GAPDHs), 131, 184, 197, 269n
  Glycosidases, 139
  Gyrases, 16
  Helicases, 16, 78, 81, 122t, 189, 191, 226
  Histidine kinase, 82
  Holoenzyme, 15, 143, 198–199
  Hydrolase, 66

Integrase, 113, 114, 115
Isomerase, 70, 177, 197, 269n
Kinase, 78, 79, 82, 138, 178
Kinesin/dynein/dynamin-ATPases, 138
Ligase, 49, 111
Malate dehydrogenase, 147
Maturases, 132, 133, 167
Mutase, 263n
NADH dehydrogenase, 132
Nucleases, 173
Oxidoreductase, 131
Peptidases, 79
Peptidyl transferase, 172
Permeases, 79, 90
Phosphatases, 79, 81. *Also see* Adenosine triphosphatase (ATPase) and Guanosine triphosphatase (GTPase)
Phosphoglucose isomerase, 131
Phosphohydrolases, 165
Phospholipase C, 185
Phosphotransferases, 165
Polymerase. *See* DNA polymerase and RNA polymerase
Proteases, 81, 177, 183, 186, 189, 191, 223
Proteinases, 113, 115, 226
Protein kinase, 78, 79
Pyrophosphatase, 162, 263n
Pyrophosphate (PPi) synthetase, 162
Pyruvate kinase (PK), 177
Replicase. *See* RNA polymerase
Ribonuclease (RNase), 78, 99, 112, 113, 115, 164, 172, 173, 202, 226, 264n
Ribonucleotide reductases, 117, 166
RNA helicases, 78
Serine protease, 81, 177
Swivel enzyme, 16
Telomerases, 16, 37, 165
Terminal transferases, 226
Thymidine kinases (TK), 226
Thymidylate synthetase, 117
Topoisomerases, 16, 41, 122t
Transcriptase. *See* Reverse transcriptase
Translocating-ATPase, 79
Transposase, 71
Triosphosphate isomerase (TPI), 269n, 275n
Triphosphatase. *See* Adenosine triphosphatase (ATPase) and Guanosine triphosphatase (GTPase)
tRNA syntheases, 78
U-exonuclease, 68
Uracil-DNA glycosylase, 173
Zinc-dependent protease, 183
Epigenesis, 23–27, 31, 241n
Episomes, 20, 223
Escherichia coli, 49, 60, 64, 67, 70–71, 79, 114, 124, 138, 143, 170, 179, 184, 187, 189, 197, 229, 246n, 249n, 250n, 253n, 264n, 273n
Eubacteria. *See* Bacteria
Eucarya (Eukaryota and eukaryotes), 34, 35, 41, 51, 59, 66, 116–117, 129–133, 149–145, 147, 151, 162, 181–183, 186, 192–193, 196–197, 201, 203, 224, 227, 261n. *Also see* Animalia, Fungi, Plantae
Amitochondriate, 87t, 145, 196
Archezoa, 98, 138–139, 196, 202, 255n
Crown clusters, 138–141
Metakaryota, 98, 138, 193
Euglena, 123, 124, 130, 164, 184, 199, 201
Euglenoids, 87t, 97, 140
Euglenophyta, 201
Eukaryota, eukaryotes. *See* Eucarya
Eustigmatophyta (eustigmatophytes), 271n
Evolutionary (maximum) parsimony analysis, 88
Evolutionary stable strategy (ESS), 194
Exons (expressed sequences), 35, 67–68, 80, 113, 117, 163, 167–168, 174–177, 190, 218, 222, 236, 267n, 273n
Exon shuffling hypothesis, 75, 164, 174–176, 204, 222, 226, 267n, 269n, 273n. *Also see* Introns-early hypothesis
Expressed sequence tags (ESTs), 51–52, 56, 63, 69, 80, 246n, 247n, 248n
Extracellular matrix (ECM), 82, 98, 223, 235

Fabricius, Hieronymus ab Aquapendente (1537–1619), 23
Fallopio, Gabriello (1523–1562), 23
Felsenstein, Joseph, 103
File transfer protocol (FTP), 57

Fisher, Ronald Aylmer (1890–1962), 11
Five Kingdoms, 97–98
Flavobacterium, 125
Flexibacterer, 125
Fluorescent *in situ* hybridization (FISH), 61, 80
Fragile X syndrome, 68
Framework map, 58–59, 62
Functional classes. *See* Protein or polypeptide functional classes
Fungi, 87t, 97–98, 114, 124, 132, 138–141, 177, 184–185, 189, 196–197, 201, 229
Fusobacterium, 125

Gaia, 212, 262n
Galen of Pergamum (131–201), 22, 241n
Galileo, Galilei (1564–1642), 28
Gamophyta, 201
Gemmules, 242n
Gene, 68
  "Ancient," 116–117, 177, 181–182, 188, 204, 216
  Capture, 226
  Conversion, 76, 176, 219, 221, 234, 252n
  Copy and copying, 75, 199, 223, 226
  Differential amplification of, 223–224
  Displacement, 229, 233
  Duplication, 13, 71, 76, 80, 142, 146, 178, 180, 184, 217, 233, 245n
  Hitchhiking, 236
  "Orphan," 71
  Rearrangement, 188, 218, 222, 252 146n
Gene clusters, 66, 130, 143, 178–180. *Also see* Clusters of orthologous groups
  Coordinated coexpression in, 178
  Duplication and, 179–180
  Facilitated coregulation of, 178
  Gene families and, 215
  Hemoglobin, 180
  Hox, 180
  in lambdoid and T4 family of bacteriophages, 268n
  in operons, 66, 67, 250n
  of paralogues, 76
  Physiological (functional) relationship in, 67, 178
  rRNA, 66, 130, 182
  Sharing a promoter, 111
  tRNA, 67, 132
Gene expression, 246n
Gene families (classes), superfamilies (superclasses), and supergene families, 63, 74–83, 82–89, 105, 118, 129, 131, 142, 144, 177, 179–186, 215–216
  Actin, α and β, 177
  Adenosine triphosphatase (ATPase), 79, 82, 118
  Adenosine triphosphate- (ATP-) binding transport proteins, 76
  Bioenergetics and intermediary metabolism. *See* Protein or polypeptide functional classes
  Chlorophyll a/b, 131, 215
  Conversion and, 252n
  Cytochromes, 244n
  Definitions of and ambiguity concerning, 215–216
  DNA replication initiation proteins, 144
  Elongation factor (EF), 142
  Gα, 185–186, 275n
  Gi, 177, 185
  Glyceraldehyde-3-phosphate dehydrogenases (GAPDHs), 184, 269n
  Guanosine triphosphatase (GTPase), 142, 177, 185
  Hemoglobin, 45, 179, 180, 244n, 275n
  Homeo-box, 82
  Homeodomain, 80
  Homeotic, 179
  HOX, 2, 180
  Immunoglobulin, 175
  Informational processing. *See* Protein or polypeptide functional classes
  Laminin, 82
  lin-12Notch, 82
  Lin proteins, 82
  MADS-box, 82
  Metabolic, 184
  Multidomain, 82
  Multigene, 177, 185–186, 188, 234
  Open reading frames (ORFs), 76–77
  Origin and evolution of, 184–185
  Orthologous, 77, 231
  Paralogous, 231
  Phylogenetic trees and, 217
  Polymerases. *See* DNA polymerase and RNA polymerase

Receptor. *See* Protein or polypeptide functional classes
Ribosomal RNA (rRNA = rDNA), 66, 181–183
RUBISCO small subunits, 198
Tubulin, 140, 177
Virogene, 223
GeneMark predictions, 70
Genes, 174–178
*al*1 and 2, 114
Arabidopsis GPA1, 177
atp, 130
cab, 131
cap, 110
cox, 114
COX, 133
CYC, 133
Delta, 82
env, 113
fat, 82
gag, 112, 113, 115
gag-like, 115
Gap, 123–124, 184, 201
gap, 184, 185, 201
Germ-line proliferation-defective-1, 82
Gnas, 177
Maternal, 230
NADH dehydrogenase subunits, 132
Notch, 82
pelota, 144
Pesticide-resistant, 229
pol, 112, 113, 114, 115
psa, 131, 199
psb, 130–131
rep, 110
rev, 113
rp, 131
sec, 79
tat, 113
Tuf, 198
tuf, 131
vif, 113
vpr, 113
vpu, 113
Genetic code(s), 35, 44, 47, 71–72, 85, 151, 168–172, 218–219, 244n, 254n, 265n
Genetic drift, 11, 37, 234–235
Genetic (linkage) maps, 59–62
Genotype, 60, 62, 230
Geoffroy Saint-Hilaire, Étienne (1772–1844), 2, 9, 83, 96
Germ line, 7, 31–32
Germ plasm, 243n
Germ theory of disease, 10, 43
Giardia, 139, 196
Gilbert, Walter (b. 1932), 54, 56, 70, 167, 175, 246n, 247n
Gilbert/Gō theory, 167
Glycolysis (Embden–Myerhof–Parnas pathway), 78, 139, 147, 184
Glycolytic enzymes, 196, 269n
Gō, Mitiko, 167
Goffeau, André, 64
Gogarten, Johann Peter, 116
Goldschmidt, Richard (1878–1958), 33, 36
Golgi, Camillo (1844[1843]–1926), 30, 242n
Golgi apparatus (GA), 98, 139, 193, 259n
Golgi transport elements, 98
Gould, Stephen Jay (b. 1941), 233, 275n
Great Chain of Being, 6, 18, 83, 96, 146, 232, 254n
Grew, Nehemiah (1641–1721), 25, 28
Griffith, Frederick (1877–1941), 179
Gu, Xun, 120
Guanosine triphosphatase (GTPase), 82, 142, 177, 185
Guanosine triphosphate (GTP), 185

Haeckel, Ernst Heinrich Philipp August (1834–1919), 10, 30, 96–97, 242n
Haeckelian recapitulation, 210, 230
Haemophilus, 64, 249n
Haldane, J(ohn) B(urdon) S(anderson) (1892–1964), 11, 261n
Haller, Albrecht von (1708–1771), 24, 27
Halobacterium, 127–128, 192
Halococcus, 128, 192
Halophiles, 192
Haplosporidians, 140
Hardy–Weinberg equilibrium, 212
Harrison, Ross Granville (1870–1959), 30
Hartl, Daniel L., 55, 85
Harvey, William (1578–1657), 19, 23–24, 241n

Health Advisory Committee, 60
Hegel, Georg Wilhelm Friedrich (1770–1831), 3
Heidegger, Martin (1889–1976), 38
Helicobacter pylori, 64
Heliozoans, 141
Helper-adenovirus, 110
Hennig, Willi, 150
Hepadnaviridae, hepadnavirus, 106–107, 112, 114
Hepatitis δ agent, 37, 165
Herophilus of Chalcedon (fl. 300 BC), 22
Hertwig, Oscar (1849–1922), 7, 25
Hippocrates (*ca.* 460–377 BC), 22
Homogeneous staining regions (HSR), 223
Homologous recombination, 221
Homologues
  Bacterial, 126, 144–145, 178, 184–185
  Cyanobacteria–chloroplast line, 184
  Genes and sequences, 104–105, 177, 230, 251n
  Human, 64
  Mouse, 68
  Orthologous and paralogous, 142, 146, 217
  RNA, 40
  Serial, 217
  Viral, 110, 187–188
  Yeast, 144
Homology, 9, 45, 61, 64, 68, 74–75, 83–85, 91, 104–105, 146, 184, 204, 230, 215, 237
Homoplasy (false homology), 89
Homosexuality, 20
Hood, Leroy, 54
Hook, Robert (1635–1703), 28
Hopeful monster, 36
Horizontal gene transfer, 71, 76, 89, 102, 119, 143, 177–180, 195, 197, 204, 223, 226, 229, 235, 243n, 257n, 274n
Hot spots in DNA, 121
  mutational (hypermutable), 37, 225
  polymerization, 161
  recombinational, 37, 60
Hot-spring mats, 123
Howard Hughes Medical Institute, 46, 56
Hoyle, (Sir) Fred (b. 1915), 156–158, 262n
Hubbard, Ruth (b. 1924), 242n
Human Genome Organization (HUGO), 56
Human Genome Project sequencing centers, 52, 63, 56–57
Human Genome Sciences, 56
Human immunodeficiency virus (HIV), 111–113, 226
Human leukocyte antigen complex, 20
Huntington's disease, 59, 68
Husserl, Edmund (1859–1938), 38
Huxley, Thomas Henry (1825–1895), 43, 210–211
Hyatt, Alpheus (1838–1902), 210
Hybrid classes, 230
Hybridization, 229–231, 235, 252n, 275n
Hydrogen hypothesis (HH), 196–197
Hydrogenosome, 98, 139, 195–197, 200
Hydrothermal vent (chimney), 117, 194

Immunodeficiency disease, 2
Immunoglobulin, 81, 175, 218
Imperial Cancer Research Fund, 56
Influenza A, Hong Kong/1968, 226
Influenzavirus. *See* Negative (–) single-stranded (ss)RNA viruses
Inheritance of acquired characteristics, 17–18, 36–37. *Also see* Lamarckism
Insertion sequences (IS), 66
In situ hybridization, 59, 61
Institute for Genomic Research, The (TIGR), 51, 56, 248n
Intervening sequences. *See* Introns
Introgression, 229–230, 252n
Introns, 35, 62, 66–68, 70, 100, 115, 120, 129–130, 132, 145, 163–164, 168, 174–176, 199, 218, 220, 222–223, 235, 236, 263n, 265n, 267n
  Group (type) I, 130, 132, 133, 164, 167–168, 172, 182, 264n
  Group (type) II, 114, 130, 132, 164, 167–168, 172, 181, 264n
  Group (type) III introns, 130, 164
  IVS intron of *Tetrahymena*, 263n
  Splicesomal, 165, 168, 174, 268n
Introns-early hypothesis, 151, 164, 167–169, 173–175. *Also see* Exon shuffling hypothesis

Introns-late hypothesis, 164, 176–178, 263n
Inversions (chromosomal), 221
Isoforms, 185, 219, 240n
Isozymes, 18, 71, 76, 105, 132
Iwabe, Naoyuki, 116

Jakobid flagellate, 271n
Johannsen, Wilhelm (1857–1927), 7, 11
Joyce, Gerald, 166
Jumping genes. *See* Transposable elements
Jussieu, Antoine Laurent de (1748–1836), 96
Jussieu, Bernard de (*ca.* 1699–1777), 95

Kandler, Otto, 99, 147
Kimura's and Ohta's principles of molecular evolution, 104
Kinetoplastid mitochondrial genes (DNA), 68, 124, 131–132
Kinetoplasts (Kinetoplastic protozoa), 132, 140, 126
Koch, Robert (1843–1910), 10, 41, 43
Köllreuter, Joseph (1733–1806), 7
Kozo-Polyanski, B.M. (1890–1957), 271n
Kuhn, Thomas Samuel (1922–1996), 44, 93, 208

Labyrinthulids, 87t, 140
Lake, James, 192–193, 257n
Lamarck, Jean-Baptiste Pierre Antoine de Monet, Chevalier de (1744–1828), 10, 17–18, 29, 96, 232
Lamarckism, 12, 18, 21, 37, 211, 237
Lambda (λ) clones, 64
Lavoisier, Antoine Laurent (1743–1794), 5
Leafhopper, 195
Lederberg, Joshua (b. 1925), 186
Leeuwenhoek, Anton van (1632–1723), 25, 28
Legionella, 124
Leibniz, Gottfried Wilhelm (1646–1716), 3, 232
Leptonema, 124, 125
Linkage homology, 61
Linnaeus, Carolus (Carl von Linné; 1707–1778), 95–96, 98
Lipoprotein membrane, 123, 258n
Lister, Joseph (1827–1912), 42, 43
Loeb, Jacques (1859–1924), 42
Long terminal repeats (LTR), 66, 113, 114–115, 181, 189–190. *Also see* Retrotransposons
Luria, Salvador Edward (1912–1991), 186
Lwoff, André Michel (1902–1994), 186
Lyell, (Sir) Charles (1795–1875), 17
Lysosomes, 98, 139

MacClade, 102, 105, 118
Madison, David and Wayne, 102
Magnetofossils, 159
Major histocompatibility complex (MHC), 234–235
Malpighi, Marcello (1628–1694), 26, 28
Mammalian egg, the, 23, 24
Manic depression, 20
Marchantia polymorpha, 171, 219
Margulis, Lynn (b. 1938), 30, 194, 202, 271n
Mars, Life on, 157, 158–160, 262n
Martin and Müller hypothesis. *See* Hydrogen hypothesis
Martin, William, 228
Martinsrieder Institut für Protein Sequenzen (MIPS), 64–65
Maternal inheritance, 32, 235, 243n
Matrix-assisted laser-desorption mass spectrometry, 47
Maxam, Alan M., 54
Maximum likelihood analysis and methods, 86, 88, 101–103, 121, 215, 248n
Maximum parsimony analysis and methods, 86, 88, 102–103, 214, 241
Mayr, Ernst Walter (b. 1904), 11, 12
Mechanists, 19, 42
Medusazoa (medusazoans), 131
Meiotic bins, 60
Mendel, Gregor Johann (1822–1884), 7–8
Mendelian factors and laws of heredity, 7–8, 11–12, 32, 80
Merck–WashU, 248n
Mereschkowdky, K.S. (1855[1865]–1921), 194
Metabacteria. *See* Archaea
Metakaryota. *See* Eucarya

Metamonada (metamonads), 138–139
Metaphyta, 97–98, 123, 141, 184
Metazoa, 40, 97, 132, 140–141, 171, 185, 219
Bilateral (Bilateria, and Eubilateria), 140–141
Radial (Radiata), 141
Meteorites, 155, 158–159, 261n
Methanococcus (Mc.), 64, 117, 127, 142, 143, 144, 183
Methanogens. *See* Archaea
Metricizing sequence data, 46, 73–74, 85, 197
Metridium senile, 132
Microbial World, The, 99
Micoplasms, 20
Microevolution, 11–12, 44, 211, 219, 234, 254n
Microsporidia, 87t, 129, 138, 139–140
Milieu interior, 4, 31
Miller, Stanley Lloyd (b. 1930), 154
Mini-exon-derived (med) RNA polynucleotide, 168
Minigenes, 75, 167, 174, 236
Mitochondria, 30, 36, 50, 54, 85, 131–133, 145, 196, 201, 202, 219–220, 223–224, 226, 228, 243n, 250n, 256n, 259n, 264n, 270n, 271n
Mobilifilum, 202
Modules (domains), motifs, and repeats, 61, 69, 75, 77–78, 80–82, 85, 107–108, 111–113, 114–115, 142, 167, 174–176, 183, 185, 187, 191, 197, 215, 221–223, 225–226, 236, 248n, 249n, 267n, 273n
Mohl, Hugo, 242n
Monera (Moneres), 97. *Also see* Bacteria
Monism or Monadism, 10
Monophyly (monophylety), 10, 117, 124, 140, 141, 151–152, 184, 189–190, 192–193, 198, 200, 201–203, 208, 224, 235, 257n
Montgomery, Hugh, 240n
Morgan, Thomas Hunt (1866–1939), 8, 10, 12, 32, 36
Morgans, 59, 248n
Motifs. *See* Modules (domains), motifs, and repeats
Motif shuffling, 222–223, 226
Muller, Herman Joseph (1890–1967), 32, 243n
Müller, Johannes Peter (1801–1858), 96
Muller's ratchet, 215
Mullis, Garry Banks (b. 1944), 52, 247n
Multicopy single-stranded DNAs- (msDNAs-) associated reverse transcriptases, 114
Multiple factors. *See* polygenes
Mus high resolution genetic linkage map, 61–62
Mutational spectrum, 76
Mutations
Amorphs, 220
Anticodon, 170
Chance or random, 12, 74, 244n
Definitions, 219
Deleterious, 11, 215, 221
Deletions, 219
Excision, 221
Frame shift, 220
Hypermorphs, 220
Hypomorphs, 220
Missense, 220
Near-neutral, 220
Neutral, 71, 220, 234
Nonsense, 220
Nonsynonymous substitution, 215
Parallel or conversions, 70
Petite (*pet*), 132
Phenotypic, 59, 220
Qualitative, 11
Rate of, 234, 272n
Recombination, 219
Single-base substitution, 220
Synonymous substitutions, 214, 218
Transitions, 220
Transversions, 220
Mychota, 97
Mycobacterium, 64, 123
Mycoplasma, 64, 121, 122, 170, 219, 249n, 259n
Myotonic dystrophy, 68
Myxobacteria, 120, 124
Myxococcus xanthus, 114
Myxozoans, 140, 242n

Nagy, Bart, 157
Nanthomonas, 124
National Advisory Council for Human Genome Research, 60
National Center for Biotechnology Information (NCBI), 51, 73, 247n, 248n
National Institutes of Health (NIH), 51, 56–57, 60–61
Naturalists, 12, 146
Natural selection, 9, 11–12, 39, 83, 146, 207, 209, 212, 214, 227, 231–237, 244n, 273n
Nature philosophers, 28–29, 96
Negative (–) single-stranded (ss)RNA viruses, 109, 191, 223
  Arenaviridae, arenavirus, 106, 109, 227
  Bipartite, 109
  Bunyaviridae, bunyavirus, 106, 109, 187, 225
  Devolution from double-stranded RNA viruses, 191
  Ebola, Zaire, 227
  Filoviridae, filovirus, 109
  Influenzavirus, 106, 109, 226
  Lyssavirus (lyssa-like), 187
  Marburg virus, 227
  Mononegrivales, 109
  Monopartite (nonsegmented, simple set), 109
  Polypartite (multipartite or multisegmented), 109
  Orthoacariviruses, 223
  Orthomyxoviridae, orthomyxovirus, 106, 109
  Paramyxoviridae, paramyxovirus, 106, 109
  Rhabdoviridae, rhabdovirus, 106, 109, 187, 225
  Segment reapportionment in, 227
  Sendai virus, 109
  Vesiculo-like, 187
Neisseria, 124
NeoDarwinism. *See* Microevolution
NeoLamarckism, 18
Neurofibromatosis, 68
Neurospora crassa, 114
Newport, George (1802–1854), 25
Nilsson-Ehle, H., 11
Nitrosovibrio, 124
Noncellular life, 105–115
Noncoding sequences, 180–181
Non-long terminal repeats (LTR-), 114–115, 181, 189–190, 236, 268n. *See* Retrotransposons
Non-Mendelian segregation, 252n
Nuclear (n)DNA, 131, 133, 145
Nucleotide-editing, 226

Oken, Lorenz (née Okenfuss; 1779–1851), 29
Oncogenes, 129, 223, 227
Oomycetes, 87t, 131, 140
Oparin, Alexander Ivanovich (1894–1980), 153–154, 261n
Open reading frames (ORFs), 67, 69–71, 74–75, 76–77, 83, 85, 89, 130, 132, 174, 218, 223–224
Operons, 66–67, 70, 76, 129, 143, 178, 250n
Oppenheimer, Mary Jane (1911–1996), 32
Organellar genes, 114, 130, 197–198. *Also see* Chloroplast and Mitochondria
Organelles, 30, 36, 65, 79, 98–99, 130, 139, 174, 193–203, 219, 254n, 271n
Organicism, 5
Orgel, Lesli, 156
Orthologues, 76–77, 89, 105, 120, 146, 177, 184, 204, 215, 217, 231
Overprinting, 223–224

Pace, Norman, 45, 121, 126
Pallas, Peter Simon (1741–1811), 95
Pandemic, 226–227
Pangenes, 242n
Panspermia, 154, 156, 158, 160
Parabasalia, 138, 139
Paracoccus, 197
Paralogues, 76–78, 116, 142, 177, 217, 231, 257n
Paramecium, 97, 132, 170
Paraphyly (paraphylety), 117, 126, 192, 208
Pararetrovirus, 106–107, 111, 112–114, 115, 191
Parasexuality, 190
Parsimony analysis. *See* Maximum parsimony analysis and methods
Parthenogenesis, 28, 227

Parvoviridae, parvovirus, 110
Pasteur, Louis (1822–1895), 10, 42, 43, 242n
Pasteur Institute, 57, 99
Pedomicrobium, 158
Pentose phosphate cycle, 78, 184
Peroxisomes, 79, 98, 139, 193, 196, 200, 202, 259n
Pfug, Hans, 158
Phaeophyta (phaeophytes; brown algae), 87t, 97, 140, 271n
Phagocytosis, 98, 133
Phenotype, 8, 11, 59, 99, 231, 233
  Archaeal, 126–128
  Cytoplasmic male sterile, 133
  Mutant, 8, 80, 85
Phospholipids, 78–79, 124, 132, 200, 261n
Photobacterium, 124
Photochemical change, 173
Photosystems, 121, 130–131, 199
Phototrophs, 121, 123, 128, 201
Phycobilins, 123, 130
Phycoerythrin, 123
Phylogenetic analysis, 86, 101, 115, 121, 125, 189, 251n
Phylogenetic analysis using parsimony (PAUP), 86, 101–102, 105
Phylogenetic inference, 103, 230
Phylogenetic trees, 88–89, 94, 102, 107, 115, 116–117, 120, 122, 138, 139–141, 146, 147, 179, 188, 225, 226, 228, 233
Phylogeny inference package (PHYLIP), 86, 88, 103
Physarum, 114, 140, 168
Physical maps, 54, 57, 59, 62–65
Placozoa, 140–141
Planctomyces, 121, 125
Plantae, 87t, 97–98. *Also see* Metaphyta
Plasmids, 49–50, 63, 65, 66, 114, 139, 190–192, 197, 221, 226, 246n, 255n
Plastids. *See* Chloroplast
Plesiomorphies, 105, 146
Polycyclic aromatic hydrocarbons (PAHs), 159, 263n
Polygenes, 11, 58
Polymerization chain reaction (PCR), 48, 52–53, 55, 60–61, 62–64, 75, 77, 80, 92
Polymorphism, 59–63, 219. *Also see* Restriction fragment length polymorphisms
Polyphylety and polyphyletism, 140, 184, 189, 193, 200–201, 203, 208, 230, 255n
Polyploidization, 12, 275
Polyproteins, 108, 189
Polytomies and resolution of, 102
Popper, Karl Raimund (1902–1994), 45, 207
Popperian hypothetical-deductive method, 16, 45, 91, 207, 237
Positive (+) single-stranded (ss)RNA viruses, 106, 107–108, 110, 189, 191, 223, 225
  Alphalpha mosaic, 187
  Alphavirus (alpha-like), 108, 189
  Animal, 225
  Anthriscus yellows, 187
  Aphthovirus, 108
  Arbovirus, 108
  Bisegmented (bipartite), 108
  Bovine viral diarrhea virus, 223
  Brome mosaic, 187
  Bromoviridae, bromovirus, 108, 109
  Caliciviridae, calicivirus, 108
  Cardiovirus, 108
  Carmovirus (carmo-like), 107–108
  Comoviridae, comovirus, 108, 187, 189, 226
  Como/nepovirus, 108
  Copy-choice in, 107, 227
  Coronaviridae, coronavirus (corona-like), 106, 108, 166, 189, 191
  Cowpea mosaic virus (CPMV), 187
  Cucumovirus, 109
  Dandelion yellow mosaic, 187
  Enterovirus, 108
  Flaviviridae, flavivirus (flavi-like), 106, 108, 189, 223
  Foot-and-mouth disease virus, 108
  Hepatovirus (hepatoviruses), 108
  Hog cholera virus, 223

Human rhinovirus, 108
Insect, 225
Luteo-like, 107
Monopartite (nonsegmented), 108, 189
Multi-component, 109
Nodaviridae, 225
Parsnip yellow flect, 187
Picornaviridae, picornavirus (picorna-like), 106, 108, 187, 189
Plant, 108, 187, 225
Poliovirus, 187
Polypartite (multipartite, multisegmented or polysegmented), 108, 189, 225
Poty/bymovirus, 108
Potyviridae, potyvirus, 108, 226
Primitive and evolution of, 191
Rhinovirus, 108
Segmented (polypartite, polysegmented), 108, 191, 225
Simple-set picornavirus, 108
Sindbis-like, 107
Sobemo-like, 107, 108
Tobamovirus, 225
Tobravirus, 109
Togavirus (toga-like), 106, 107, 108
Tymoviruses, 224
Western equine encephalomyelitis virus, 108

Posttranslational modifications, 47
Pouchet, Félix-Archimede (1800–1872), 10
Preformation, 25–28, 31
Prions, 3, 16, 20, 200, 204, 240n
Proctolaelaps regalis, 229
Progenote, 170, 176–177, 181–182, 192, 257n
Prokaryotes (Prokaryota), 87t, 34, 40–41, 50, 53, 64, 67, 70, 91, 97–98, 114–115, 117, 119–120, 126, 129–130, 138–139, 142–143, 164, 176, 177, 181–182, 191, 192–193, 194, 201, 215, 216, 224, 246n, 250n, 255n, 257n, 268n. *Also see* Archaea and Bacteria
Promoter elements, 67, 70, 74, 110–111, 113, 129, 130, 132, 168, 178, 250n
Protein families (pfam), 67, 69. *Also see* Gene families
Protein ladder sequencing, 47
Protein or polypeptide functional classes, 64, 75–77, 78–80, 83, 88. *Also see* Gene families
Bioenergetics and intermediary metabolism, 78–79, 90–91, 99, 106, 118, 182, 183–185
Cell rescue/disease/defense, 80
Informational processing, 78–80, 90,182,
Receptor proteins, 79, 185–186
Protein or polypeptide structural families, 64, 75, 80–83. *Also see* Gene families
Proteins
"Ancient," 70, 177, 185, 189
Binding, 79
Cdc54, 144
Cell structure, 79
CrtK, 131
CS-RBD-type RNA-binding, 198
Fibronectin, 82, 197
Fibrinogen, 177
G-proteins, 79
Heat shock protein (HSP), 140, 183
Histone and histone-like, 40, 78, 81, 111, 116, 129, 145, 275n
Homeodomain, 81
"Hypothetical," 75
Intracellular, 79
Lipo-, 75
Mammalian interleukin-1 beta, 77
Mediator, 79
Membrane, 79
Morphogenesis, 79
Nutritional response, 79
Oligopeptide ABC transporters, 79
Osmosensing, 79
Quorum sensing, 79
Pheromone response, 79
Profilins, 86
Protease-inhibitor, 81
Retrotransposons/plasmid, 251n
Ribosomal, 78, 116, 130–131, 144, 145,
RNA-binding, 131, 198
Second messenger formation, 79
Signal-transduction pathway, 79
Single-stranded (DNA) binding, 16
Solute transporters, 79
Soybean trypsin-inhibitor, 77

Stabilizing, 16
Transporter, 76, 79
Transport facilitator, 79
"Unclassified," 75
VPg, 108
Protein sequence space, 12
Protista (protists), 97–98, 140, 202
Protobacteria, 144, 145, 179, 185, 196, 197, 202
Protoctista (protoctistans), 87t, 97–98, 140–141
Proto-exons, 167
Protogenes, 236
Protophyta, 97
Protozoa (protozoans), 68, 97–98, 114, 131, 140, 170, 189, 195, 219
Punctuated equilibrium, 233
Purkynjê, Evangelista (1782–1869), 29
Purple bacteria. *See* Bacteria
Pyrhotite, 159
Pyrodictium occultum, 129

Radiation hybrid (RH) panels, 61, 80
Radiolarians, 90, 194
Raff, Rudolf, A., 117, 141, 228, 251n, 254n
Ramón y Cajal, Santiago (1852–1934), 30, 242n
Ray, John (1627–1705), 95
Reading frame, 111, 168. *Also see* Open reading frames
Reclinomonas americana, 202
Recombination, 59, 60, 66, 75, 76, 78, 80, 107, 111, 113, 132, 167, 172, 173, 176, 178, 191, 199, 215, 219, 221–224, 225, 226–227, 267n
Regulation, 14, 16, 17–18, 19, 20, 21, 31, 33, 36
of cell division, 144
by guanosine triphosphate subunits, 185
of protein processing, 106
by retrotransposon integration, 181
Remak, Robert (1815–1865), 30
Renaissance, 5, 18, 23, 28
Repeats. *See* Modules (domains), motifs, and repeats
Replicase. *See* RNA polymerase
Replication, 13, 15–16, 22, 34, 41, 66, 82, 90, 111, 142, 144, 171, 172, 182, 247n, 252n
and copy choice, 191, 199, 267n
and differential gene amplification, 223
of DNA viruses, 110, 188
and duplication of LTRs, 113
Error prone, 113, 172, 225
and gene copying, 222
Initiation of, 78, 266n
Initiation-specific factors, 144, 78, 266n
Interruption by dideoxiribonucleotides, 54
Origin of, 66, 111, 132
Polarity of, 67, 249n
Primers for, 165, 243n
Reiterative, 52–53
and repair and recombination, 78
Reverse, 37, 51–53, 167, 176
RNA-dependent, 166
of RNA viruses, 108, 166, 189, 191
Termination factor, 78
Replication factor complex (rfc), 113, 144
Reservoir species, 226
Restriction enzymes, 48–50, 54, 246n
Restriction fragment length polymorphisms (RFLPs), 59–60
Reticulate evolution, 229–231, 252n, 268n
Retortamonad, 196, 271n
Retroelements, 114–115, 181, 189–190, 256n
Retroids, 181
Retroposons (nonLTR retrotransposons or LINEs), 181
Retrotransposons, 66, 114–115, 130, 181, 189–191, 199, 226, 236
Retrovirus, 51, 65, 106, 107, 111–112, 113–114, 115, 165–166, 181, 190–191, 216, 223, 226. *Also see* Pararetrovirus and Switching viruses
Retrovirus-like elements, 114, 181, 190, 221
Reverse transcriptase, 37, 51, 106, 111–115, 130, 150, 166, 173–174, 188–190, 199, 246n, 256n, 263n, 267n
Reverse transcriptase-like (RTL) elements of organellar genes, 114

RFLP. *See* Restriction fragment length polymorphisms
Rhodobacter, 124, 131, 197
Rhodocyclus, 124
Rhodophyta (rhodophytes; red algae), 87t, 140–141, 184
Rhodospirillum rubrum, 263n
Ribonuclease (RNase), 99, 113, 164, 166, 173, 226, 264n, 265n
Ribonucleic acid (RNA). *Also see* Cistrons, Gene families, Ribosomal RNA, RNA virus families, and Transfer RNA
  Antisense (nonreading frame) strands of RNA viruses, 108, 224
  Editing, 68, 170, 219
  Guide RNA (gRNA), 68
  Posttranslational (transcript) processing, 34, 175, 249n, 263n
  Pre-edited mRNA, 68
  Small-cytoplasmic RNA (scRNA), 66
  Small-nuclear RNA (snRNA), 66
  Sense strand, 256n
Ribonucleoprotein particles (RNPs) and complexes, 34
  Chloroplast, 198
  Cytoplasmic, 133
  Small nuclear (snRNPs; "snurps"), 165, 249n, 264n
Ribosomal proteins, 78, 116, 130–131, 144, 145, 182–183, 200, 268n
Ribosomal RNA (rRNA = rDNA), 66, 99, 117–118, 119, 120–121, 125–127, 130, 138–141, 181–183, 190, 192–193, 202, 236, 255n, 266n, 268n
  Large subunit (LS), 34, 130, 132, 138, 140, 182, 192, 202
  Organizing centers, 66, 223
  Small subunit (SS), 34, 99, 117, 120–121, 126, 130, 132, 138, 140–142, 182, 192
Ribozymes, 164–168, 171–172, 263–264n, 265n
Ribulose bisphosphate carboxylase/oxidase (RUBISCO), 143, 179, 198–199
RNA polymerase, 74, 78, 116, 131, 133, 142, 144–145, 147, 160, 166, 172, 191–193, 202, 224, 226
  Binding sequence, 74
  DNA-directed (dependent) RNA polymerase, 122t, 129, 144, 147, 166, 226, 243n, 267n
  Error-prone replicases, 113, 265n
  Eukaryotic (pol1–3; polymerase I–III), 34, 98, 129, 138, 142, 144, 145, 147, 182, 202, 243n
  Prokaryotic, 34
  RNA-dependent (directed) RNA polymerase (replicase), 107–109, 164–166, 173, 186, 224, 226, 256n, 265n, 267n
  Subunits, 138, 144, 193, 202,
  Viral (replicase), 107–109, 165, 187, 190, 224, 226
RNA virus families, 107–109
RNA world hypothesis, 166–165, 173
Rockefeller Foundation, 46, 47
Roux, Wilhelm (1850–1924), 27
Royal Society (London), 26
Ruse, Michael, 210–211

Saccharomyces cerevisiae, 64, 65, 67, 114–115
Salmonella, 124, 179
Sanger, Frederick (b. 1918), 47, 54, 246n, 247n
Sanger Center, 57
Saprospira, 125
Schizophrenia, 20, 241n
Schizophyta, 97. *Also see* Bacteria
Schleiden, Mathias Jakob (1804–1881), 29–30, 242n
Schwann, Theodore (1810–1882), 29–30
Segmented (virus) genomes, 107–109, 110, 189, 191, 222, 225
Segment reapportionment in segmented viruses, 227
Selfish DNA, 35, 235
Selfish genes, 212, 271n
Selfish operon model, 178
Self-replication, 166, 172
Senapathy, Periannan, 163, 167
Sequence length polymorphism (SSLPs), 60
Sequence recognition particle (7S SRP), 182
Sequence skimming, 69

Sequence specific tags (SSTs) or sequence tagged sites (STSs), 62
Sequential endosymbiosis theory (SET), 194–203
Sex, 5, 60, 192, 227–228, 243n, 252n, 272n, 274n
Sexual reproduction, 11, 217, 227–228
Shine–Dalgarno sequences, 40, 70, 85, 139, 147, 250n
Short tandem repeat polymorphisms (STRPs), 62
Signal recognition particles (SRP), 66, 144, 182, 249n
Signature elements, 66
Single-stranded (ss)DNA viruses, 110, 191
Single-stranded (ss)RNA viruses, 21, 107–109, 110, 187, 189–191, 214
Sinsheimer, Robert, 56
Site-directed mutagenesis, 81, 224
Sober, Elliott, 233
Sogin, Mitchell, 117, 138, 141, 255n
Somatic cell line, 7
Southern, Edward M., 246n
Southern transfer or blotting, 48, 52, 197
Spallanzani, Lazaro (1729–1799), 25
Spemann, Hans (1869–1941), 27, 242n
Spirillum, 124
Spirochaeta, 202
Spirochetes. *See* Bacteria
Spliceosome complex, 165, 168, 264n
Split genes, 98, 117, 163, 173
Spontaneous generation, 10, 31, 190
Stainier, Robert Yate, 99
Staphylococcus, 122
Stebbins, George Ledyard (1906–1993), 11, 99
Sticky ends, 49, 111
Stramenopiles, 87t, 140
Streptococcus, 122
Structured Query Language (SQL), 57
Sturtevant, Alfred Henry (1891–1970), 8
Sulfolobus, 116, 120, 129
Sutton, William (1877–1916), 8
Svedberg, Theodor (1884–1817), 47, 245n, 249n
Swammerdam, Jan (1637–1680), 19
Switching viruses, 111–115
Swofford, David, 101
Symbiogenesis, 194, 203
Symbiont ball, 195
Symplesiomorphies, 105, 146
Synapomorphies, 105, 146, 192
Synechocystis, 64, 123, 184, 201
Systema Naturae, 95

Taq, 53
Taylor, "Max," 202
Telomeres, 16, 37, 60, 65, 71, 165, 173, 174, 180, 221
Temin, Howard Martin (b. 1934), 186, 246n, 256n
Template-switching. *See* Copy-choice
Tetrahymena, 86, 132, 165, 170, 263n, 264n
TIGR. *See* Institute for Genomic Research, The
Thermococcales celer, 128
Thermofilum pendens, 129
Thermoproteus tenax, 129
Thermoplasm celer, 128
Thermotoga, 121, 125, 258n
Thermus aquaticus, 53
Tiselius, Arne (1902–1971), 47, 245
Transcription, 16, 34, 67–68, 70, 78, 81, 90, 110–112, 114, 117, 129, 132–133, 142, 168, 191, 223, 233, 244n, 246n, 247n, 250n, 252n, 265n
  Differential, 108, 225
  Factors (TFs), 81, 82, 129
  Initiation of, 34, 75, 90, 129, 144
  Reverse, 165, 173, 181, 268n
  Termination of, 70, 257n
Transcriptome, 69. *Also see* Open reading frames
Transfer RNA (tRNA), 34–35, 66–67, 69, 72, 78, 84, 99, 113, 118, 120, 126, 129–130, 132, 142, 145, 168, 169–172, 182, 200, 218–219, 222, 224, 243n, 264n, 266n, 273n
  Anticodon loop, 168, 169
  D arm, 169
  Genes, 66–67, 120, 129, 132, 145, 167, 170, 190, 224
  -like template, 165
  Recruitment of, 169
  Short introns in, 145
  TψCG arm, 169, 258n, 266n

Translation, 34, 90–91, 142, 218, 250n
Translocation of genes, 199
Translocation of protein through membranes, 79, 142
Transposable elements (TEs), 115, 116, 180–181, 190, 221, 236
Transposon-integration sites, 66
Transposons, 66, 71, 78, 168, 181
Treeing methods and software, 70, 72, 86, 88, 101–103, 105, 118, 192
Tree of life, 96–97, 118. *Also see* Universal tree of life
Tricarboxylic (citric) acid cycle, 78, 125, 132, 184
Trichomonads, 139
Trichomonas, 139, 196
Trypanosoma, 115, 168
Trypanosomes (trypanosomids), 68, 123, 131, 133, 201
Tuberculosis bacillus, 57
3′ untranslated regions (3′ UTRs), 62

Ubiquitin, 223
Ultraviolet (UV) irradiation, 173
Undulipodia, 194, 200, 202
Uniformitarianism, 17, 212
Unity of life, 5–6, 10, 15, 16
Universal tree of life, 116, 182
Upstream activation sites (UAS), 129
Urey, Harold Clayton (1893–1981), 154
Ur-kingdoms, 116, 118
Use and disuse, 17–18, 37, 211. *Also see* Lamarckism

Vahlkampfia, 140
Vairimorpha necatrix, 139
Van Niel, Cornelis Bernardus Kees, 99
Variable number of tandem repeats (VNTRs), 62
Venter, Craig, 51
Vesalius, Andreas (1523–1562), 24
Virchow, Rudolf (1821–1902), 30
Virogenes, 197
Viroids, 16, 166, 172, 204
Viruses. *See* Double-stranded (ds)DNA viruses, Double-stranded (ds)RNA viruses, Negative (–) single-stranded (ss)RNA viruses, Positive (+) single-stranded (ss)RNA viruses, Pararetroviruses, Retroviruses, Single-stranded (ss)DNA viruses, and Switching viruses
Virusoids, 165
Vitalism, 5, 10, 31
Von Baer, Karl Ernst (1792–1876), 9, 24, 242n

Wallace, Alfred Russel (1823–1913), 232
Watson, James Dewey (b. 1928), 13–15, 21, 33, 47, 56, 93, 146, 149, 217, 244n, 245n, 272n
Watson/Crick pairs, 13, 15, 33
Weismann, August (1834–1914), 31–33, 36, 217, 242n, 243n
Weldon, (W.F.) Raphael (1860–1906), 210
Wellcome Trust, 46, 57
Whittaker, Robert Harding (1920–1980), 210
Wickramasinghe, Chandra, 156
Williamson, Donald I., 230–231
Wilson, E[dmund].B. (1856–1939), 8, 210
Woese, Carl R. (b. 1928), 45, 99–100, 104, 115–117, 122, 126, 145, 169, 192, 251n, 255n, 256n, 257n
Wolfe, Ralph, 99
Wolff, Caspar Friedrich (1738–1794), 28
Wright, Sewall (1889–1988), 11–12, 211, 234

Xanthophylls, 97
Xanthophyta (xanthophytes), 87t, 140, 194, 271n

Yarus, Michael, 169
Yeast, 28, 49, 59, 168, 170, 226. *Also see* *Saccharomyces cerevisiae*
Yeast artificial chromosomes (YACs), 61, 64

Zillig, Wolfram, 99
Zoochlorella, 194–195
Zooxanthella, 194